ASHEVILLE-BUNCOMBE TECHNICAL INSTITUTE

CHEMICAL EXPERIMENTATION

A Series of Books in Chemistry

Editors: Linus Pauling
Harden M. McConnell

CHEMICAL EXPERIMENTATION

An Integrated Course in Inorganic, Analytical, and Physical Chemistry

Ursula A. Hofacker

W. H. FREEMAN AND COMPANY
San Francisco

Printed in the United States of America

International Standard Book Number: 0-7167-0166-9

1 2 3 4 5 6 7 8 9

Contents

Preface

Today's research in inorganic, physical, and analytical chemistry shows that these fields overlap considerably. In recent years, the development of new instrumental techniques, which had been mainly the territory of physical chemistry, has become an important area of analytical chemistry. Now chemists of all fields of specialization use almost identical instrumentation. Inorganic chemistry has received a great deal of impetus from concepts of physical chemistry. Structure determination methods and reaction kinetics, subjects under investigation by physical chemists for many years, have become established parts of inorganic chemistry. The solid state, a classical domain of physics, is now being researched by physical and inorganic chemists. Countless other examples attest to the permeability of the boundaries between the classical chemical disciplines. One result is the use of the same laboratory experiments in different chemistry courses; another is uncertainty about the appropriate course in which to include such subjects as vacuum techniques, basic electronics, chemical instrumentation, and the use of computers. In view of these circumstances, a program for a combined "advanced practices" laboratory was inaugurated at Northwestern University in 1964. The experiments presented in this manual originated in that course.

As the program developed, it became apparent that exercises concerned mainly with the principles and application of certain techniques were necessary to provide the tools with which to perform integrated experiments. The primary criterion for the preparation of integrated experiments was that inorganic, physical, and analytical chemistry should all come to bear on the procedures. Obviously, such a requirement limits the choice of topics considerably; therefore no claim is made here for the complete coverage of any field. Nevertheless, experience at Northwestern University indicates that what is gained in coherence more than makes up for the lack of breadth.

Much of the success of the combined-practices laboratory was due to the emphasis placed upon the interrelationships between various experiments. These interrelationships were subjects of discussions between students and teachers, which took the place of conventional written reports on individual experiments. The course finally emerged as an interwoven pattern of learning activities; consequently, its presentation in a manual creates special problems.

To capture the spirit of the course while clearly presenting its content, the original experiments have been organized into five major groups, although it is intended that they be reassembled into the integrated laboratory projects described in "Projects to be Constructed from Experiments in the Manual" (page ix). Interrelations between the projects are manifested by the inclusion of some experiments in several different projects. This does not mean that an experiment must be performed twice if it is included in two different projects; the intention is to emphasize the various aspects of the experiment within the contexts of the projects.

Some of the information required for the interpretation of many of these experiments can be obtained only by searching the literature and textbooks. To reduce the amount of time spent by the student on research in the library, discussions of particular features of such experiments are incorporated into the exercises. The manual thus contains a measure of basic information in excess of that normally found in laboratory manuals.

Not all of the experiments in this manual are entirely new, nor are they all inventions of the author. In the development of a new course, commonly used experiments are invariably drawn upon: sources have been acknowledged wherever possible. However, for advice and criticism, I am indebted to all of my former colleagues in the Department of Chemistry at Northwestern University. In particular, I wish to thank Duward Shriver and Fred Stafford, who participated in teaching the course, and whose contributions are too numerous to list here. Special thanks also go to Donald Smith, who shared the joys and burdens of the course with me during most of my stay at Northwestern, for his contribution of Sections B–D of Exercise 12. It is also a pleasure to thank Donald DeFord, then Chairman of the Department of Chemistry, for his active support and personal encouragement. A great deal of inspiration, critical evaluation, and enthusiasm was furnished by Northwestern University's graduate students. Mark Ratner, who was once one of them and is now an assistant professor at New York University, has my sincere gratitude for reading the entire manuscript and for help in my struggle with the English language.

March 1972 *Ursula A. Hofacker*
Munich, Germany

Projects to be Constructed from Experiments in the Manual

1. Preparation of tris(ethylenediamine)cobalt(III) chloride followed by resolution of the optical isomers and analysis by optical rotation. Observation of the visible spectrum and measurement of the magnetic susceptibility. Discussion of optical and magnetic properties on the basis of crystal field theory. Measurement of O.R.D. and C.D. in relation to the absolute configuration of the complex.
(Exercises: 1A, 7A and C, 9A, 8A and B)

2. Preparation of nitro- and nitritopentaamminecobalt(III) chloride. Identification by ligand group frequencies. Discussion of the U.V., visible, and I.R. spectra, observation of crystal field, charge transfer, and ligand bands. Investigation of the kinetics of isomerization and formation of the nitro complex from the aquopentaammine complex. Discussion of the underlying reaction mechanism. Auxiliary preparations: aquopentaamminecobalt(III) nitrate and chloropentaamminecobalt(III) chloride.
(Exercises 1B, 7A and C, 14B, 1A and D)

3. Preparation of *cis-* and *trans*-dichlorobis(ethylenediamine)cobalt(III) chloride. Identification of structure by comparison of spectra with those of tris (ethylenediamine)cobalt(III) chloride. Investigation of the kinetics of *cis-trans* isomerization in methanol.
(Exercises 1C, 7A and C, 14A)

4. Conductivity measurements on chloropentaamminecobalt(III) chloride and tris(ethylenediamine)cobalt(III) chloride. Debye-Hückel-Onsager treatment to obtain limiting equivalent conductance. Experiments prove the existence of the complex ion.
(Exercises 1A and B, 12G)

5. Potentiometric titration of copper(II) with ethylenediamine. Use of Bjerrum's method in the calculation of formation constants. Observation of visible spectra of hexaaquo- and bis(ethylenediamine)diaquo complexes. Data reduction by digital computer.
(Exercises 12E, 7A and C, 16B)

6. Preparation of cobalt ferrite. Analysis by EDTA titration (direct and back titration) and by spectrophotometry following ion-exchange chromatographic separation. Measurement of x-ray powder pattern and density of the solid. On the basis of the above measurements, spinel structure, crystal field stabilization energies, and the existence of normal and inverse spinels are discussed. Estimation of residual entropy at absolute zero and the possibility of connecting ferrimagnetism to stoichiometry and structure are considered.
(Exercises 2A, 13A and B, 10A and D).

7. Preparation of sodium tripolyphosphate. Potentiometric titration to obtain average chain length, following conversion to acid by ion exchange. Hydrolysis of the salt and identification of products and reactant material by paper chromatography.
(Exercises 2B, 11C, 12F)

8. Preparation of chromous acetate in Schlenck apparatus. Measurement of magnetic susceptibility as a

means of structure determination and as a criterion of purity.
(Exercises 3A and B, 9A)

9. Vacuum line preparation and purification of DC1. Identification by vapor pressure. Rotational-vibrational spectrum as a tool in detection of impurities and in the determination of molecular parameters and thermodynamic state functions. Observation of isotope effect (H/D, and Cl^{35}/Cl^{37}). Calculation of vibrational frequency, bond length, approximate dissociation energy, first anharmonicity constant, and rotational temperature. Construction of potential curve (Morse) and calculation of thermodynamic properties.
(Exercises 4A and C, 6B)

10. Vacuum line preparation and purification of ND_3. Analysis of vibrational-rotational spectrum. Isotope effect (H/D). Determination of impurities. Calculation of bond angles and bond lengths. Observation of inversion line splitting. Interpretation of tunneling in a double minimum potential. Isotope effect on splitting.
(Exercises 4A and B, 6C)

11. Electrochemical methods: Introduction to use of operational amplifiers. Construction of a polarograph from operational amplifiers used for (a) observation and interpretation of current-voltage curves (unified approach to electrometric titrations); selection of end-point methods from current-voltage curves; comparison of predictions with actually performed titrations; (b) investigation of electrode processes in direct-current polarography with the D.M.E.; their influence on the appearance of the polarographic wave; analytical and kinetic applications; and (c) investigation of electrode processes by cyclic voltammetry; introduction to the concept of time scales for reactions.
(Exercise 12A, B, C, and D)

12. Chromatography: Gas-liquid chromatography and ion-exchange chromatography. Optimization procedures and performance characteristics.
(Exercises 11A and B)

13. X-ray powder diffraction: Analytical application—analysis of a binary mixture of crystalline solids. Structure investigations—determination of lattice parameters for cubic or tetragonal substances and of unit cell characteristics for a more complicated structure (spinel).
(Exercise 10A, B, C, and D)

14. Spectra and behavior of rare earth and transition metal ions: Determination of magnetic susceptibilities of rare earth and transition metal compounds with a Gouy balance. Interpretation of coordination compound spectra and susceptibilities in terms of crystal field theory as opposed to the free-ion interpretation for rare earth compounds.
(Exercises 9A and B, 7A, B, and C)

15. Spectroscopic investigations of atoms: The free atom—observation of the visible spectrum of sodium; determination of an energy diagram; selection rules; determination of Rydberg corrections and ionization energy; discussion of Aufbau principle and periodic table on the basis of the observed energy level diagram. The weakly interacting atom—spectrum of Eu^{3+}; nomenclature and selection rules; quenching of fluorescence by transfer of energy to the solvent.
(Exercises 5A, 7A and B)

16. Molecular information from spectroscopy: Observation of the vibronic spectrum of I_2. Potential curves and vibrational states. Different electronic states, their dissociation energies, and the states of the products. Selection rules and the Franck-Condon principle. Birge-Sponer treatment, data evaluation, and curve fitting by digital computer. The vibration-rotation spectrum of a heteronuclear diatomic molecule (HCl or DCl). Information derived from resolved rotational structure—rotational temperature and bond length; construction of a potential (Morse) curve; calculation of thermodynamic functions; infrared spectrum of a symmetric top molecule (ND_3) under high resolution; normal modes and nomenclature; rotational structure and calculation of bond angles and bond lengths; observation of inversion splitting; discussion in terms of a double minimum potential.
(Exercises 6A, B, and C, 16A)

17. Optical activity: Discussion of optical rotatory dispersion and circular dichroism. Optical rotation as an analytical tool. O.R.D. and C.D. as a means of improving resolution and identification of bands. Absolute configuration and optical activity of Co(III) complexes.
(Exercise 8A and B)

18. Kinetics of homogeneous gas-phase reactions: Thermal cyclization of hexafluorobutadiene. Analysis by gas chromatography. Since the reaction does not go to completion, the rate of forward and reverse reactions is determined from rate and equilibrium measurements. Calculation of activation energy for both reactions and for reaction enthalpy and entropy. The experiment is performed by a team of several groups.
(Exercises 14C, 4A, 11A)

19. Data evaluation by digital computer: Curve fitting by least-squares method of a large data pool to a linear and a quadratic equation is performed in the Birge-Sponer treatment of the observed I_2 spectrum. The standard deviation is used as the criterion for best fit. The second example involves the calculation of stepwise formation constants from a titration curve. The computer allows the use of more of the data in a recorded titration curve.
(Exercises 6A, 12E, 16A and B)

CHEMICAL EXPERIMENTATION

PART I

PREPARATIVE METHODS

Preparation of Coordination Compounds

A. OPTICAL ISOMERS

Apparatus

500-ml Erlenmeyer flask
100-ml graduated cylinder
4-inch diameter Büchner funnel

Chemicals

30% aqueous ethylenediamine
Concentrated hydrochloric acid
Hexaaquocobalt(II) chloride
Ethyl alcohol
Ether
30% aqueous hydrogen peroxide
Sodium potassium D-tartrate-tetrahydrate
Sodium iodide

Introduction

This experiment demonstrates the resolution of optical isomers. The octahedral complex $[Co(en)_3]Cl_3$ exists in two optically isomeric forms. In the first part of the experiment the racemic mixture of the two is prepared. Resolution of the D and L isomers is achieved by crystallization with sodium potassium tartrate, which occurs naturally in the D form. The D-D salt precipitates while the L-D compound stays in solution, from which it can be recovered in the form of the (−)-$[Co(en)_3]I_3$. Ultimately, the (+)-$[Co(en)_3]I_3$ is obtained from the D-D salt and the optical rotation of both iodides is measured. Part of the racemic mixture is set aside for the measurement of its magnetic susceptibility (see Exercise 9B) and the observation of its spectrum, together with those of the geometric isomers *cis*- and *trans*-$[Co(en)_2Cl_2]Cl$ (see Exercise 7C).

Preparation of Tris(ethylenediamine)-cobalt(III) Chloride

To a solution of 75 ml of 30% aqueous ethylene diamine, 50 ml of H_2O, and 11.5 ml of concentrated hydrochloric acid in a 250-ml beaker, add 15 g of ground hexaaquocobalt(II) chloride. Oxidize with

16 ml of 30% H_2O_2, added slowly while cooling the solution in an ice bath. Evaporate the solution until a thin crust at the surface is observed, at which point the volume will be approximately 40 ml. Cool, then add 7 ml of concentrated hydrochloric acid while stirring, and then 15 ml of ethyl alcohol. Cool, and filter, using a Büchner funnel with suction. Wash the crystals with ethyl alcohol until no color appears in the wash solution. The product should be yellow-orange. If the crystals are green, they should be dissolved in water and treated with ethylenediamine again. Dry the crystals by washing once with ether or by heating at a temperature above 100°C. Calculate the percent yield. Bottle and label the part not required in the next step, and give it to the instructor to retain for measuring magnetic susceptibility and observing spectra later. Record the visible spectrum of a 0.01 *M* solution in water (see Exercise 7C).

Resolution of the (+) and (−) Isomers

Heat to boiling a solution of 12.2 g (4×10^{-2} mole) of tris(ethylenediamine)cobalt(III) chloride and 50 ml of water in a 500-ml Erlenmeyer flask. After the solution is completely dissolved and still boiling, add 10.0 g (4×10^{-2} mole) of sodium potassium D-tartrate-tetrahydrate. Crystallization of (+)-tris(ethylenediamine)cobalt(III)-D-tartrate-pentahydrate begins upon cooling with ice and is completed by allowing to stand overnight, although it may take longer. Collect the crystals by filtration and reserve the filtrate for the isolation of the levo isomer. Wash the crystals with 40% ethanol in water and recrystallize by dissolving them in 6 ml of hot water. Allow them to cool to room temperature before cooling in an ice bath. After filtration, wash the crystals with 40% ethanol in water and then with absolute ethanol, and air-dry. Calculate the percent yield of the optically pure isomer. Determine the content of the optically pure isomer by measuring the optical rotation: $[\alpha]_D = +102°$ (see Exercise 8A). If only a small amount of the product is obtained, proceed to the next step and omit the measurement of optical rotation.

Dextro Iodide

Dissolve the (+)-tris(ethylenediamine)cobalt(III)-D-tartrate in 9 ml of hot water. While stirring the solution, add 2 ml of concentrated ammonia solution and then 10.2 g of sodium iodide dissolved in 5 ml of hot water. The iodide crystallizes into reddish-brown needles upon cooling. Complete crystallization is effected by cooling in an ice bath for about 15 minutes. After vacuum filtration, wash the crystals with 10 ml of ice-cold 30% sodium iodide solution to remove any tartrate, and then with ethanol and acetone. Calculate the percent yield from the measurement of optical rotation: $[\alpha]_D = +90°$ (see Exercise 8A).

Levo Iodide

Treat the (−)-tris(ethylenediamine)cobalt(III)-D-tartrate remaining in the solution from which the (+)-tris(ethylenediamine)cobalt(III)-D-tartrate-pentahydrate has crystallized with 0.2 ml of concentrated ammonia solution and heat the mixture to about 80°C. Stir in 10.2 g of solid sodium iodide and cool in an ice bath. Collect the impure levo iodide by filtration and wash with 13 ml of ice-cold 30% sodium iodide; then wash with alcohol and air-dry. To purify this crude material, stir it into 20 ml of water that has been heated to 50°C. Remove the undissolved racemate by filtration. Add 3 g of sodium iodide to the warm filtrate (50°C) and allow crystallization to take place. After cooling the solution in an ice bath, collect the solid by filtration, wash it with ethanol and then acetone, and air-dry. Calculate the percent yield: $[\alpha]_D = -90°$.

B. LINKAGE ISOMERS

Apparatus

125-ml Erlenmeyer flask
100-ml graduated cylinder
Büchner funnel

Chemicals

Hexaaquocobalt(II) chloride
30% aqueous ethylenediamine solution
30% aqueous hydrogen peroxide
6 *N* ammonium hydroxide
Dilute hydrochloric acid
Sodium nitrite
Ammonium chloride
Ethyl alcohol
Ether

Introduction

This study of nitro- and nitritopentaamminecobalt(III) chloride begins with the preparation of chloropentaamminecobalt(III) chloride, in which one chlorine is then replaced by either the nitro or the nitrito group. Since the nitrito compound is very unstable and converts to the nitro compound, it will be necessary to keep the nitrito compound cold during the entire preparation. Measurements must be taken as soon as possible after preparation. Store in the freezing compartment of a refrigerator if storage is necessary.

Preparation of Chloropentaammine-cobalt(III) Chloride

To a cold solution of 10 g of hexaaquocobalt(II) chloride and 20 g of NH_4Cl in 60 ml of concentrated ammonia, add 5 ml of 30% H_2O_2 solution at once. Then add 2–3 ml at a time, until 10 ml have been added. After oxidation is complete, heat the solution to boiling and allow it to cool. Collect the purple precipitate by filtration.

Preparation of Nitropentaammine-cobalt(III) Chloride

To a 125-ml Erlenmeyer flask, add 15 ml of water, 6 ml of 6 *N* ammonium hydroxide, and 1.5 g of $[Co(NH_3)_5Cl]Cl_2$. Warm to dissolve the salt. Filter, cool, and acidify slightly with dilute HCl. Add 2 g of sodium nitrite and heat gently for complete dissociation. Cool, and cautiously add 20 ml of concentrated HCl. Cool in an ice bath, filter, wash with ethanol, and dry by washing with ether. Calculate the percent yield. Bottle and label. The compound is yellow.

Preparation of Nitritopentaammine-cobalt(III) Chloride

Dissolve 1.5 g of $[Co(NH_3)_5Cl]Cl_2$ in 25 ml of water plus 5 ml of concentrated ammonium hydroxide. Warm slightly, if necessary, and filter, if any of the starting material has not been dissolved. Carefully add 6 *M* HCl until the solution is just neutral when tested with litmus. Cool in an icebath; add 1.5 g of sodium nitrite and allow to remain in the icebath for 1–2 hours. Filter the precipitate, and wash with ice water, then alcohol, and finally ether. Calculate the percent yield (based on the dissolved starting material!). Record the infrared spectrum as soon as possible and keep the material cold at all times.

Note: Ethanol may be added to obtain precipitation. Both nitro- and nitritopentaamminecobalt(III) chloride are formed in this preparation. The nitrito compound, however, is less soluble and the precipitate will, with proper cooling, consist mainly of the nitrito complex. The addition of too much ethanol, however, precipitates the nitro compound also. The pure compound is salmon-pink.

Characterization

Record the infrared spectrum of each substance in a KBr pellet. Consult your instructor for the experimental details. Observe and record the spectrum of the nitrito compound immediately after its preparation. Retain your preparation and record the spectrum again after one or two weeks have elapsed. Check whether your second observation of the spectrum is that of the nitro compound.

Both complexes absorb at 1595, 1315, and 850 cm^{-1}. These absorptions are due to NH_3 vibrations, as can be seen by checking the spectrum of $[Co(NH_3)_6]Cl_3$. Additional absorptions at 1460 and 1065 cm^{-1} are due to the ONO group; those at 1430 and 825 cm^{-1} characterize the NO_2 group. The latter, again, can be checked against the spectrum of $[Co(NO_2)_6]^{3-}$. Record the visible spectrum of a 0.01 *M* solution of chloropentaamminecobalt(III) chloride (see Exercise 7C).

Additional Work

This experiment is related to Exercise 14B, in which some of the steps of the preparation are investigated with regard to kinetics and the mechanism of the reaction.

Questions

The following questions are more easily answered if Exercises 1B and 14B have been performed. Otherwise consult Burmeister and Basolo (1968).

1. Why are Co(II) salts invariably used as starting materials in preparations of Co(III)

complexes? Why is oxidation always performed in the presence of the ligands?
2. $[Co(NH_3)_5Cl]^{2+}$ is a comparatively stable complex ion (stable enough to measure its conductivity in Exercise 12G) in a neutral solution. In spite of this, $Co(NH_3)_5Cl_3$ is used as the starting material in this experiment and the implication is that $[Co(NH_3)_5H_2O]^{3+}$ is formed. In which step of the preparation do you actually obtain this ion?
3. Why is it necessary to allow much more time for the preparation of the nitrito than for the nitro compound?

C. GEOMETRIC ISOMERS

Apparatus

Büchner filter
100-ml graduated cylinder

Chemicals

Hexaaquocobalt(II) chloride
10% aqueous ethylenediamine solution
30% aqueous hydrogen peroxide
Concentrated hydrochloric acid
Acetone
Ethyl alcohol
Ether

Introduction

In this experiment *cis*- and *trans*-dichlorobis(ethylenediamine)cobalt(III) chloride are studied as examples of geometric isomers. Generally, *cis* compounds are prepared by replacing a bidentate ligand with two monodentate ligands, which, because of the nature of the bidentate bond, leads to *cis* substitution exclusively. For the preparation of each *trans* compound a suitable method must be devised. In this experiment we will not follow the general procedure, but will instead exploit the difference in solubility of the *cis* and *trans* compounds in different media.

The green *trans*-$[Co(en)_2Cl_2]Cl$ is less soluble than the *cis* compound in an aqueous solution containing hydrochloric acid. The *trans* compound can therefore be readily prepared in a highly pure form. The *cis* compound, on the other hand, is the less soluble compound in a neutral solution. It can be prepared by the recrystallization of the *trans* complex from a neutral aqueous solution. However, substitution of the chloro for the aquo ligand in the coordination sphere is easily facilitated and crystallization from an even slightly basic solution, especially after extensive evaporation of water, will yield the reddish aquo complex.

In equilibrium, only the *trans* complex is present in methanol solution. The *cis* complex is converted to *trans* in a reaction of easily measurable speed. (See Exercise 14A).

Cis and *trans* compounds are characterized by their visible spectra, which are compared with the spectrum of tris(ethylenediamine)cobalt(III) chloride. (See Exercise 7C).

Preparation of *Trans*-dichlorobis(ethylenediamine)cobalt(III) Chloride

Add 60 ml of a 10% solution of ethylenediamine, while stirring, to a solution of 16 g of cobalt(II) chloride hexahydrate in 50 ml of water in a 400-ml beaker. While the solution is cooling, add 16 ml of a 30% solution of H_2O_2.

Add 35 ml of concentrated HCl and allow the solution to evaporate over a steam bath until a crust forms on the surface (the volume is reduced to approximately 75 ml). Allow the solution to cool overnight before collecting the bright green square plates of the *trans* form (to which some HCl still adheres) by filtration. Wash the plates with alcohol and ether and dry them at 60°–70°C. At this temperature the hydrogen chloride is lost, and the crystals crumble to a dull green powder.

Preparation of *Cis*-dichlorobis(ethylenediamine)cobalt(III) Chloride

Dissolve 5 g of *trans*-$[Co(en)_2Cl_2]Cl$ in a minimum amount (10 ml) of hot water contained in a 50-ml beaker. Place this solution over a steam bath and evaporate until crystals begin to separate. Break up the crust of crystals with a glass rod and allow the mixture to remain over the steam bath for another half hour. Remove the mixture before it has evaporated to dryness and cool it in an ice bath. Collect the product by vacuum filtration and wash it carefully with a minimum amount of water. Finally, wash with acetone and dry at room temperature.

Characterization

Observe and record the spectra of saturated solutions of the *cis* and the *trans* compounds in methanol between 300 mμ and 700 mμ. Record the visible spectra of both isomers in 0.01 *M* solutions in water (see Exercise 7C). How can the difference between these spectra and the spectrum of the tris(ethylenediamine) compound be explained in terms of the distortion originated by the introduction of Cl^- into the octahedral coordination sphere of Co(III)? The magnetic susceptibility will be measured in Exercise 9A.

Questions

1. What other methods can be used to identify *cis* and *trans* compounds?

D. PREPARATION OF AQUOPENTA-AMMINECOBALT(III) NITRATE

Apparatus

Büchner funnel
100-ml graduated cylinder

Chemicals

Hexaaquocobalt(II) nitrate
Ammonium carbonate
Concentrated ammonium hydroxide
30% aqueous hydrogen peroxide
Alcohol
Ether

Introduction

To investigate the kinetics of the formation of nitro- and nitritopentaamminecobalt(III) ions (see Exercise 14B) from the aquo complex and NO_2^- ion in solution, the pure aquo complex must be prepared. The aquo complex is formed readily in solution by the substitution of water for chloride in the coordination sphere of the chloropenta-amminecobalt(III) ion at increased temperatures. This is the first step in the preparation of nitro and nitrito complexes performed in Exercise 1B. However, this method is unsatisfactory for the purposes of this experiment since it does not readily yield a pure precipitate. Instead, carbonato-pentaamminecobalt(III) nitrate is prepared first—a compound that is easily prepared in a highly pure form. Then the carbonato complex is converted into the aquo complex by the acidification of the solution. The free carbonate is driven off as CO_2. The reaction proceeds quantitatively (because of the removal of CO_2) in so short a time that practically no other exchange reactions occur. The pure aquo complex is obtained by precipitation upon the addition of methanol.

Preparation of Carbonatopenta-amminecobalt(III) Nitrate

The reaction proceeds according to:

$$4Co(NO_3)_2 + 4(NH_4)_2CO_3 + 16NH_3 + O_2 \rightarrow 4[Co(NH_3)_5CO_3]NO_3 + 4NH_4NO_3 + 2H_2O$$

Mix 30 g of hexaaquocobalt(II) nitrate (in 15 ml of water) with 45 g of ammonium carbonate (in 45 ml of water) and 75 ml of concentrated ammonia (28% or specific gravity 0.9). Add 25 ml of 30% H_2O_2 solution. When the reaction has stopped, heat over a steam bath for 30 minutes. Bubble air through for another hour and cool the solution in an ice salt bath for several hours or overnight. Filter. Wash with 5 ml of ice-cold water, then with 5 ml of alcohol, and then with ether. A 50%–60% yield of deep red crystals should be obtained.

Recrystallization

Dissolve 12 g of the carbonatopentaammine salt in 30 ml of water at 90°C. Filter, and cool the filtrate in an ice salt bath for several hours. Filter the precipitate and wash with 3 ml of ice-cold water, followed by 3 ml of ice-cold alcohol, and then ether. Dry at 50°C.

Preparation of the Aquopentaammine Salt

To 5 g of the carbonato complex suspended in 12.5 ml of water, add 10 ml of colorless HNO_3 prepared from equal amounts of concentrated acid and water. After the evolution of CO_2 has stopped (approximately 10 minutes), add 50 ml of pure methanol. Collect the aquopentaammine-cobalt(III) nitrate on a filter and wash with alcohol and ether. A quantitative yield should be obtained. The crystals are a much lighter red than those of the carbonato complex.

Characterization

Record the visible spectrum of an 0.01 *M* solution of aquopentaamminecobalt(III) nitrate in water (see Exercise 7C).

REFERENCES

Basolo, F., and Pearson, R. G. 1958. *Mechanism of Inorganic Reactions*. New York: Wiley, pp. 254–256. 1967. 2d ed., pp. 291–292.

Broomhead, J. A., Dwyer, F. P., and Hogarth, J. W. 1957. In *Inorganic Syntheses*. Vol. 5. Ed. T. Moeller. New York: McGraw-Hill, p. 183.

Burmeister, J. L., and Basolo, F. 1968. In *Inorganic Preparative Reactions*. Ed. W. L. Jolly. New York: Wiley.

Faust, J. P., and Quagliano, J. V. 1954. *J. Am. Chem. Soc.* **76**, 5346. Spectra.

Walton, F. H. 1948. *Inorganic Preparations*. New York: Prentice-Hall, p. 92. $[Co(NH_3)_5ONO]Cl_2$ and $[Co(NH_3)_5NO_2]Cl_2$.

Willard, H. H., and Hall, D. 1922. *J. Am. Chem. Soc.* **44**, 2220. $[Co(NH_3)_5Cl]Cl_2$.

Work, J. B. 1946. In *Inorganic Syntheses*. Vol. 2. Ed. Y. C. Fernelius. New York: McGraw-Hill, pp. 221–222.

EXERCISE 2

High-Temperature Preparations

A. PREPARATION OF COBALT FERRITE

Apparatus

Magnetic stirrer and hot plate combination
Teflon-covered stirring bar
N_2 tank
Aspirator
Schlenk apparatus:
  500-ml three-necked flask
  Fritte
  Dropping funnel
  Transfer cross

Chemicals

Iron powder, reagent grade
Cobalt acetate, reagent grade
Acetic acid, 1:3 dilute
Oxalic acid (dihydrate), reagent grade

Introduction

Very pure ferrites are prepared in large quantities by sintering the well-mixed metal oxides in stoichiometric proportions at a temperature of approximately 1000°C. Ferrites are technically important substances because of their magnetic properties. They are ferrimagnetic and their susceptibilities depend decisively on their stoichiometry and the oxidation state of the metals. Sintering, therefore, has to be conducted in an atmosphere that matches the composition of the equilibrium gas phase at sintering temperature to avoid reduction or oxidation of the metals. The intimacy of mixing is also of importance if a single-phase product of desired uniformity is to be obtained. Ferrites can be prepared by the much more convenient method described herein [see also Tyree (1967)] if small quantities suffice for the purpose of further investigation. In this experiment the method for the preparation of cobalt ferrite is described. Nickel ferrite can be obtained in exactly the same way.

The method is altered slightly, however, for the preparation of many other ferrites (Tyree 1967).

Cobalt oxalate and ferrous oxalate are coprecipitated in correct stoichiometric proportions from acetic acid solution. The mixed oxalates are ignited at temperatures exceeding 800°C in the presence of air. The intimate mixing necessary in the technical process is replaced by coprecipitation in this procedure. Stoichiometry is achieved by the addition of exactly the correct amounts of cobalt and iron ions. Care should be taken that no iron is oxidized to the ferric state while in contact with water since ferric oxalate is soluble and might alter the stoichiometry. Thus, the wet reaction must take place in a protective nitrogen atmosphere.

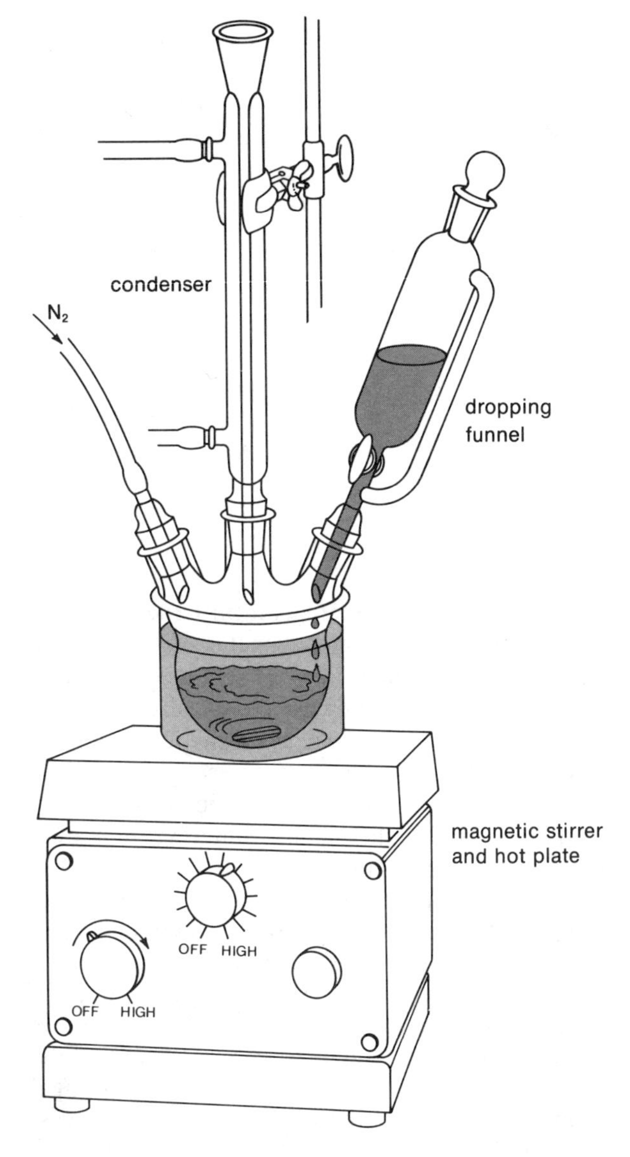

FIGURE 2-1
Apparatus for preparation of cobalt ferrite. A flat-bottomed flask placed directly on the hot plate may be used instead.

Procedure

Transfer exactly 0.0600 mole of iron powder and 0.0300 mole of cobalt acetate into a three-necked flask. Add 300 ml of acetic acid (1:3 dilute) and bubble nitrogen through the solution. Assemble the apparatus in Figure 2-1. Fill the dropping funnel with a solution of 11.7 g of oxalic acid (dihydrate) in 75 ml of distilled water through which nitrogen has been bubbled previously. Heat and stir while admitting nitrogen until all of the iron powder is dissolved. This step takes at least 4 hours. Keep the solution boiling during this time. After the iron powder has been dissolved, rapidly add the oxalic acid solution to the boiling contents of the flask, allow it to digest for 10 minutes, and replace the water bath with an ice bath to cool the solution. Maintain sufficient nitrogen flow at all times to prevent the admission of air through the condenser. When the solution is cold, quickly replace the funnel with a Schlenk transfer cross to which the fritte has been connected. Stopper the open ends of the transfer cross and replace the condenser with a stopper. Open the sidearm of the fritte and meticulously transfer the precipitate under nitrogen flow into the fritte. Figure 2-2 illustrates the transfer process. Connect the nitrogen to the sidearm of the fritte and set up the fritte for filtering as shown in Figure 3-7. Filter. Wash the filtrate with three 25-ml portions of cold water, and rinse with acetone.

Gather the precipitate in a flat dish (preferably platinum), and ignite it in a muffle oven with the door somewhat ajar at 800°C or slightly higher. Weigh the product and calculate the yield.

Characterization

Cobalt ferrite is soluble in hot concentrated HCl and can be analyzed chemically for its Co^{2+} and Fe^{3+} content. The analysis can be performed either by complexation titration (see Exercise 13A) or by spectrophotometry following separation of

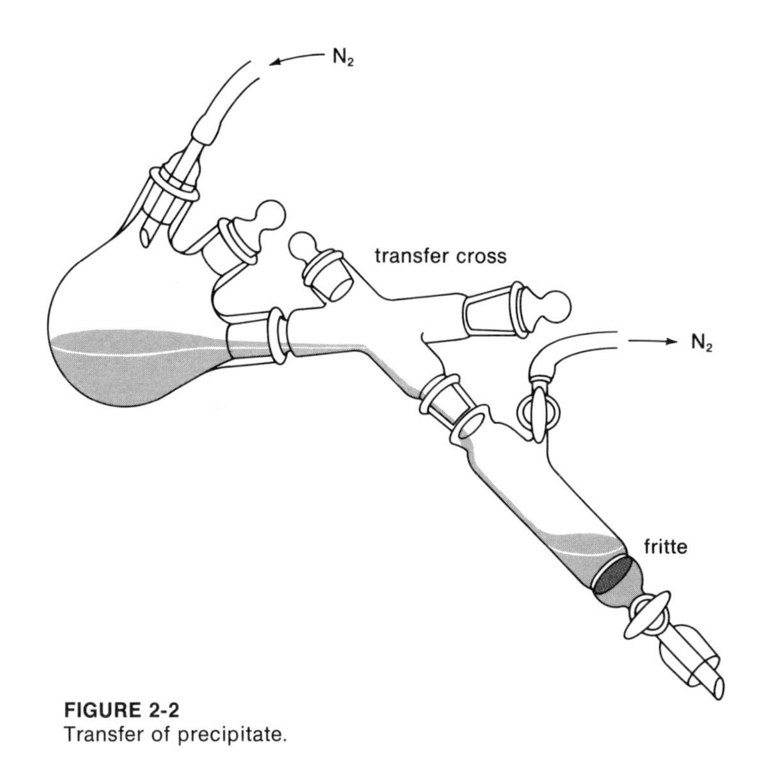

FIGURE 2-2
Transfer of precipitate.

the ions by ion exchange chromatography (see Exercise 13B).

Ferrites crystallize in the spinel structure. Obtain the X-ray powder pattern (see Exercise 10). Determine the unit cell dimensions, measure the density, and calculate the atomic distances in the cell (see Exercise 10C and D).

You may also compare the X-ray powder pattern of cobalt ferrite with those of Fe_2O_3 and CoO. Judging from the powder patterns would you describe $CoFe_2O_4$ as a mixture of oxides or as a unique substance?

Additional Work

The measurement of saturation magnetization enables us to distinguish between normal and inverse spinels (Gorter 1954), if we accept Néel's theory that ferrimagnetic substances possess two sublattices. Néel's theory postulates that all spins within one sublattice are parallel at saturation magnetization and that they are antiparallel to those in the other sublattice. The difference of spins in the two sublattices determines the measured saturation magnetization. The terms "normal" and "inverted" spinels are defined in Exercise 10D.

B. PREPARATION AND HYDROLYSIS OF A TRIPOLYPHOSPHATE

Apparatus

Platinum crucible
Mortar and pestle
Oven at 580°C
Drying oven at 80°C
*p*H meter with glass and standard calomel electrodes

Chemicals

Dibasic sodium phosphate
Monobasic sodium phosphate
Concentrated nitric acid

Introduction

Structures of polyphosphates can be visualized as chains or rings of phosphate tetrahedra in which one or more corners (oxygen atoms) are shared

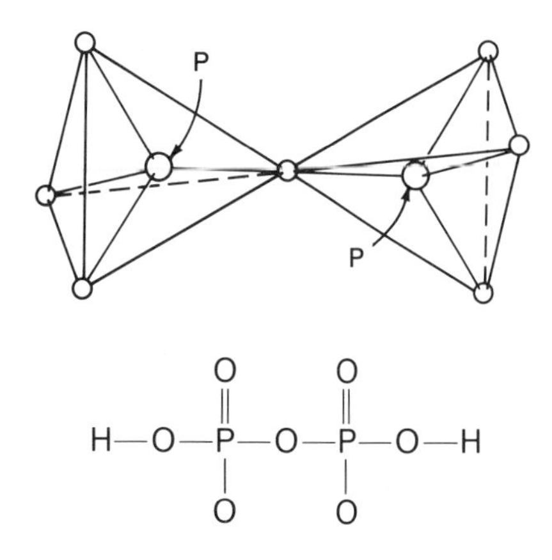

FIGURE 2-3
Two PO_4^- tetrahedra linked at one corner to form the $P_2O_7^{2-}$ ion.

between two phosphorus centers. In pyrophosphoric acid, $H_4P_2O_7$, only one corner is shared (see Figure 2-3).

Branched polyphosphates, which contain at least one phosphate tetrahedron with three shared corners, are extremely unstable in aqueous solution and hydrolyze immediately at one of the three corner bonds. Chain and ring phosphates also hydrolyze easily. This causes the scission of adenosine triphosphate, which forms adenosine diphosphate and orthophosphate, to be a major source of free energy in biological processes [see Klotz (1967)].

Many polyphosphates can be prepared by the dehydration of solid phosphates at an increased temperature, with the nature of the main product depending on the stoichiometry of the polyphosphate mixture. Any corner of the tetrahedron to which a sodium is affixed is nonreactive. Dehydration and fusion take place at those corners which carry hydrogen. Thus sodium tripolyphosphate is a product of the following reaction:

Since the starting materials may react somewhat differently, one of the tasks of this experiment is to find out as much as possible about the reactions of the products obtained.

The most common methods for the analysis of polyphosphate mixtures are:

1. Determination of average chain length by *p*H titration
2. Paper chromatography, a powerful method for the separation of polyphosphates, which is especially useful because it has been found that the logarithm of the R_f value is linearly related to the chain length

Polyphosphates are known for their chelating abilities and are used as water softeners because of their strong complex formation with metal ions such as Ca^{2+}. Even the sodium salts are not completely dissociated in aqueous solution. The industrial and biological importance of polyphosphates has provided a stimulus for research in this area of chemistry.

Preparation

Prepare an intimate mixture of monobasic sodium phosphate (NaH_2PO_4) and dibasic sodium phosphate (Na_2HPO_4) in an exact molar ratio of 1:2 by grinding the constituents together in a mortar. Start with enough material to yield 0.01 mole of tripolyphosphate. The purity of the product depends decisively on the molar ratio of the reactants and their degree of mixing. Put the mixture into a platinum crucible and heat at 540°–580°C for 2 hours. Allow the product to cool at room temperature.

What are the possible byproducts of the reaction? Assume that chains or rings of more than five phosphorus atoms are not formed in any measurable quantity. What is the stoichiometry for each of the side reactions, taking into account the stoichiometry of the initial mixture?

$$\text{Na—O—}\underset{\substack{|\\ \text{O}\\ |\\ \text{Na}}}{\overset{\substack{\text{O}\\ \|}}{\text{P}}}\text{—O—H} + \text{H—O—}\underset{\substack{|\\ \text{O}\\ |\\ \text{Na}}}{\overset{\substack{\text{O}\\ \|}}{\text{P}}}\text{—O—H} + \text{H—O—}\underset{\substack{|\\ \text{O}\\ |\\ \text{Na}}}{\overset{\substack{\text{O}\\ \|}}{\text{P}}}\text{—O—Na} \rightarrow \text{Na—O—}\underset{\substack{|\\ \text{O}\\ |\\ \text{Na}}}{\overset{\substack{\text{O}\\ \|}}{\text{P}}}\text{—O—}\underset{\substack{|\\ \text{O}\\ |\\ \text{Na}}}{\overset{\substack{\text{O}\\ \|}}{\text{P}}}\text{—O—}\underset{\substack{|\\ \text{O}\\ |\\ \text{Na}}}{\overset{\substack{\text{O}\\ \|}}{\text{P}}}\text{—O—Na} + 2H_2O$$

The purity of the products will be analyzed later by end-group titration and paper chromatography (see Exercises 12F and 11C).

Hydrolysis

The hydrolysis of tripolyphosphate proceeds according to

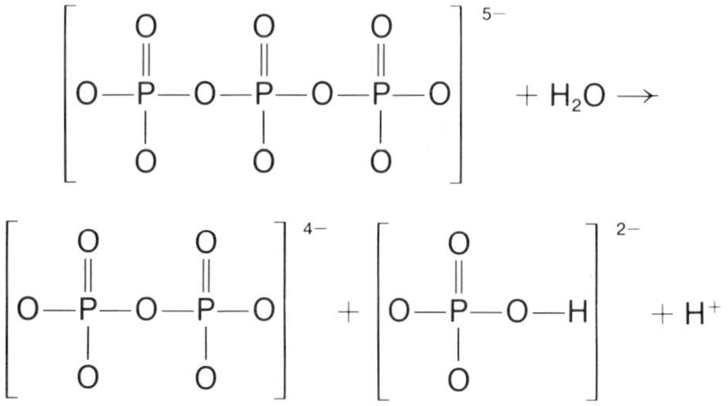

$$\left[\begin{array}{ccccccc} & O & & O & & O & \\ & \| & & \| & & \| & \\ O- & P & -O- & P & -O- & P & -O \\ & | & & | & & | & \\ & O & & O & & O & \end{array}\right]^{5-} + H_2O \rightarrow$$

$$\left[\begin{array}{ccccc} & O & & O & \\ & \| & & \| & \\ O- & P & -O- & P & -O \\ & | & & | & \\ & O & & O & \end{array}\right]^{4-} + \left[\begin{array}{ccc} & O & \\ & \| & \\ O- & P & -O-H \\ & | & \\ & O & \end{array}\right]^{2-} + H^+$$

The reaction possesses a half-life ($\tau_{1/2}$) of 106 minutes at pH 2 and 85°C. Simultaneously, hydrolysis of pyrophosphate and higher poly- and metaphosphates[1] occurs. Fortunately, the rates of these reactions are slower than that of tripolyphosphate; that of pyrophosphate is similar to the rate of tripolyphosphate hydrolysis only at higher temperatures.

It is desirable to hydrolyze enough tripolyphosphate to obtain almost stoichiometric amounts of all phosphate species. Since prolonged reaction leads to excessive phosphate production, the reaction time should be limited to 80 minutes at pH 2 and 80°C.

The reaction can take place in the oven, set at 80°C, to be used later for drying paper chromatograms. Prepare approximately 20 ml of 0.05 M reaction product, calculating the molarity under the assumption that the starting material was pure tripolyphosphate. Adjust the pH by adding several drops of concentrated HNO_3, and follow the change in pH on a pH meter. The final pH should be between 2.0 and 2.5. Hydrolysis seems to be accelerated by a further drop in pH; therefore, adjust the pH carefully. Stop the reaction after 80 minutes by putting the solution in an ice bath and/or in the refrigerator.

The presence of tripolyphosphate, its hydrolysis products, and possible by-products of the reaction can be demonstrated by paper chromatography (see Exercise 11C).

If less hydrolysis of pyrophosphates is desired, the reaction must take place at a lower temperature for a longer period of time. The required time may be estimated by assuming that the reaction rate doubles for every 10°C.

[1]Ring polyphosphates are called metaphosphates.

$$\text{Trimetaphosphate} - \left[\begin{array}{ccccccc} & & O & & O & & \\ & & \| & & \| & & \\ O & - & P & -O- & P & - & O \\ & & \backslash O & & O/ & & \\ & & & P & & & \\ & & O // & & \backslash O & & \end{array}\right]^{3-}$$

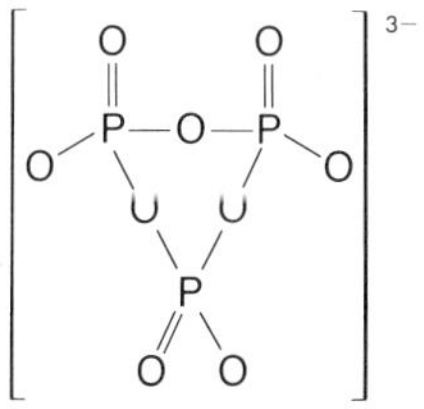

REFERENCES

Cotton, F. A., and Wilkinson, G. 1966. *Advanced Inorganic Chemistry*. New York: Wiley, pp. 396–400.

Gorter, C. I. 1954. *Philips Res. Rept.* 9, 241–320.

Jolly, W. L. 1960. *Synthetic Inorganic Chemistry*. Englewood Cliffs, New Jersey: Prentice-Hall.

Klotz, I. M. 1967. *Energy Changes in Biochemical Reactions*. New York: Academic Press.

Thilo, E. 1962. In *Advance in Inorganic Chemistry and Radiochemistry*. Vol. 4. New York: Academic Press.

Tyree, S. Y., Ed. 1967. *Inorganic Syntheses*. Vol. 9. New York: McGraw-Hill.

van Wazer, J. R. 1958. *Phosphorus and Its Compounds*. Vol. 1. New York: Wiley.

EXERCISE 3

Preparation of Air-Sensitive Substances

A. INERT-ATMOSPHERE TECHNIQUES

Manipulation of air- or moisture-sensitive solids or liquids is frequently and conveniently done in a glove box. A glove box consists essentially of a gastight box with a window, through which operations can be observed, and rubber gloves, which allow the handling of substances and apparatus in the interior (see Figure 3-1). The box is flushed with an inert gas after all necessary equipment and substances have been introduced. Reactive samples can then be manipulated in the customary way. Since it takes considerable time to flush a large box, most of them are fitted with a transfer lock, which can be rapidly evacuated and refilled with inert gas. Thus, materials can be introduced into the box without flushing the entire box.

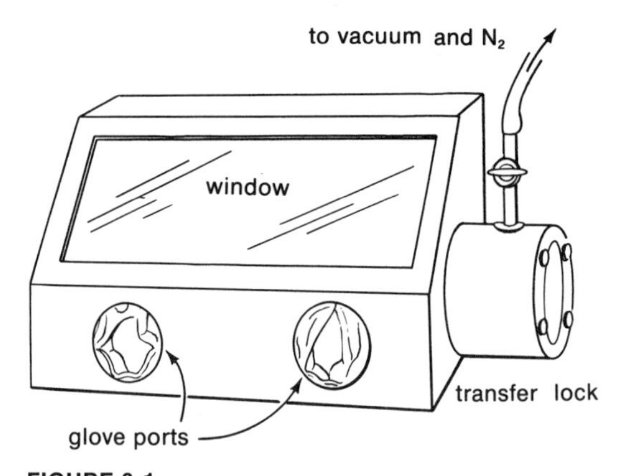

FIGURE 3-1
Schematic drawing of a glove box.

Boxes for very exact work are quite expensive and time-consuming to use. The best ones contain a recirculating purification system for the elimination of oxygen and moisture that enter by diffusion through the rubber gloves, even though the box is completely airtight. Thus, for moderately sensitive compounds it is much more economical and less time-consuming to use a glove bag, which consists of a polyethylene bag with a nitrogen inlet and built-in gloves. One end of the bag is open for the introduction of substances and apparatus. The open end is rolled up and clamped while work is being done, and a small flow of nitrogen maintains the inert atmosphere. The bag is flushed by flattening it, rolling it up tightly, and then inflating it with nitrogen. This procedure is repeated several times.

In many preparations the use of special glassware is preferred to the manipulation of normal apparatus in a glove box, although some operations can be performed only by using the latter. Standard glassware fitted with a nitrogen inlet and bubbler can be used for such operations as refluxing and distillation, but specialized apparatus is needed if crystallization, filtration, or solid and liquid transfer are included in the preparation. Apparatus and techniques have largely been developed by organometallic chemists like Schlenk, Hein, and Herzog.

The essential feature of any apparatus used in working with air- or moisture-sensitive solids or liquids is a sidearm with a stopcock, by which the apparatus can be evacuated and filled with inert gas. The suction does not need to be very strong since the flushing procedure is repeated several times. In the examples given in this manual (see Exercises 2A and 3B), an aspirator will do the job. A slight flow of nitrogen is maintained during some operations. The glassware required for performing the next part of this exercise is shown in Figures 3-2–3-8.

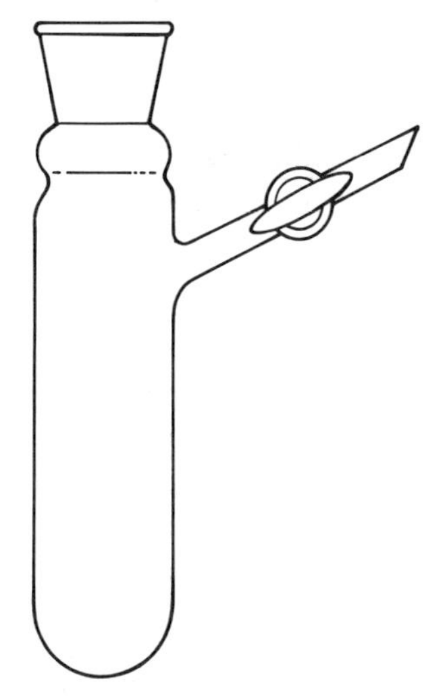

FIGURE 3-2
Schlenk tube serves as a reaction vessel and filtrate receiver. [From Shriver (1969).]

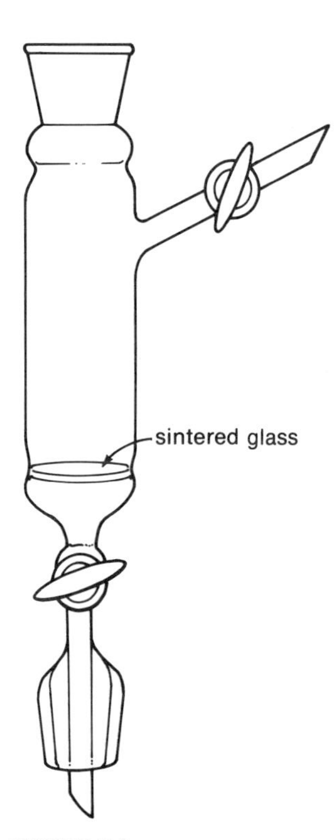

FIGURE 3-3
Fritte used for filtrations. [From Shriver (1969).]

B. PREPARATION OF CHROMOUS ACETATE

Apparatus

1000-ml beaker (tall type for ice bath)
N_2 tank
Aspirator
Schlenk apparatus:
- 3 reaction tubes
- Dropping funnel
- Fritte
- Transfer cross
- Solids container and receiver

Chemicals

Potassium dichromate, reagent grade
Mossy zinc, reagent grade
Concentrated hydrochloric acid
Sodium acetate, reagent grade
Alcohol (pure ethanol or methanol)
Ethyl ether

Procedure

Because the Cr(II) ion is easily oxidized by air while in solution, the reactions should take place in a Schlenk apparatus. However, if the product

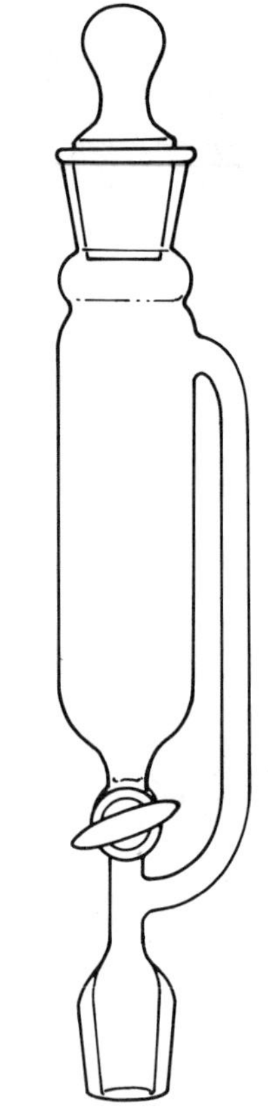

FIGURE 3-4
Dropping funnel.

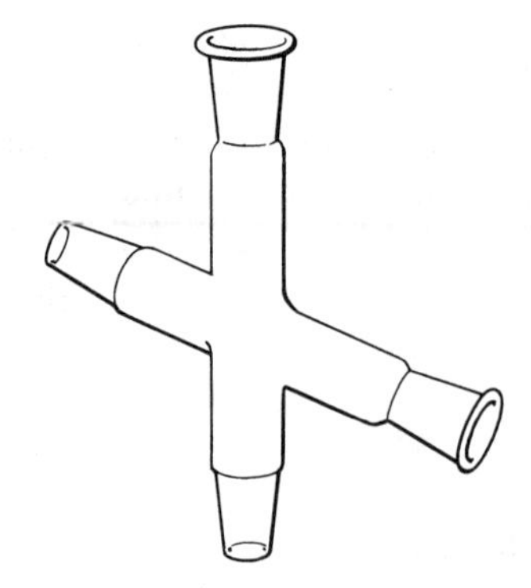

FIGURE 3-6
Transfer cross. [From Shriver (1969).]

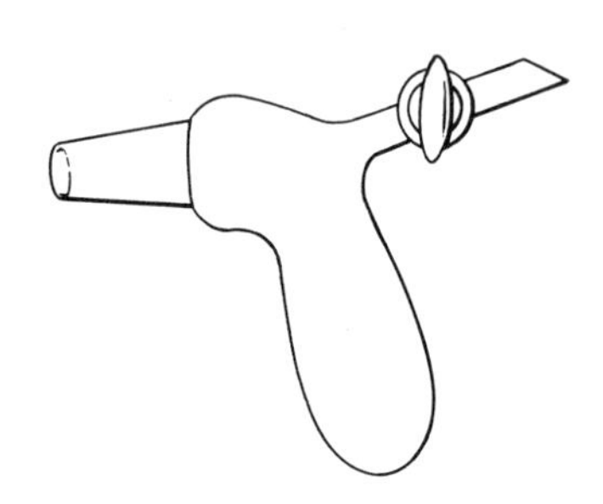

FIGURE 3-5
Solids container and receiver. [From Shriver (1969).]

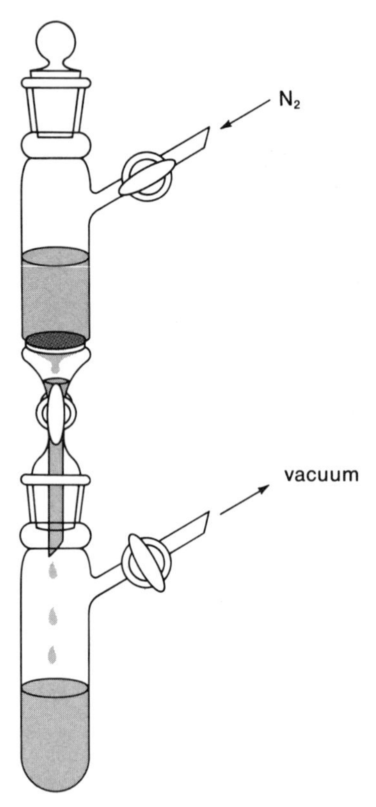

FIGURE 3-7
Filtration. [From Shriver (1969).]

is pure and kept dry, it will remain stable for months in a moisture-free atmosphere without undergoing oxidation.

Put 100 ml of distilled water and 50 ml of alcohol in an icebath to use as cold washing materials at the end of the experiment. Place 30 g of reagent grade, mossy zinc (anything else contains too much carbon) and 15 g of $K_2Cr_2O_7$ in reaction tube *C* and set up the apparatus as shown in Figure 3-9. Grease all joints and stopcocks well, and use rubber bands to secure the joints. Fill the dropping funnel with 200 ml of concentrated HCl.

Attach the nitrogen cylinder to stopcock B' and the aspirator to stopcock D'. Close all stopcocks except D' and evacuate the system until gas is evolved from the hydrochloric acid solution.

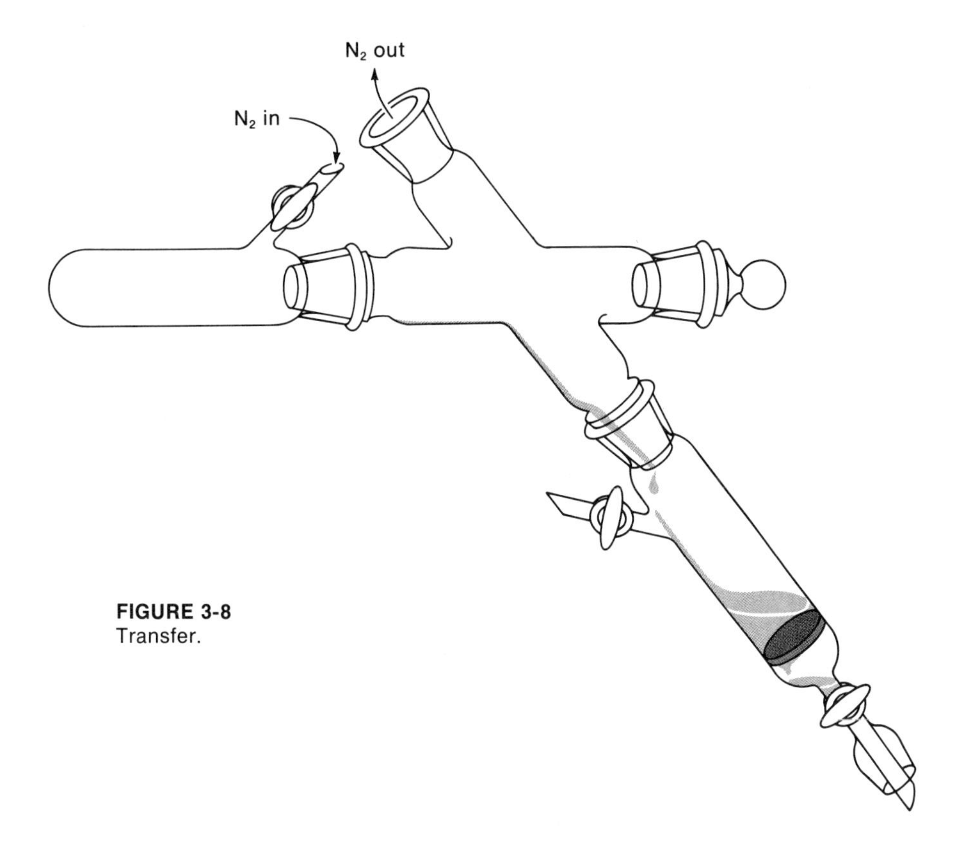

FIGURE 3-8
Transfer.

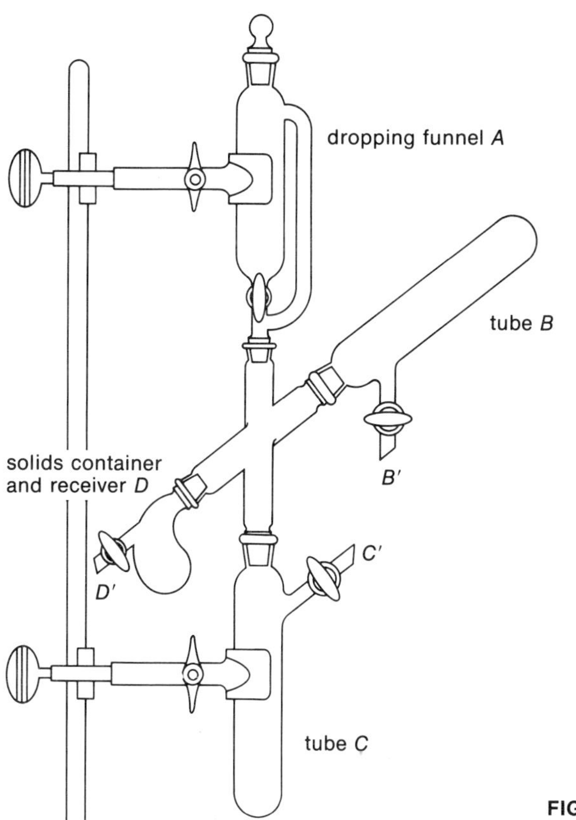

Open B' and flush with N_2; repeat this procedure three times. Fill the apparatus with nitrogen by closing stopcock D'. Quickly remove the hose leading to the aspirator from closed stopcock D'. Cover the opening with your finger, open stopcock D', and feel (by a reduction of the suction on your finger) the increase in pressure caused by the admission of N_2 through stopcock B'. Release your finger as soon as there is a positive pressure difference. Then attach a hose from a beaker of water to D' so that the gas flow from D' can be controlled according to the rate at which bubbles appear in the water.

Add HCl from the dropping funnel a little at a time until the reaction is underway. Continue the addition of HCl until the solution is clear blue, without a tinge of green. The source of nitrogen may be removed as the reaction proceeds, since the H_2 evolved protects the solution.

Meanwhile, prepare a solution of 100 g of sodium acetate in 90 ml of distilled water in another

FIGURE 3-9
Schlenk apparatus set up for preparation of chromous acetate.

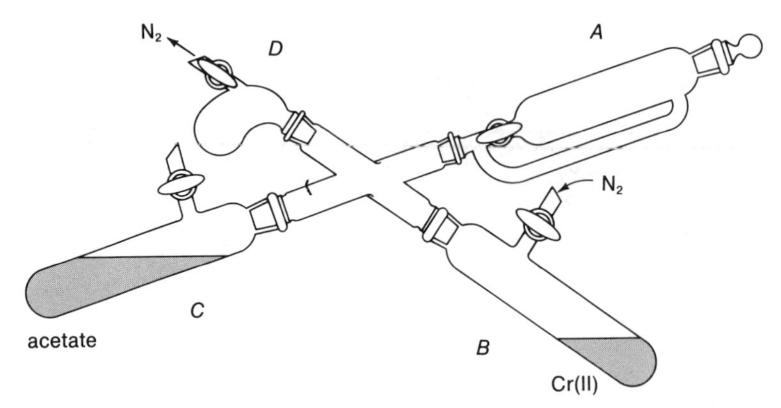

FIGURE 3-10
Position of Schlenk apparatus ready for precipitation.

Schlenk tube *C*. Bubble nitrogen through the solution vigorously for at least 5 minutes. Cover or stopper the solution and retain for later use.

When the reaction in the Schlenk apparatus nears completion, reattach the nitrogen tank to *B′* and let the gas escape at *D′*. Pour the Cr(II) solution into tube *B* under the nitrogen flow so that zinc that has not reacted and possibly carbon that may be present are trapped in sidearm *C′*. Increase the flow of nitrogen and remove tube *C*, replacing it with the tube containing the sodium acetate solution.

Keep the apparatus in the position shown in Figure 3-10, and aspirate and flush with nitrogen as in the beginning of the experiment. Repeat three times.

Tilt the apparatus back into the starting position, thereby causing the Cr(II) solution to flow into the acetate solution. Do this fast enough to make the two solutions mix well. The brick-red chromous acetate precipitates immediately. Cool the precipitate in the mixed solutions in an ice bath. While it is cooling, pass N_2 through *C′* at an increased rate, replace tube *B* with the fritte, and remove the dropping funnel and stopper. Connect the nitrogen flow to sidearm *B′* of the fritte. Aspirate through *D′* and rinse with nitrogen through *B′* three times as before. Remove the solids container and stopper. Let the N_2 escape through *C′*. After the precipitate is completely cold, tilt the apparatus so that the precipitate and liquid run into the fritte. Remove the fritte and clamp it in an upright position to a ring stand. Connect a Schlenk tube to the bottom and aspirate through the tube while continuing to add nitrogen at the sidearm of the fritte. Put a tissue into the opening of the fritte to prevent air convection. After the precipitate is dry, wash with three 25-ml portions of cold water, adding the new water only after the previous portion has disappeared. Then add two 25-ml portions of alcohol and, finally, 25 ml of ether. Dry the precipitate using suction. Any precipitate that has gathered on the wall of the fritte must be tamped down with a glass rod so that all of the solid is completely dry at the end of the procedure.

Remove the product from the fritte by using a glove bag filled with nitrogen. As a test of the quality of the product determine the magnetic susceptibility directly after preparation and again a few weeks later (see Exercise 9A). Keep the product in a dry, tightly closed bottle with a threaded cap that can be sealed with plastic adhesive tape.

Questions

1. Could iron be substituted for zinc as the reducing agent in the chromous acetate preparation? Could tin? Give reasons for your answers, referring to a table of standard electrode potentials.
2. How would you prepare a specimen of chromous chloride? What would be the chief difficulty of this preparation?
3. Since the hydrated chromium(II) is blue, it is somewhat surprising that the acetate is brick red. Refer to Van Niekerk (1953) and Van Niekerk, Schoening, and de Wet (1953), and draw the structure of chromium(II) acetate.

Knowing its structure, is the color still as much of a surprise? How does this structure compare to that of copper(II) acetate? How does knowledge of these structures permit an understanding of the low magnetic susceptibilities of the compounds?

REFERENCES

Audrieth, L. F., Ed. 1950. *Inorganic Syntheses*. Vol. 3. New York: McGraw-Hill, p. 148.

Booth, H. S., Ed. 1939. *Inorganic Syntheses*. Vol. 1. New York: McGraw-Hill, p. 122.

Rochow, E. G., Ed. 1960. *Inorganic Syntheses*. Vol. 6. New York: McGraw-Hill, p. 145.

Shriver, D. F. 1969. *The Manipulation of Air-Sensitive Compounds*. New York: McGraw-Hill.

Van Niekerk, J. N. 1953. *Acta Cryst*. **6**, 501.

Van Niekerk, J. N., Schoening, F. R. L., and de Wet, J. F. 1953. *Nature*. **171**, 36.

EXERCISE 4

Vacuum-Line Preparations

A. VACUUM-LINE TECHNIQUES

Vacuum techniques are applicable to chemistry mainly in two areas: (1) the preparation and characterization of gaseous or air-sensitive, volatile compounds and (2) gas-phase kinetics or gas-surface reactions. Although the purposes for which vacuum lines are constructed may be quite different, they are made from the same material and a few general principles of construction apply to all of them.

Construction

A good vacuum line should meet the following requirements:

1. Air or other forms of contamination entering one part of the system should not affect other parts. Therefore, the vacuum line should be constructed in separate manifolds, each suited to its purpose and easily separated from the rest of the system by stopcocks.
2. Each part of the system must have direct access to the pump or high-vacuum manifold. There should be as little dead volume as possible between each part and the device for measuring pressure. This requirement is imperative for effective leak detection.
3. U-bends or traps should be provided wherever necessary to prevent the loss of a substance by accidental access to the wrong section of the system. The substance is recovered by trap-to-trap transfer.
4. Since the speed with which parts of the apparatus are evacuated affects the amount of time it takes to carry out a procedure, the following conditions should be met:
   a. The various parts should be as close to one another as possible.
   b. Tubing and stopcock bores should have large inner diameters.
   c. The number of stopcocks should be minimal (they are the most common source of leaks).
   d. Joints that are permanent or infrequently removed should be sealed with wax.
5. Pressure gauges should be placed so that the

vacuum attained in the system is measured, rather than the pressure adjacent to a very efficient pump. A mercury manometer for measuring higher pressures (~1 mm) should be accessible to all parts of the system that may contain the substance to be worked with.

Requirements 1 through 5 are easily met by the construction of a system around a single section of large-diameter tubing, directly connected to the pump, from which all other sections can be evacuated. The vacuum line used for the preparation of gaseous hydrides in Sections B and C of this exercise is shown in Figure 4-1. The line is held under the best vacuum the pump system can offer between stopcocks *B* and *K*. Stopcock *K* remains open at all times, unless air should enter the main vacuum line, which is very unlikely since stopcocks *B*, *H*, and *J* separate it from the working manifolds. These stopcocks remain closed after the total system has been evacuated, unless a separate section needs evacuation during the experiment. A large trap to the right prevents vapors from the diffusion pump from entering the system and condensable gases from the system from getting into the diffusion pump while the experiment is in progress. The working manifold below the main vacuum line consists of two sections. Section 1, between stopcocks *A*, *B*, and *C*, contains the reaction vessel and can be flushed with N_2 while the reactants are introduced. It has access to a bubbler manometer through stopcock *A*. A bubbler manometer is a device that limits the pressure to about atmospheric pressure. It is evacuated with the entire system. The mercury in the right (permanently evacuated) reference arm falls until it reaches the level of the mercury in the reservoir during evacuation. When N_2 is admitted through *A* (a three-way stopcock) to the reaction section and the bubbler manometer, the mercury in the reference arm rises until the pressure in the system has reached atmospheric pressure (plus an additional pressure caused by the mercury through which the gas escapes—see Figure 4-1), at which time N_2 escapes through the reservoir.

Section 2, between stopcocks *B*, *H*, and *J*, is designed for the purification of the substance by means of trap-to-trap distillation, vapor-pressure measurement at temperatures controlled by slush baths, and the transfer of a desired quantity to a storage vessel or infrared cell by means of three-way stopcock *G*. The mercury manometer is placed in this section because it is used for vapor-pressure determinations and filling the infrared cell.

Elements

Pumps

Mechanical and diffusion pumps are usually employed together. A rotary oil pump (mechanical) and a mercury diffusion pump are shown in Figure 4-2. The section of the rotary pump shown in Figure 4-2A is filled with oil. A vane is held on top of the eccentric rotor, separating the inlet and outlet at all times. As the rotor moves, the volume connected to the inlet expands and air is drawn in. At the same time the volume containing the outlet is compressed and air, which was previously captured in it, escapes through the outlet. The ball in the outlet prevents oil from flowing back. This kind of pump becomes inefficient with falling pressure because it depends on compression and expansion to function.

The diffusion pump, on the other hand, works only if the pressure is sufficiently low to allow free passage for the mercury atoms that are boiled off the reservoir. Under these circumstances it is possible to obtain a dense stream of mercury atoms directed downward in the nozzle section (*N*). The mercury atoms in the nozzle section collide with the gas molecules from the vacuum line and transfer their momentum to the latter, which, consequently, collect in the lower part of the pump and are transported out of the system by the forepump.

The diffusion pump, then, takes over when the rotary pump becomes inefficient. A cooled trap between the rotary and diffusion pumps prevents mercury vapors from entering the forepump. Another trap separates the diffusion pump from the main vacuum line.

Figure 4-3 shows the pumping speed of both types of pump as a function of pressure.

Pressure gauges

Three types of gauges used to measure pressures below the range of a normal mercury manometer are: a McLeod gauge, a thermocouple gauge, and an ionization gauge. The last two are very convenient to use but must be calibrated for each gas in precision work, whereas the first provides a means of absolute measurement, but only for noncondensable gases.

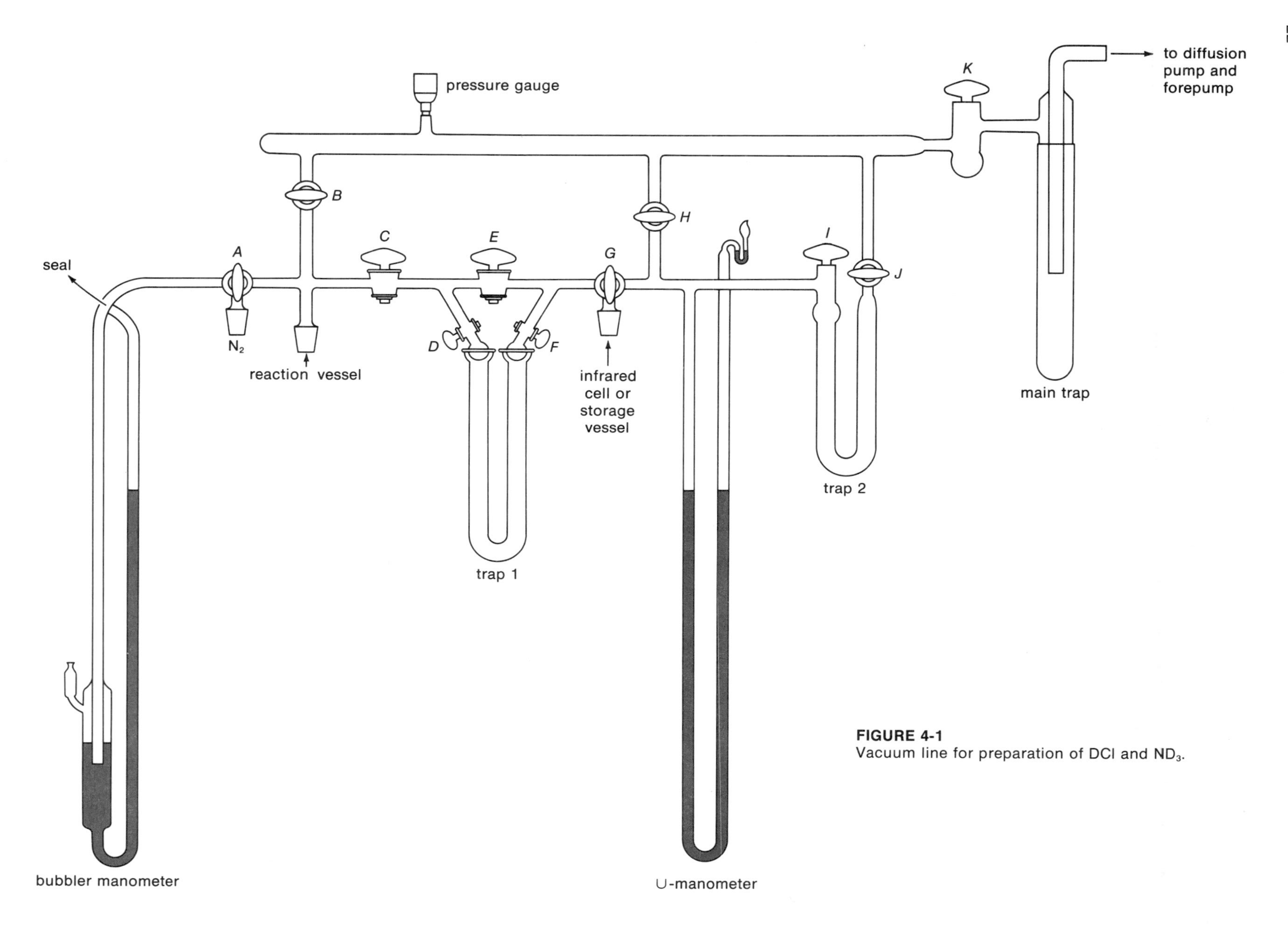

FIGURE 4-1
Vacuum line for preparation of DCl and ND_3.

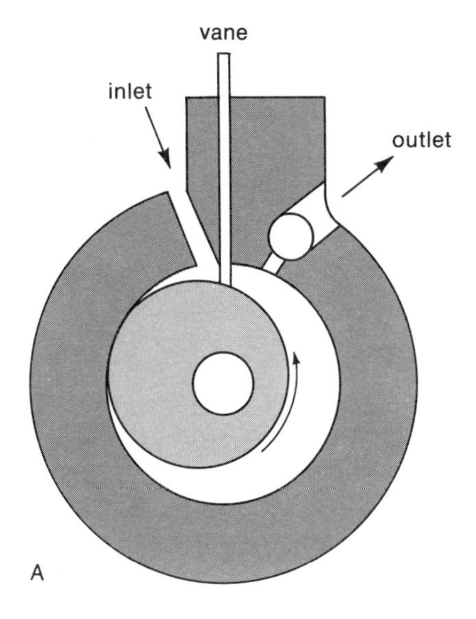

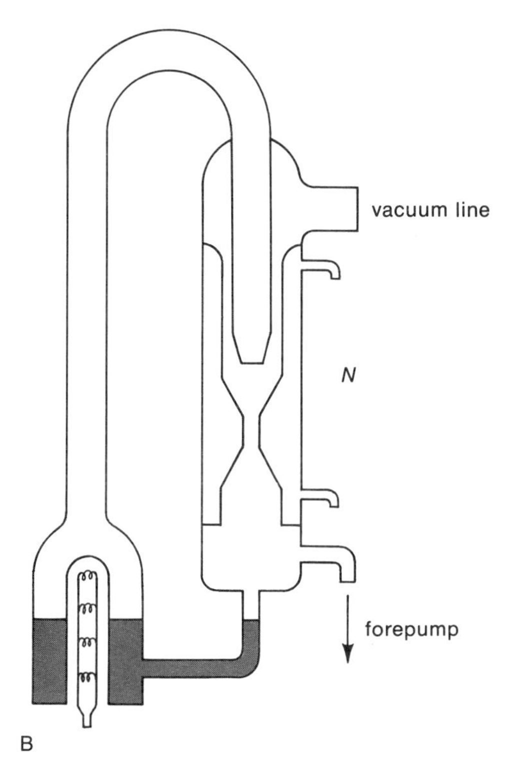

FIGURE 4-2
A. Rotary oil pump. B. Mercury diffusion pump.

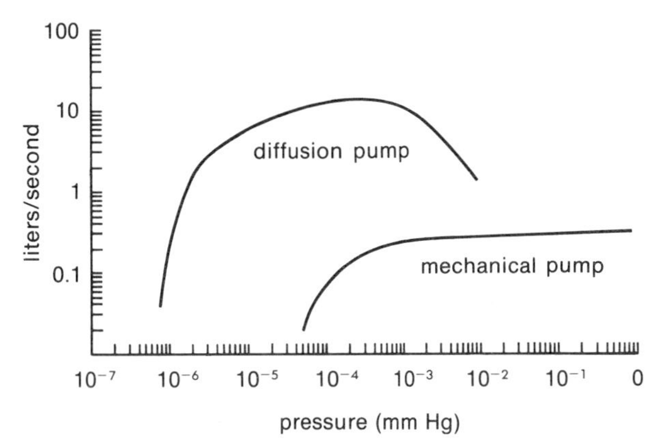

FIGURE 4-3
Pumping speed of a typical two-stage mechanical pump and a single-stage diffusion pump. [From Shriver (1969).]

Operation of the McLeod gauge is based on the compression of a large volume of gas of very low pressure, resulting in a small volume of gas of measurable pressure (see Figure 4-4). The mercury in the reservoir can be regulated by pumping or by the addition of air through the air inlet under the control of three-way stopcock S. As long as the mercury stays below level A, the volume (V) of the bulb connected to the closed capillary is filled with gas at pressure P_i, the pressure in the vacuum line. As soon as the mercury rises above A, the gas is trapped and finally compressed into the closed capillary. When the mercury in the reference capillary reaches point O (the height at which the bore of the closed capillary ends—see Figure 4-4), the mercury level in the closed capillary is read at point B. Thus the volume, V_e, of trapped gas is given by the inner cross section, C_i, of the closed capillary and the length filled with gas, which is equal to B, since the scale begins at O. The pressure of trapped gas is also equal to B in millimeters of mercury.

$$V_e = BC_i$$

and

$$P_e = B$$

Under the assumption of ideal behavior of the gas

$$P_i V = P_e V_e = B^2 C_i$$

and the initial pressure in the vacuum line is obtained as

$$P_i = \frac{B^2 C_i}{V}$$

in milliliters of mercury. Since C_i/V is known from previous calibration, the pressure in the vacuum line can be calculated.

The thermocouple and the Pirani gauges employ the pressure dependence of thermal conductivity.

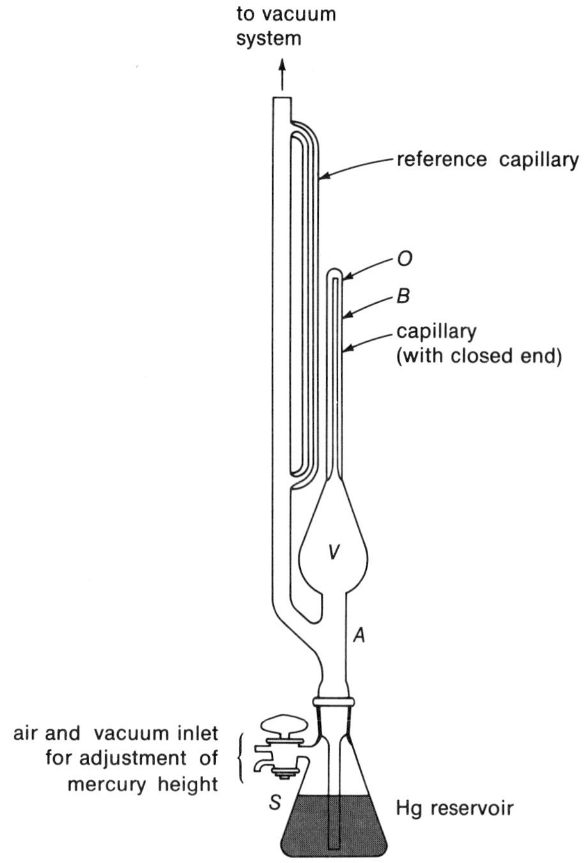

FIGURE 4-4
McLeod gauge. [From Shriver (1969).]

Since thermal conductivity is independent of pressure unless the free path length of molecules approaches the distance between cold and hot surfaces, it is only suitable for low pressures.

The thermocouple gauge measures the temperature of a heated wire by means of a thermocouple attached to it. The Pirani gauge uses a bridge arrangement to determine the temperature by means of the resistance of the wire. The temperature of the wire in turn depends on the thermal conductivity of the gas.

Ionization gauges are of two types: thermionic (hot cathode) or Phillips (cold cathode). In both types gas molecules are ionized. The measured ionization current depends on the gas pressure. In thermionic gauges electrons are emitted from a hot cathode and accelerated in an electric field. The ionized molecules are collected on a plate anode. The Phillips gauge employs a high voltage discharge between a cold cathode and a ring anode. Whereas the thermionic gauge has to be kept under reduced pressure while operating to prevent it from burning out, the Phillips gauge is quite rugged and is therefore used more frequently.

Stopcocks and Joints

As mentioned previously, the design of the glass apparatus determines the speed with which a system can be evacuated. At low pressures the free path length is of the order of the smallest dimension of the glass assembly so that collisions with the walls limit the pumping speed. Small-diameter tubing and small-bore stopcocks seriously interfere with the flow of gas. The same considerations hold for the transfer of condensable gases between traps. Vacuum stopcocks, therefore, have large bores and hollow plugs. When mounted in the correct way, the hollow plug helps to keep the stopcock leakproof by means of suction. Plugs and shells of stopcocks are ground to match precisely and usually both are numbered to avoid mismatching. Figure 4-6 shows two types of vacuum stopcocks. Needle-valve stopcocks are sometimes employed in cases where the substance in the vacuum line must not come in contact with grease.

Joints in vacuum lines should match the diameter of the tubing. Figure 4-7 shows two commonly used types. In addition to these, O-ring joints, metal valves, and fittings are used in special cases.

Joints and stopcocks must be greased, and three types of grease are commonly used: hydrocarbons, halocarbons, and silicones. The halocarbon greases are the most inert, but their vapor pressure is higher than that of the other two. Hydrocarbon greases have lower vapor pressures than silicones; they are also more inert but do not possess temperature stability and resistance to flow, which are characteristic of the silicones.

Vacuum-Line Operations

Leaks

Most leaks are caused by the faulty greasing of joints and stopcocks. A great deal of time is saved if stopcocks and joints are carefully checked before evacuation. All seals must be completely clear, and stopcocks should turn easily. Should there be the slightest doubt, the joint or stopcock must be taken apart and regreased. In this case, all of the old grease must be removed, and the plug and shell must be completely clean. Stopcock bores can be emptied by the use of a pipe cleaner. After applying thin, longitudinal strips of grease

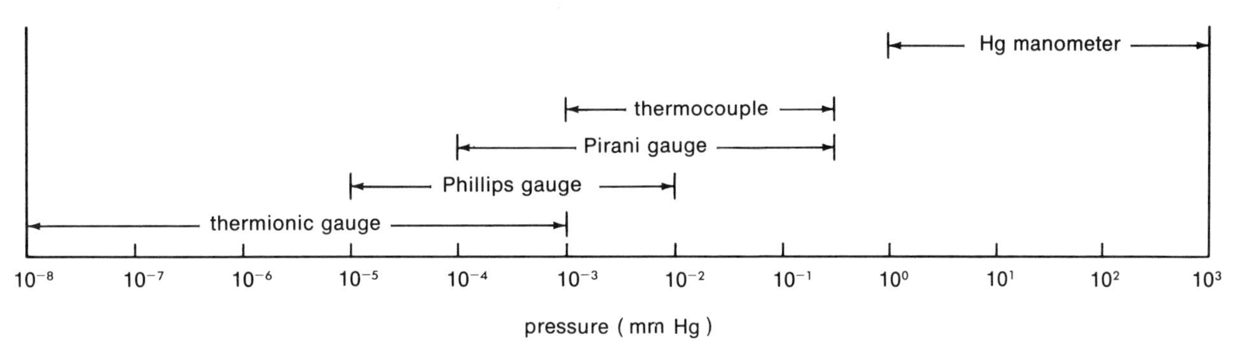

FIGURE 4-5
Operating ranges of different pressure gauges.

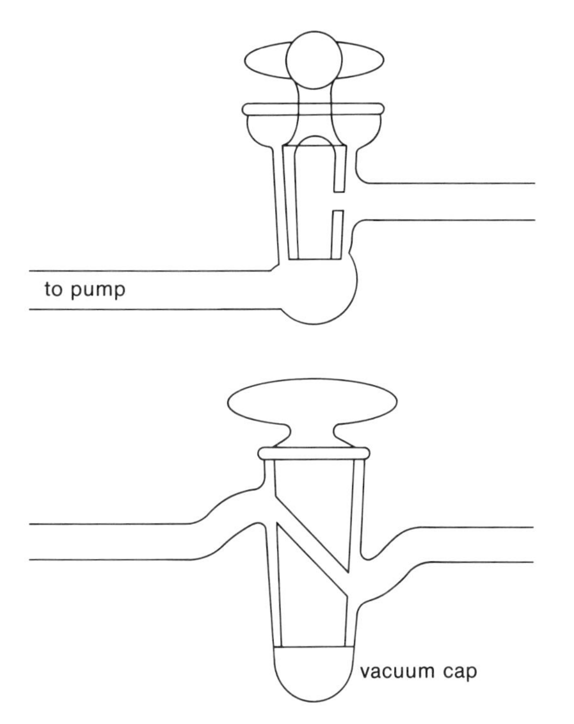

FIGURE 4-6
Hollow-plug vacuum stopcocks.

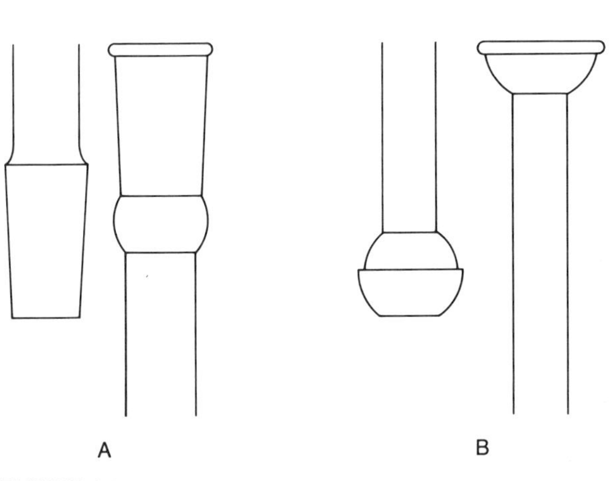

FIGURE 4-7
A. Taper joint. B. Ball joint.

to the plug, it is inserted with slight pressure into the shell. The grease should flow together and form a clear seal before the stopcock is turned. Air escapes through the channels between strips as pressure is applied and would be trapped and cause leaks if the plug were turned prematurely. Both hands should be used when applying pressure, one to press the plug and one to support the shell, so that no stress is put on the glass system. A frequent mistake is the use of too much grease, which then partially clogs bores, adsorbs vapors, and, in general, produces a bad seal because channels are sucked into the grease when vacuum is applied. Dendritic patterns on the stopcock are an indication of excessive grease. A freshly greased system should be evacuated for several hours before it is used. Desorption of moisture from the line can be achieved by flaming the tubing under vacuum. Once the system is airtight and dry it should be left under vacuum between laboratory periods. The general location of leaks can be found by methodically closing off sections of the line and reopening them to the pressure gauge after some time has elapsed. If slow leaks occur, the whole system should be evacuated, all sections closed off, and, after a few hours or overnight, each section should be opened separately while reading the gauge. Sections close to the gauge should be opened first. Pinholes are sometimes found in newly constructed lines and can be located by using a Tesla coil, which provides a high voltage discharge. A blue spark appears in the hole when the coil is passed over it. Suspected areas can be squirted with acetone, and if a hole exists, the thermocouple or discharge gauge will suddenly indicate a lower pressure.

Transfer of Condensable Gases

Condensable gases are transferred by means of evaporation and condensation at the desired place. The presence of noncondensable gases greatly impedes the gas-phase diffusion by which molecules reach their destination. The initial pressure should be maintained below 10^{-3} mm Hg for reasonably quick transfers. The pressure above the condensed substance should also be kept low for the same reason; therefore efficient cooling at the trapping location is necessary. The most commonly used refrigerant is liquid nitrogen (boiling point: −196°C).

As an example, consider the transfer of DCl or ND_3 from the vacuum line to a storage bulb, where it can be retained for later use. Figure 4-8 illustrates a part of a vacuum system shown in Figure 4-1, together with the attached storage vessel. It is necessary to know the amount of gas in trap 2, as well as the volume of the storage bulb, so that the transfer of more gas than corresponds to one atmosphere pressure in the bulb can be avoided. The bulb is attached to the line by means of a carefully but lightly greased taper joint and is held in place by springs or rubber bands. The substance to be transferred is frozen in trap 2. The section shown in Figure 4-8 is then evacuated through stopcocks *J*, *I*, and *H*, with *G* open to include the storage bulb. Stopcocks *E* and *F* are closed. It is perfectly safe to pump on trap 2 while the substance is held at liquid nitrogen temperature. Neither DCl nor ND_3 has an appreciable vapor pressure at this temperature. Then *H* and *J* are closed and the liquid nitrogen bath removed from trap 2. A small Dewar flask containing liquid nitrogen is placed under the tip of the storage vessel. The contents of trap 2 are then condensed into the tip of the storage bulb. The bulb can then be closed and taken off for storage after the liquid nitrogen has been removed.

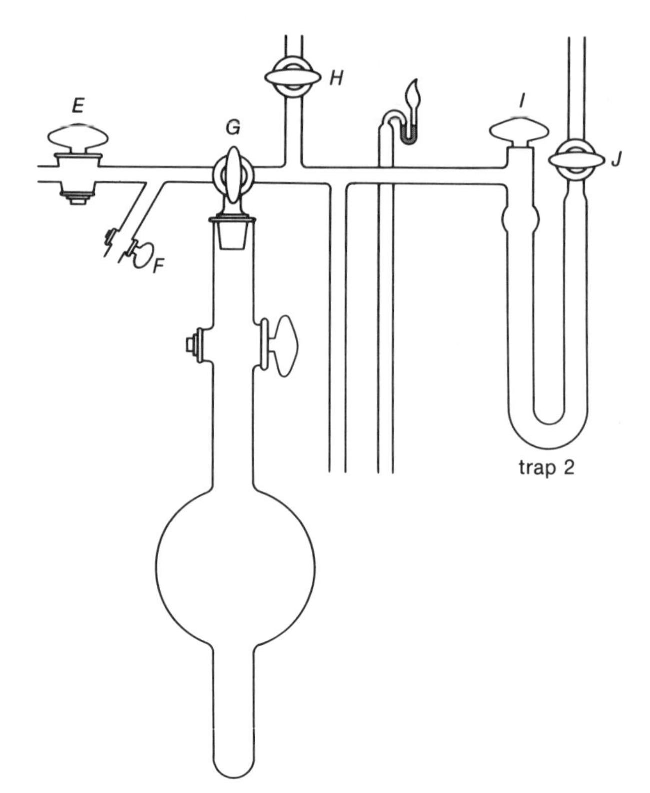

FIGURE 4-8
Transfer of DCl or ND_3 to storage bulb.

Trap-to-Trap Distillation

The separation of volatile compounds in a vacuum line is achieved by trap-to-trap distillation at temperatures at which their vapor pressures are sufficiently different. The procedure is as follows: First, the trap containing the mixture is cooled with liquid nitrogen so that the system can be evacuated (if this had not been done previously) without loss of substance. Then the liquid nitrogen bath is replaced by one of a temperature at which the vapor pressure of the more volatile compound is a few centimeters of mercury, if possible, whereas the pressure of the other compound(s) is negligible. The second trap is cooled so that the volatile compound is trapped. If the volatilities of the compounds are not very different, the process has to be repeated. Trap-to-trap distillations are used in the preparation of DCl or ND_3 to purify the products. The section between stopcocks *C*, *H*, and *J* in Figure 4-1 is used for this purpose.

Vapor-Pressure Measurements

Vapor pressures are characteristics of a substance and are used for identification and as a criterion of purity. They are easily measurable in a vacuum line simply by keeping the trap containing the substance at a constant temperature by means of a slush bath. The trap is opened to a mercury manometer. The temperature of the slush bath can be conveniently changed to obtain the vapor pressure as a function of temperature. To check the purity of a substance, the vapor pressure at constant temperature should be measured before and after the vaporization of an appreciable portion of the substance. If contaminants are present, they will be enriched in the remaining portion of the substance, and the two measurements of vapor pressure will be different. Trap 2 (in Figure 4-1) can be used conveniently in connection with the mercury manometer to measure the vapor pressures of ND_3 or DCl.

Yield Determination

The amount of gaseous substance present is usually calculated on the basis of the Ideal Gas Law. Simultaneous measurements of pressure, temperature, and volume are necessary. The temperature is normally determined by room temperature; pressure is measured on the mercury manometer; and the volume must be known. Since the volume includes one arm of the manometer, in which the height of the mercury column varies with pressure, the volume has to be known as a function of pressure. Suppose that we want to calibrate the sections of the vacuum line in Figure 4-1 between E and I, with stopcocks F and H closed to the rest of the apparatus. Figure 4-9 shows this section.

A bulb of known volume is attached to G, after which the section is evacuated by opening J and H. After the system has been closed to the pumps, gaseous CO_2 is introduced through F from a tank connected to the line at the left, which is not shown in Figure 4-9. After a suitable pressure, P', has been reached, F is closed, the pressure noted, and all of the CO_2 is distilled into trap 2 by means of a liquid nitrogen bath. Then G is closed and the temperature bath removed from trap 2. All of the CO_2 is then allowed to fill the section without the bulb. The new pressure P'' is measured after the gas has reached room temperature. Since pressures P' and P'', as well as the volume of the bulb, V_{bulb}, are known, the volume of the section can be calculated at any given pressure if several pairs of P' and P'' are used.

The volume of the section at $P = 0$ is $V_{\text{sect},0}$. Accordingly, the volume of the section at a particular pressure P is $V_{\text{sect},P}$ and can be expressed as

$$V_{\text{sect},P} = V_{\text{sect},0} + a \cdot P$$

where $a \approx (\pi/2)r^2$ and r is the inner diameter of the manometer tubing. The term $a \cdot P$ is due to the depression of the mercury level in the left side of the manometer (see Figure 4-9) caused by the rise in pressure from 0 to P.

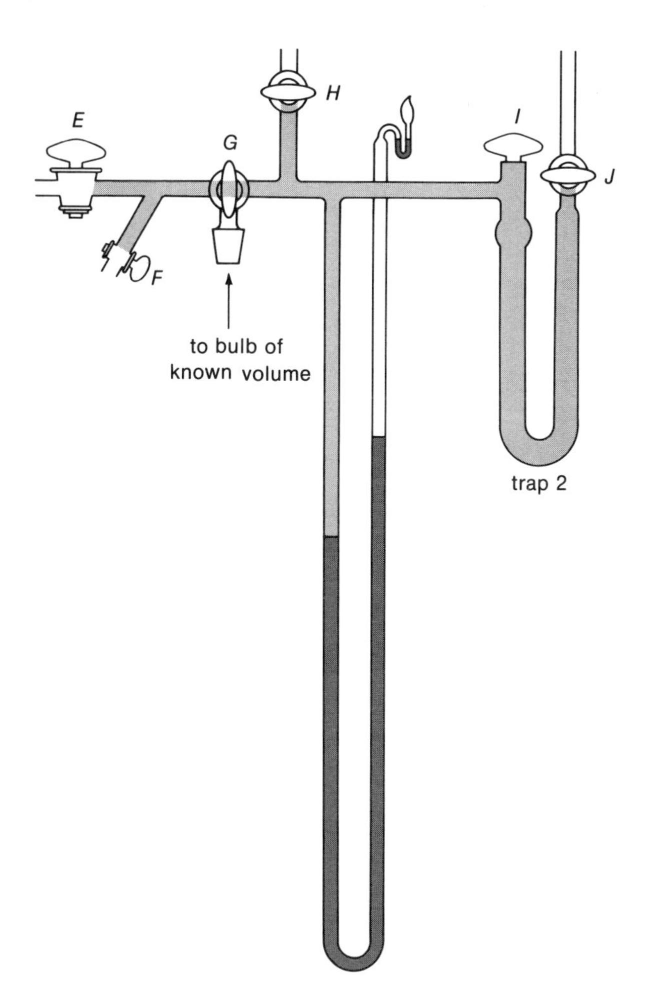

FIGURE 4-9
Section of vacuum line suitable for yield determination. The light grey indicates V_{sect}.

If ideal behavior of CO_2 gas is assumed, the products of volume and pressure for the two measurements (with and without the bulb) must be equal.

$$P'(V_{\text{sect},0} + a \cdot P' + V_{\text{bulb}}) = P''(V_{\text{sect},0} + a \cdot P'')$$

or

$$P'V_{\text{bulb}} = (P'' - P')[V_{\text{sect},0} + a(P'' + P')]$$

Thus the volume of the section at pressure $P'' + P'$ can be calculated from

$$V_{\text{sect},P''+P'} = V_{\text{sect},0} + a(P'' + P') = \frac{P'V_{\text{bulb}}}{P'' - P'}$$

A plot of $V_{\text{sect},P''+P'}$ against $P'' + P'$ for different sets of pressures P' and P'' yields the constants a and $V_{\text{sect},0}$, from which the volume of the section at any given pressure can be calculated. Alternatively, the volume may be read directly from the plot.

Low-Temperature Baths

Constant-temperature baths, or "slush baths," for vapor-pressure measurements or trap-to-trap distillation are prepared by partially freezing a liquid by the addition of liquid nitrogen. Since a constant freezing point depends on the purity of the liquid, clean Dewar flasks and pure substances should be used. Small amounts of liquid nitrogen are added to a Dewar flask that is partially filled with a liquid of the proper freezing point. The liquid must be stirred rapidly, and the procedure should take place in a hood. Because of the danger of explosion or fire, slush baths should not be prepared with air or liquid nitrogen contaminated by oxygen through prolonged exposure to air (boiling point of N_2: −196°C; of O_2: −183°C). Substances frequently used for slush baths and their melting points are indicated in Table 4-1. For more information refer to Dodd and Robinson (1954).

TABLE 4-1
Substances used for slush baths.

Compound	Melting point (°C)
Benzene	+ 5.5
Ice	0.0
Carbon tetrachloride	− 22.9
1,2-dichloroethane	− 35.3
Chlorobenzene	− 45.2
Chloroform*	− 63.5
Ethyl acetate	− 83.6
Toluene	− 95.0
Carbon disulfide	−111.9
Methylcyclohexane	−126.6
Isopentane	−160.0

Source: Shriver (1969).
*Reagent grade chloroform contains a significant amount of alcohol, which lowers the melting point.

B. PREPARATION OF DEUTEROAMMONIA

Apparatus

Vacuum line with two U-traps
Vacuum line reaction tube with fitting joint and side inlet
Magnetic stirrer and teflon-covered stirring bar
2-ml hypodermic syringe with needle
Infrared gas cell (10 cm long)
Pyrex glass wool

Chemicals

Magnesium nitride
Heavy water
Dry ice
Liquid nitrogen
Acetone or isopropyl alcohol for dry-ice baths
Chloroform (1 lb)
Chlorobenzene (1 lb)

This experiment demonstrates the use of a vacuum line in the preparation and characterization of a gaseous compound. ND_3 is prepared, rather than NH_3, because the amounts of partly deuterated compounds (NH_2D, NHD_2) obtained serve as an indication of experimental skill, provided pure D_2O has been used as starting material. Figure 4-1 illustrates the vacuum line to be used.

Deuteroammonia is prepared by the hydrolysis of magnesium nitride (Jolly 1960).

$$Mg_3N_2 + 6D_2O \longrightarrow 3Mg(OD)_2 + 2ND_3$$

Because heavy water is the most expensive item in the preparation, the experiment is designed so that D_2O is the limiting reagent.

Procedure

Before starting the procedure put a glass-wool plug in the large, inner standard taper joint below *B*. Do not allow wisps of the glass wool to protrude because they may ruin the seal when the reaction tube is attached. The purpose of this glass-wool plug is to prevent small particles of magnesium nitride from flying into the vacuum line.

Next, carefully grease the joint and attach the reaction tube shown in Figure 4-10. Hold it in place with rubber bands or springs. Connect an N_2 tank to *A*, attach all traps, and start the mechanical pump. Evacuate the entire system by opening all stopcocks to the pump. Support all large stopcocks as you turn them slowly, or they may break off. When the rotating pump has quieted down, put a Dewar flask containing liquid nitrogen under the main trap (the trap on the right in Figure 4-1), and start the diffusion pump, turning on the heat

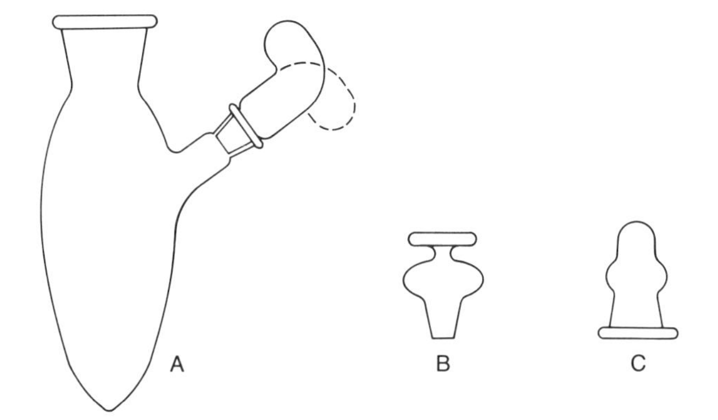

FIGURE 4-10
A. Vacuum-line reaction tube with side inlet and magnesium nitride container (dotted line indicates arrangement before the addition of Mg_3N_2). B. Stopper for side inlet. C. Cap for magnesium nitride container.

and cooling water. Close stopcock *A* to the N_2 tank. It will take about one-half hour to evacuate the system. While it is being evacuated, calculate the volume of D_2O (density = 1.1 g/cm^3 at 25°C) that will be needed to prepare 200 ml of ND_3 gas at 25°C and 760 mm Hg. Next, calculate the amount of Mg_3N_2 necessary for a threefold excess. Weigh out this quantity of Mg_3N_2 (somewhat more than the threefold excess is acceptable). Handle Mg_3N_2 in a glove bag only (see Exercise 3A) and keep the magnesium nitride container closed at all times! The time required for this operation should be held to a minimum.

When the diffusion pump has been operating for about 15 minutes, turn on the thermocouple vacuum gauge. If the vacuum line is free of leaks the gauge should indicate a pressure of less than 10^{-3} mm Hg. If this is not the case, examine the line for leaks. When the correct pressure is attained isolate the reaction section from the system (i.e., close *B* and *C*). Admit N_2 through *A* until the section is at one atmosphere pressure. Watch the bubbler manometer to check the rate of admission of gas to the system. Then remove the stopper from the sidearm on the reaction tube and allow N_2 to flush through the opening. Use the syringe to inject the required quantity of D_2O into the sidearm. (Avoid prolonged exposure of D_2O to the atmosphere, since atmospheric H_2O will contaminate the heavy water.)

Freeze the D_2O by placing a small Dewar flask containing liquid nitrogen under the tip of the reaction vessel. Next, quickly turn off the nitrogen flow through the reaction tube and attach the magnesium nitride container to the sidearm. Keep the D_2O frozen with liquid nitrogen and slowly open stopcock *B*. Thoroughly evacuate the reaction tube; then close *B*. The vacuum line is now ready for the reaction to take place. The procedure to be followed is to be determined by the student; however, the following general remarks should serve as guidelines.

1. The hydrolysis of Mg_3N_2 is vigorous, so moderate the reaction by keeping most of the D_2O frozen and/or by very carefully adding the Mg_3N_2 from the tipped-up container.
2. Halfway through the reaction, cool the tip of the reaction tube with a dry-ice bath at −78°C (acetone or isopropyl alcohol) and distill the crude ND_3 into trap 1 (which should be maintained at liquid nitrogen temperature).
3. When the reaction is finished, devise a means of separating the ND_3 from traces of D_2O that may be present by using trap-to-trap distillation (see Exercise 4A).

Write a step-by-step procedure for the experiment and obtain the approval of the instructor before carrying it out. (*Note:* One of the most common errors is to leave a stopcock open and consequently pump the ND_3 out of the system. Never allow the ND_3 to go through stopcocks *B*, *H*, or *J*.)

When ND_3 is frozen at liquid nitrogen temperature its vapor pressure is very low so it may be pumped on at this temperature without loss of substance.

Characterization

Vapor pressures of a volatile compound serve as a valuable verification of its authenticity. Two convenient temperatures at which to check ND_3 are −63.5°C (chloroform slush) and −45.2°C (chlorobenzene slush). These slush baths should be made from pure materials in a clean half-liter Dewar flask (See section A of this exercise for directions).

The infrared spectrum of a substance can be used for its identification with only a minimal understanding of the nature of the absorption process. A frequently used method is to "fingerprint" molecules, comparing identified spectra with the spectrum of the molecule under investigation. The only additional information necessary is the state of aggregation of the molecule. Gas-

TABLE 4-2
Useful vapor pressures.

Vapor pressure (mm Hg)	Temperature (°C)		
	ND_3	NH_3	D_2O
3.8			5.05
60	−74.0		44.0
100	−64.72	−67.4	50.3
150	−58.88	−61.53	
300	−47.95	−50.52	
400	−43.02	−45.54	

phase infrared spectra look very different from those in liquid phase because of the fine structure due to rotational transitions superimposed on the vibrational bands. Frequently, the fine structure is not completely resolved, but the envelope of the band traced by the spectrophotometer reveals its presence. Spectra of liquids may, on the other hand, be quite different for different concentrations or solvents because of the influence of intermolecular interactions in the denser medium.

A somewhat more sophisticated, but nevertheless empirical approach is that of explaining the spectrum by "group frequencies" and "stretching" and "bending" modes—often the only reasonable method for complex molecules. Group frequencies can be identified experimentally by the observation of isotopic effects. The process of rationalizing a spectrum begins with the identification of fundamentals, overtones, and combination bands. Fundamentals correspond to absorptions due to one normal vibration, that is, a change of $\Delta v = 1$, where v is the vibrational quantum number. Overtones are also transitions in one normal mode but with $\Delta v = 2,3, \ldots$ The observed frequencies are very nearly $2,3, \ldots$ times the fundamental frequency. Combination bands are due to transitions in two normal modes at the same time, and their frequencies can be expressed as the sum or difference of frequencies of fundamentals or overtones. Fundamentals are usually characterized by high absorption coefficients, but combination bands and some overtones may possess higher absorptivities than some fundamentals. In some normal modes, the amplitude of a particular group may be especially large, so that the frequency of this absoprtion is characteristic of that group. Absorption of such a mode may then be said to be due to a "group frequency." Group frequencies depend somewhat on the nature of the rest of the molecule, but experimental evidence demonstrates the applicability of the concept. For ND_3 only ND frequencies are observed. If the motion of vibration is mainly in the direction of the bond, the absorption is said to be a stretching mode. If it is mainly perpendicular to the bond, the result is a bending mode. Stretching and bending modes may be symmetric or asymmetric depending on whether all or only some of the groups bend or stretch in the same way. Symmetric stretching modes possess the most clearly explainable isotopic effects. The normal modes of ND_3 are shown in Figure 6-12. Mode ν_1 would clearly be called a symmetric N-D stretching mode.

The motions of all deuterium atoms are very nearly in the direction of the bond. There is a strong analogy between these vibrations and those of a fictitious N-D molecule, if just one particular deuterium atom is considered. The relation of frequency to mass may be expressed in the simple classical approach used in Exercise 6B, equation (2):

$$\nu \propto \sqrt{\frac{k}{\mu}}$$

where μ is the reduced mass. The force constant, k, is determined by the electronic configuration of the molecule, which defines the potential for nuclear motion. Thus k should be the same, regardless of how many D are exchanged for H. The ratio of frequencies of the same mode upon isotopic substitution becomes

$$\frac{\nu_D}{\nu_H} = \sqrt{\frac{\mu_H}{\mu_D}}$$

where

$$\mu_H = \frac{m_H m_N}{m_H + m_N}$$

and

$$\mu_D = \frac{m_D m_N}{m_D + m_N}$$

If the structure of the absorbing molecules is known and simple enough, μ_D and μ_H can be expressed accurately (see, e.g., Exercise 6C), and the isotopic shift can be predicted more precisely. For general purposes, $\mu_H \approx 1$ and $\mu_D \approx 2$, so that the isotopic shift for a symmetric stretching mode can be approximated by:

$$\frac{\nu_D}{\nu_H} = \sqrt{\frac{1}{2}}$$

Asymmetric stretching modes and bending modes are shifted much less than symmetric ones. Consequently, bands can be categorized by means of isotopic substitution. The isotopic shift of partly substituted molecules can be estimated by using an average atomic weight; for XY_3, where Y_3 is H_2D and $X = N$, $m_Y = (1 + 1 + 2)/3$.

Identify the impurities in the ND_3 you prepared by their spectra, using the concept of isotopic shift. Try to find stretching and bending modes, as well as the fundamentals. Bands that you may use for "fingerprinting" and orientation in the spectrum are given in Table 4-3. The procedure for obtaining a gas-phase infrared spectrum is quite simple. Attach the 10-cm gas cell to the vacuum line (stopcock and joint G in Figure 4-1), making sure it is held securely. (Infrared cells are extremely costly! Although it is generally bad practice to touch the windows of any infrared cell, those manufactured from polished NaCl plates must be handled with extreme care, or they may require repolishing. Infrared cells with alkali halide windows are stored in desiccators). Evacuate the cell, then admit ND_3 until a pressure of 100 mm Hg has been reached. Remove the cell and record the infrared spectrum on a Beckman I. R. 5 spectrophotometer or a similar instrument. Return the ND_3 to the line and evacuate the cell. Place it in a desiccator if one is provided.

TABLE 4-3
Infrared absorptions of isotopically labelled ammonia molecules.

NH_3 bands (cm^{-1})	NH_2D bands (cm^{-1})	NHD_2 bands (cm^{-1})	ND_3 bands (cm^{-1})
932	874	808	749
968	894	818	1191
1627			2419
3336			2555
3338			

The relationship between the spectrum and the structure of ND_3 is the subject of Exercise 6C.

C. PREPARATION OF DEUTERIUM CHLORIDE

Apparatus

Vacuum line with two U-traps
Vacuum line reaction tube with fitting joint and side inlet
Magnetic stirrer and teflon-covered stirring bar
2-ml hypodermic syringe with needle
Pyrex glass wool
10-cm gas cell for infrared spectroscopy
10-cm Pyrex gas cell for use at wavelengths below 25000 Å

Chemicals

Succinyl chloride
Deuterium oxide
Liquid nitrogen
Methylcyclohexane (1 lb)
Toluene (1 lb)

Preparation

Deuterium chloride is produced by the hydrolysis of succinyl chloride:

$$2D_2O + \begin{array}{c} H_2C{-}C(=O)Cl \\ | \\ H_2C{-}C(=O)Cl \end{array} \longrightarrow \begin{array}{c} H{-}CH{-}C(=O){-}OD \\ | \\ H{-}CH{-}C(=O){-}OD \end{array} + 2DCl$$

Prepare 0.75 moles of DCl for spectroscopic investigation. Since D_2O is the most expensive chemical, it should be the limiting reagent. Slightly more than a twofold excess of succinyl chloride is sufficient. The reaction is slow and more than one laboratory period is required for completion. Employ the techniques used to prepare deuteroammonia (see Section B of this exercise) with the following exceptions. (For all other information refer to Sections A and B of this exercise.)

1. Since the reactants are immiscible, stir the reaction mixture.
2. Keep the reactants at room temperature since the reaction is slow.
3. Use methylcyclohexane slush baths for trap-to-trap distillation.
4. Use toluene and methylcyclohexane in slush baths for purity determination.

Table 4-4 lists some temperatures and corresponding vapor pressures for DCl and HCl to serve as a guide in trap-to-trap distillations and purity determination.

TABLE 4-4
Vapor pressures of DCl and HCl.

Vapor pressures (mm Hg)	Temperature (°K) DCl	Temperature (°K) HCl
1	119.7	121.4
10	136.4	137.1
100	159.4	158.7
760	188.9	188.1

Spectroscopy

Record the infrared spectrum on a Beckman I.R. 5 or similar spectrophotometer. The cell must be filled with DCl at a pressure of at least 300 mm Hg.

The calculation of molecular parameters is described in Exercise 6B. This experiment requires a spectrum between 23300 Å and 25400 Å. An instrument comparable in resolution to the Cary 14 should be used, and the required pressure of the DCl in a 10-cm Pyrex cell is from 400 mm to 500 mm Hg.

REFERENCES

Dodd, R. E., and Robinson, P. L. 1954. *Experimental Inorganic Chemistry*. New York: Elsevier.

Jolly, W. L. 1960. *Synthetic Inorganic Chemistry*. Englewood Cliffs, New Jersey: Prentice-Hall.

Sanderson, R. T. 1948. *Vacuum Manipulation of Volatile Compounds*. New York: Wiley.

Shriver, D. F. 1969. *The Manipulation of Air-Sensitive Compounds*. New York: McGraw-Hill.

PART II

STRUCTURE DETERMINATION METHODS

Atomic Spectroscopy

A. SPECTRUM OF SODIUM

Atomic spectroscopy has been crucial to the development of quantum mechanics and presents striking evidence of quantization. In this experiment the spectra of alkali atoms will be analyzed to provide a better understanding of the arrangement of elements in the periodic table with respect to chemical properties and behavior. The spectra of alkali atoms are particularly suited to this purpose since they are sufficiently simple for analysis in the laboratory, yet complex enough for the study of the general structure of atomic energy levels. In addition, the experiment offers an introduction to high-resolution spectroscopy.

Procedure

Record the spectrum emitted by a General Electric sodium Na I polarimeter lamp between 6400 Å and 3000 Å, employing a Jarrel-Ash spectrometer with an Ebert-mount plane diffraction grating of 30,000 lines per inch. The detector is a 1 P 28 photomultiplier tube, the output of which is read from a strip-chart recorder. For calibration, record the spectrum of a 4-watt germicidal mercury lamp, together with the sodium spectrum. (See Figures 5-1 and 5-2 for schematic drawings of the arrangement.) The half-silvered mirror in Figure 5-2 reflects the mercury emission, without focusing, into the monochromator and reduces the intensity of the sodium lines, so that both spectra can be recorded simultaneously.

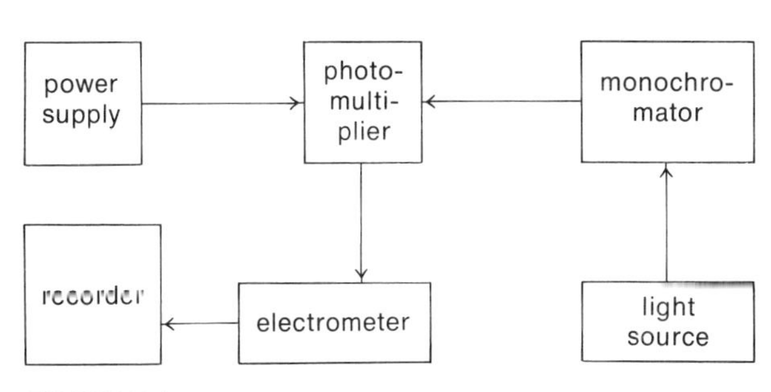

FIGURE 5-1
Diagram of spectrophotometer arrangement.

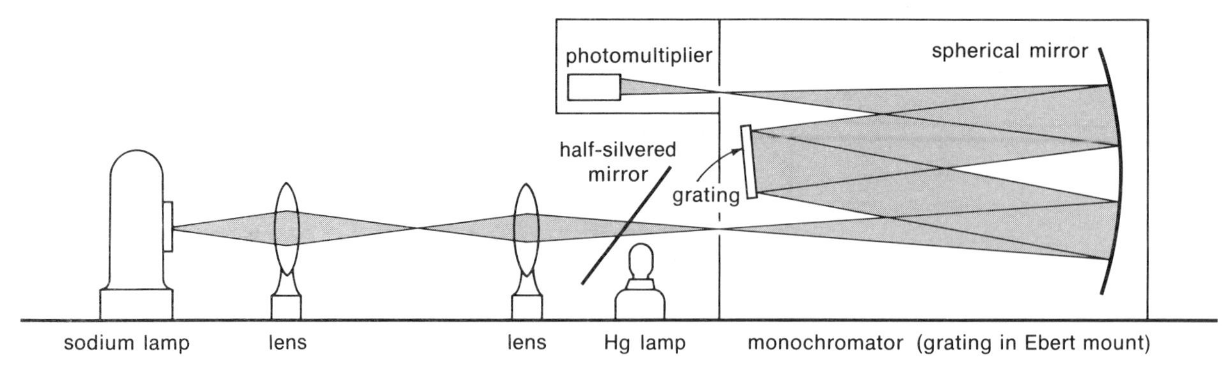

FIGURE 5-2
Optical arrangement for the simultaneous measurement of the sodium and mercury spectra. Wavelength selection is facilitated by turning the reflection grating as if it were on an axis perpendicular to the plane of this page of the manual. The spherical mirror is constructed and situated in such a way that a collimated beam of light hits the grating. The beam is focussed at the entrance and exit slits to minimize light loss and to create a well-defined beam.

Adjust the sensitivity of the recording system to accommodate the difference in intensity between the sodium and mercury lines. A manual or high-speed automatic scan (from which the settings for different wavelengths are obtained) is necessary before the actual recording is made. Between 4100 Å and 4700 Å the intensity of sodium lines is so weak (the limits of the diffuse and the sharp series are observed here) that the half-silvered mirror, together with the mercury lamp, must be removed from the optical path. (Leave the mercury lamp switched on during this time to avoid warming it up again.) One or two sodium lines may be used for internal calibration in this region [for values see Table 5-1, Herzberg (1944), and Landolt-Börnstein (1950)]. Several potassium lines, originated by potassium impurities in the sodium lamp, lend themselves to calibration in the various regions of the spectrum (see Table 5-2 and Figure 5-5 for K lines).

Record the spectrum with a chart speed of 4 inches/minute and a scan speed of 50 Å/minute. Scan in one direction only to prevent backlash.

Theoretical Background

The spectra of atoms or ions with a single electron outside a closed shell can be interpreted in terms of the transitions of this electron between allowed states. The spectra of such atoms resemble the spectrum of hydrogen. The discovery that lines in the hydrogen spectrum occur in series was made as early as the 1880's. The Rydberg equation holds for any line in the hydrogen spectrum.

$$\tilde{\nu} = R\left(\frac{1}{n_1^2} - \frac{1}{n_2^2}\right) \tag{1}$$

where R is the Rydberg constant, n_1 and n_2 are

TABLE 5-1
Sodium spectral lines.

Transition	Wavelength (Å)				
	$n=3$	$n=4$	$n=5$	$n=6$	$n=7$
$3\,^2S_{1/2}$–$n\,^2P_{3/2}$	5889.950	3302.34	2852.828	2680.335	2593.828
$3\,^2S_{1/2}$–$n\,^2P_{1/2}$	5895.924	3302.94	2853.031	2680.443	2593.927
$3\,^2P_{3/2}$–$n\,^2S_{1/2}$	5889.950	11403.96	6160.747	5153.402	4751.822
$3\,^2P_{1/2}$–$n\,^2S_{1/2}$	5895.924	11381.62	6154.225	5148.838	4747.941
$3\,^2P_{3/2}$–$n\,^2D_{5/2}$	8194.828	5688.205	4982.813	4668.560	4497.657
$3\,^2P_{3/2}$–$n\,^2D_{3/2}$	8194.791	5688.193	.....	.....	.....
$3\,^2P_{1/2}$–$n\,^2D_{3/2}$	8183.256	5682.633	4978.541	4664.811	4494.180
$3\,^2D$–$n\,^2F$	.....	18459.5	12677.6	.....	.....

Source: Landolt-Börnstein (1950).

TABLE 5-2
Potassium spectral lines for calibration.

Transition	Wavelength (Å)	Transition	Wavelength (Å)
$4\,^2S_{1/2}$–$5\,^2P_{3/2}$	4044.14	$4\,^2P_{3/2}$–$5\,^2D_{5/2}$	5831.89
$4\,^2S_{1/2}$–$5\,^2P_{1/2}$	4047.20	$4\,^2P_{3/2}$–$5\,^2D_{3/2}$	5831.72
$4\,^2S_{1/2}$–$6\,^2P_{3/2}$	3446.72	$4\,^2P_{1/2}$–$5\,^2D_{3/2}$	5812.15
$4\,^2S_{1/2}$–$6\,^2P_{1/2}$	3447.70	$4\,^2P_{3/2}$–$6\,^2D_{5/2}$	5359.66
		$4\,^2P_{1/2}$–$6\,^2D_{3/2}$	5343.07

Source: Landolt-Börnstein (1950).

integers, and $n_1 < n_2$. The symbol $\tilde{\nu}$ indicates wave number, the reciprocal of the wavelength λ, measured in centimeters, for a particular line. In each series of lines n_1 is constant so that equation (1) can be written as

$$\tilde{\nu} = T - \frac{R}{n_2^2} \qquad (2)$$

The interpretation of this equation becomes obvious if one considers Bohr's frequency relation, connecting the energy difference E of two atomic states with the frequency ν of the light absorbed or emitted by the atoms in the course of transition from one state to the other: $E = h \cdot \nu$, where h is Planck's constant (6.626×10^{-27} erg sec). Frequency ν and wavelength λ of emitted or absorbed light are related by the light velocity c. Light velocity $c = \lambda \cdot \nu$, and the energy of this quantum of radiation can be written as $E = hc \cdot \tilde{\nu}$.

Equation (2) expresses the energy of light quanta absorbed or emitted as the difference in energy of the initial and the final state of the system undergoing the transition. All lines emitted in a series possess a common final state, which is the common initial state for the same series observed in absorption. In spectroscopy the energies of allowed states are expressed in wave numbers. The n's in equation (2) can be identified as principal quantum numbers, but each state of a one-electron system is characterized by three more quantum numbers, that is, the quantum number of orbital angular momentum l, the magnetic quantum number m_l, and the spin quantum number m_s.

Yet, in hydrogen the energy of the electronic states is almost degenerate with respect to the last three quantum numbers and depends on the principal quantum number only.[1] For other atoms possessing one optical electron this is not true; the energy of the spectroscopic states depends strongly on n, l, and m_s. The main feature of this splitting of levels can be explained by the difference between the effective potential experienced by a hydrogen electron and the optical electron of an alkali atom. In the hydrogen atom the potential is a simple coulombic type:

$$U = -\frac{e^2}{r}$$

because of the electrostatic attraction between nucleus and electron. Quantum mechanics implies that electrons cannot be visualized as mass points of charge e in confined orbits. For the spectroscopist who measures their energies, electrons appear to be spread out in space in a form determined by their wave function Ψ, which is a function of the coordinates of all electrons; that is, for a single electron, Ψ is a function in three dimensional space.

The function $\rho(x,y,z) = |\Psi(x,y,z)|^2$ is interpreted as the probability distribution in the space of all coordinates, since

$$\int_{-\infty}^{+\infty} \rho(x,y,z) \cdot dx \cdot dy \cdot dz = 1$$

Of greater interest is the radial probability distribution, $\rho(r)$, where r is the distance from the center (the nucleus). The probability of finding the optical electron a distance r from the nucleus (i.e., on a sphere of radius r, with surface $4\pi r^2$) is

$$\rho(r) = 4\pi r^2 |\Psi|^2$$

A glance at functions $\rho(r)$ in standard textbooks (e.g., Moore 1962) shows that the maximum density of the electron cloud is found further from the nucleus as n becomes higher. On the other hand, the density near the nucleus depends decisively on the quantum number of the orbital angular momen-

[1] A very minimal dependence of energy on the other quantum numbers can only be detected in spectroscopes of very high resolution.

tum, l. The lower l is, the greater the probability of finding the electron near the nucleus for a given "radial" quantum number n. The nucleus of an atom bears a number of unit charges equal to its atomic number Z. If the optical electron is in a state where it has little chance of penetrating the closed shells of electrons around the nucleus, it will effectively experience a coulomb potential, because the inner-shell (core) electrons shield $Z - 1$ positive charges of the nucleus. If the optical electron is close to the nucleus where it is unaffected by shielding, its effective potential is

$$U = -\frac{Ze^2}{r}$$

The effective potential acting on the optical electron, called the "core potential," must exhibit the correct asymptotic behavior for short and long distances r. Figure 5-3 shows the construction of a core potential. Obviously, orbitals with low n and l, which penetrate the core electrons, have a much lower energy than the corresponding hydrogenic orbitals with the same quantum numbers. It is also plausible that for the same n, orbitals of different values for l must have different energies and become more hydrogenlike as l becomes larger. In addition to the splitting of energy levels described previously, a small splitting between levels possessing equal values of n and l but different values of m_s occurs because of magnetic interactions.

Preliminary to a discussion of the lines that may arise from transitions between so many levels, the allowable combinations of quantum numbers and selection rules for transitions should be reviewed. For a given n, the only allowed values are $l = 0, 1, \ldots, n - 1$. The spin quantum number m_s can only be $+\frac{1}{2}$ or $-\frac{1}{2}$ and m_l is not important for this experiment. Since orbital and spin angular momenta may couple, an "inner quantum number" j must be defined: $j = l + m_s$, that is, $j = l \pm \frac{1}{2}$ with $j > 0$.

For information on allowed energy levels, their characteristics, and transitions between them, refer to Kuhn (1962) and Davis (1965).

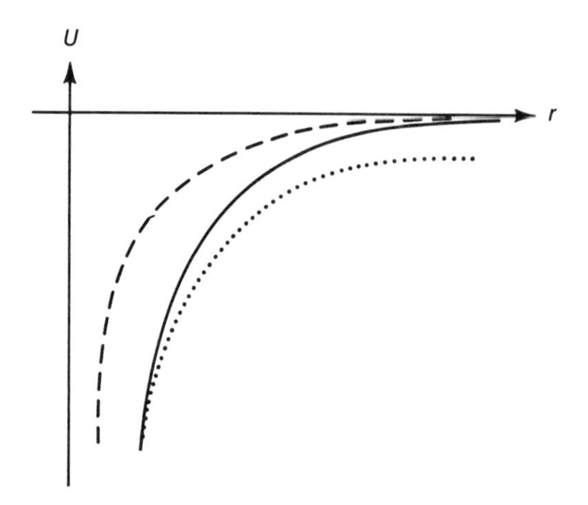

FIGURE 5-3
Construction of a core potential (solid line) from coulomb potentials with $Z = 1$ (broken line) and $Z = 3$ (dotted line).

Nomenclature

Each energy level or "term" is given a symbol as follows:

$$n^M L_j$$

The letters n and j have their usual meaning, L is given the letter symbol, familiar to chemists, that belongs to each numerical value of l, but is written as a capital letter,[2] and M is the multiplicity of the term $M = 2S + 1$; S is the total spin of the system. For one electron S always equals $\frac{1}{2}$. An electron with $n = 3$ and $l = 2$ can possess the values of $\frac{3}{2}$ or $\frac{5}{2}$ for j, and M is 2, since $S = \frac{1}{2}$. The two states bear the symbols:

$$3\,^2D_{3/2}$$

and

$$3\,^2D_{5/2}$$

which are read as 3 doublet D threehalves, and 3 doublet D fivehalves.

Selection Rules

Between all of the states mentioned here, only transitions for which $\Delta l = \mp 1$ and $\Delta j = 0$ or ± 1 are allowed. Transitions between terms of all different values of n, including equal values of n, are allowed.

---

[2]Capital letters indicate the total orbital angular momentum L. For atoms with a single electron, aside from those in filled shells, L must be equal to the orbital angular momentum quantum number of the electron, l, since filled shells possess zero orbital angular momentum. Do not confuse S for the total spin and the spectroscopic notation S, which stands for $L = 0$.

Spectrum

In the range of wavelengths observed in this experiment only those lines can be found that involve either the lowest S or the lowest P state as a lower state. (The lowest P state is split into two, the $3P_{1/2}$ and $3P_{3/2}$ states.) All other lines lie in regions of longer wavelengths. According to the selection rules the following types of series can be observed.

(a) $3\,^2S_{1/2}\text{–}n\,^2P_{1/2}$ $\quad n = 3,4,5, \ldots$

$3\,^2S_{1/2}\text{–}n\,^2P_{3/2}$

(b) $3\,^2P_{1/2}\text{–}n\,^2S_{1/2}$ $\quad n = 4,5,6, \ldots$

$3\,^2P_{3/2}\text{–}n\,^2S_{1/2}$

(c) $3\,^2P_{1/2}\text{–}n\,^2D_{3/2}$ $\quad n = 3,4,5, \ldots$

$3\,^2P_{3/2}\text{–}n\,^2D_{3/2}$

$3\,^2P_{3/2}\text{–}n\,^2D_{5/2}$

The series under (a) and (b) give rise to two lines very close together since the difference in energy of states $P_{1/2}$ and $P_{3/2}$ is slight. These two lines are called a doublet. The series under (c) predict three lines, called a "composite doublet," not a triplet! The difference in energy for the two D states is even smaller than for the P states. For Na the second series under (c) is practically indistinguishable from the third and only two lines will be observed in this experiment.

Series types (a), (b), and (c) have been given names that reflect their properties. The series under (a) involve the ground state and therefore exhibit lines of the greatest intensity in absorption; these series are called the "principal series." The series under (b) are called the "sharp series," and under (c) the "diffuse series." Because of the composite doublets, lines in the latter series appear diffused in spectroscopes not having a high enough resolution. Transitions in the principal series have a P level as the variable term, those in the sharp series an S, and those in the diffuse series a D. This is the origin of the nomenclature for s, p, and d electrons.

Wave numbers of lines in a series are expressed in equations similar to equation (2). The first term is again a constant T to which the suffix P, S, or D is added to indicate the principal, sharp, or diffuse series. For the principal series:

$$\tilde{\nu} = T_P - n^2P_{1/2} \tag{3}$$

or

$$\tilde{\nu} = T_P - n^2P_{3/2}$$

with T_P being the energy of the ground state $3\,^2S_{1/2}$ in wave numbers. Experimentally it has been found that each series can be expressed well by term differences written as

$$\tilde{\nu} = T - \frac{R}{(n^*)^2} \tag{4}$$

where n^* is not the principal quantum number but a noninteger quantum number related to it, which is always smaller than or equal to the principal quantum number. The difference δ between n and n^*, called the "quantum defect" or "Rydberg correction," is strongly dependent on L and decreases in going from s to p to d electrons, and so forth. The Rydberg correction therefore needs a suffix indicating the orbital angular momentum L of the electron. With $n^* = n - \delta_L$ a valid wave-number expression can be written for any series:

$$\tilde{\nu} = T_{\text{Series}} - \frac{R}{(n - \delta_L)^2} \tag{5}$$

From the previous discussion of the core potential it is known that δ_L is an empirical parameter that describes the reduction of energy due to the core potential, which in turn depends on the degree to which the orbitals penetrate the core. (Remember that the binding energy of the electron is negative by definition whereas wave numbers are positive quantities; the conversion factor between them is $-hc$.) The small energy differences between states of different values of j do not appear in this equation. Depending on the accuracy of the measurement one can either introduce wave numbers for each line of a doublet into equation (5) of the appropriate series, or substitute the average measured value for each doublet. The first method leads to two series under types (a), (b), and (c), the second to only one per type. (See the energy-level diagram in Figure 5-4.)

Calculations

Calculate the Rydberg constant R for Na, the ionization potential I, the Rydberg correction δ_S for the sharp series, (δ_P cannot be calculated since only two lines are observed, δ_D is taken from the literature as $\delta_D = 0.01$), and the splitting of the $P_{1/2}$, $P_{3/2}$ levels. Include in your report tables

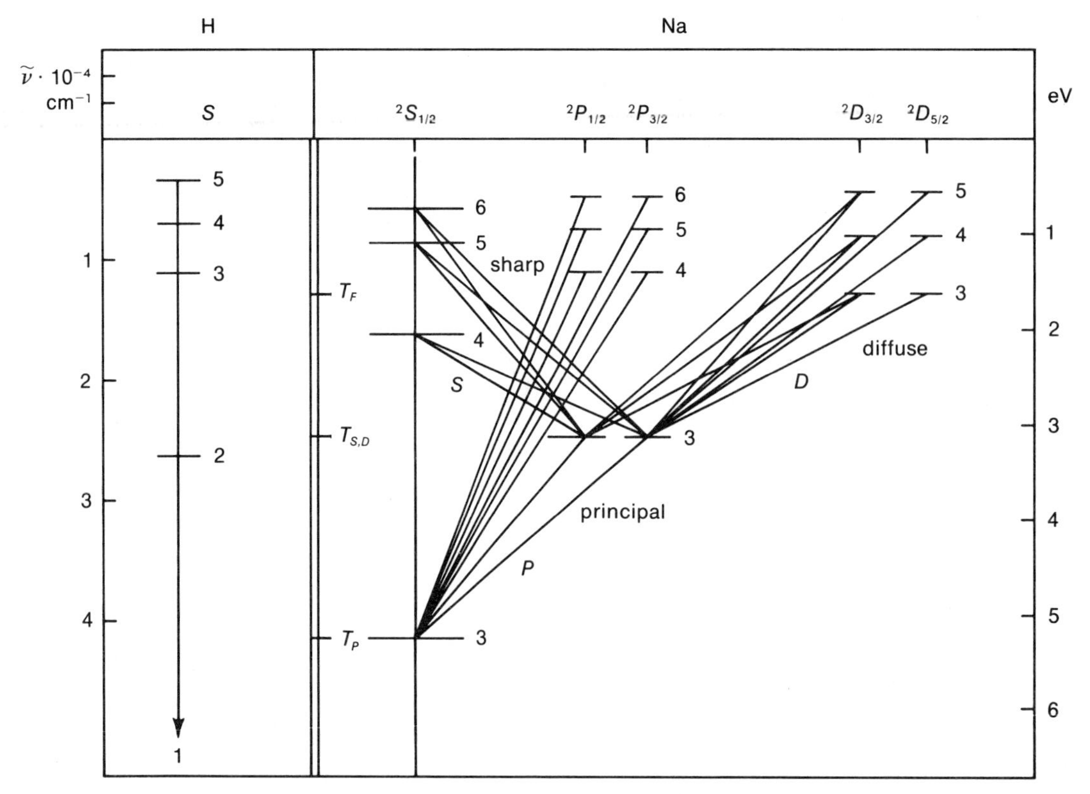

FIGURE 5-4
Energy levels of sodium compared with hydrogen. Numbers to the right of the term indicate the principal quantum number. The energy scale is given in wave numbers $\times 10^{-4}$ on the left and in eV on the right. The series limits are indicated to the right of the division line.

of wave numbers for all states between which transitions have been observed. Using these tables, construct an energy-level diagram and add it to the report. Predict the first line and the series limit for the fundamental series ($3\,^2D$–$n\,^2F$). The first step is to calibrate the spectrum by the identification of lines belonging to the Hg, Na, and, in some cases, the K spectra. This can probably be done during one laboratory period. Discuss the results with your instructor before you proceed. (Refer to Tables 5-1–5-3 for spectral lines of sodium, potassium, and mercury. Consult Figure 5-5 as well.) Plot (λ observed $-$ λ literature) against (λ observed) to obtain a calibration curve. All necessary corrections for Na wavelengths may be obtained from this graph. Assign as many Na lines as possible to the series to which they belong and label them. Make a crude sketch of the Na spectrum, obtaining wavelengths from Table 5-1.

According to equation (5), plotting the wave number against $(n - 0.01)^{-2}$ for each line of the diffuse series yields R as the slope and T_D as the intercept.[3] On the same graph a plot of wave numbers against n^2 for the sharp series is not straight but converges toward the same series limit

[3]If all doublets are resolved, prepare two plots, one for the lines at the higher wave number of each doublet, the other for the lines at the lower wave number.

TABLE 5-3
Mercury spectral lines for calibration (wavelength in Å).

3650.15	4046.50	5460.74	The following lines may also be observed in second order:
3654.83	4347.50	5790.66	3025.62, 3023.47, 3021.50, 3131.56, 3125.66,
3662.88	4348.34		and 2536.52.

Source: Landolt-Börnstein (1950).

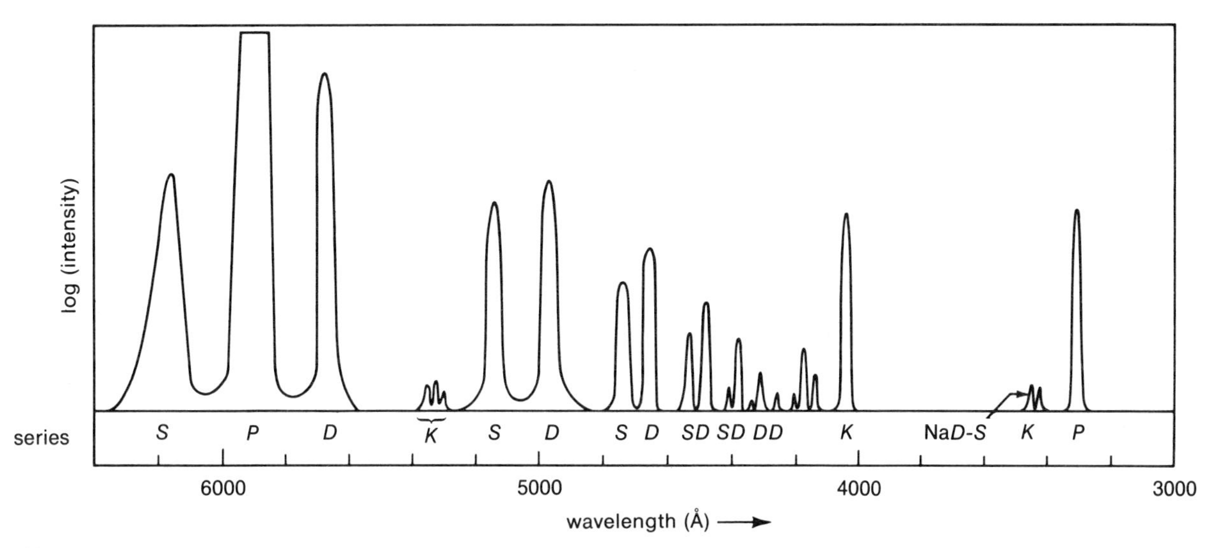

FIGURE 5-5
Spectrum of the sodium NA-I polarimeter lamp: log intensity versus wavelength in angstroms. The letters *S*, *P*, *D*, and *K* indicate whether the line belongs to the sodium sharp, principal, or diffuse series, or is due to a potassium impurity. Doublets are not resolved. [From Stafford and Wortman (1962).]

if all lines are correctly assigned. Obtain the Rydberg correction for the sharp series by plotting $R/(\tilde{\nu} - T_D)$ against n. To avoid unnecessary work this should be done for the average wave number of each doublet. The corrections for both lines in a doublet are the same. Calculate the ionization energy. The energy-level diagram (Figure 5-4) demonstrates that the ionization energy can be obtained by the addition of the term limit for the sharp series and for the diffuse series to the first line of the principal series.

$$I = T_D + \tilde{\nu}(3\,^2S_{1/2}\text{–}3\,^2P_{\text{av}})$$
$$= T_D + T_P - T_D$$

(Report in wave numbers as well as in eV.) The last part of the equation shows that I could have been obtained directly as the series limit of the principal series, which is impossible in the experiment since only two members of this series are observed.

Since S levels are not split and the splitting of D levels is too small to be observed in this experiment, the energy difference for the two $3P$ levels can be calculated from all the observed doublets. Check for constant splitting. In addition, construct an energy diagram for Na from the lines you have observed and tabulate the energy levels. The method is straightforward if you begin with the ionization potential, then go to the series limit for the sharp and diffuse series, and then use your knowledge of measured lines. This energy diagram should be used to illustrate, with the help of the "Aufbau Prinzip," or "building principle," the existence of the transition elements. To check for accuracy and internal consistency, calculate the series limit for the fundamental series and the first line, using your values for R and the fact that $\delta_F = 0$.

REFERENCES

Daniels, F., Williams, J., Bender, P., Alberty, R. A., and Cornwell, C. D. 1962. *Experimental Physical Chemistry*. 6th ed. New York: McGraw-Hill, p. 503. Contains mercury spectrum in convenient form.

Davis, J. C. 1965. *Advanced Physical Chemistry: Molecules, Structure and Spectra*. New York: Ronald Press.

Finkelnburg, W. 1950. *Atomic Physics*. New York: McGraw-Hill.

Herzberg, G. 1944. *Atomic Spectra and Atomic Structure*. New York: Dover.

Kuhn, H. 1962. *Atomic Spectra*. New York: Academic Press.

Landolt-Börnstein. 1950. *Zahlenwerte und Funktionen*. Vol. 1, Part 1. Ed. A. Eucken and K. H. Hellwege. Berlin-Göttingen-Heidelberg: Springer.

Moore, W. J. 1967. *Physical Chemistry*. 3d ed. Englewood Cliffs, New Jersey: Prentice-Hall.

Stafford, F. E., and Wortman, J. H. 1962. *J. Chem. Ed.* **39**, 630.

EXERCISE 6

Molecular Spectroscopy

A. BAND SPECTRUM OF I_2 IN THE VISIBLE REGION

Procedure

In this experiment the vibronic (vibrational and electronic) absorption spectrum of gaseous I_2 in the wavelength range of 6100 Å to 5000 Å is observed, using a Jarrel-Ash spectrophotometer containing an Ebert-mount plane diffraction grating of 30,000 lines per inch with attached recorder. Figure 6-1 shows the arrangement of a tungsten light source, an absorption cell filled with I_2, and a mercury lamp. (The mercury spectrum is necessary to obtain wavelength calibration.) For the rest of the optical and electronic arrangement see Figures 5-1 and 5-2.

The I_2 absorption cell consists of a conventional 10-cm Pyrex cell with two joints, one fitted with a small tube containing solid iodine. The vapor pressure of iodine can be increased by heating the cell to obtain sufficient absorption. Since temperature and spectral-line intensities are not important for this experiment (except that bands should be conveniently measurable), heat the cell by means of irradiation, placing a heat lamp perpendicular to the optical path. Otherwise, wind a small heating coil around the cell and regulate the temperature by the input voltage.

To obtain the wavelength calibration record the mercury and the iodine spectra simultaneously, with the unfocussed light from a germicidal mercury lamp reflecting on a half-silvered mirror into the light path as shown in Figure 6-1. Mercury lines situated between 6100 Å and 5000 Å are very different in intensity (some are observed in second order). To avoid changes in the recorder or photomultiplier settings, which also affect the underlying iodine spectrum, lower the mercury lamp sufficiently before approaching a strong line, thereby decreasing the recorded intensity. For the observation of lines in the second order (for example, around 6050 Å), the lamp must be relatively close to the entrance slit. A high speed and/or manual scan of the region under investigation is necessary before the actual recording is started to find the correct settings for the photomultiplier and recorder, as well as convenient

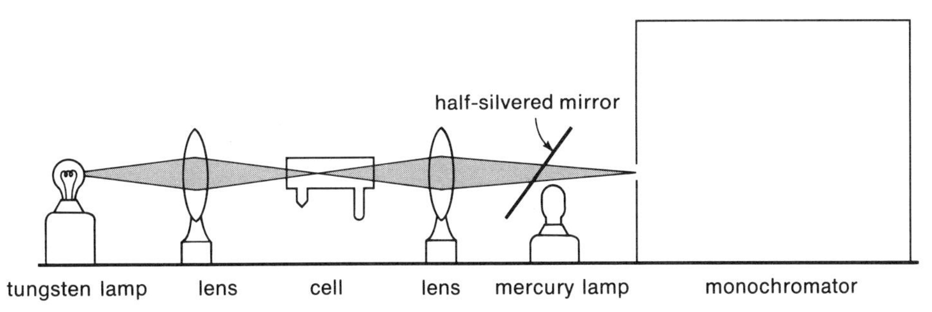

FIGURE 6-1
Optical arrangement for the simultaneous measurement of I_2 absorption and Hg emission.

positions for the mercury lamp. When all of the conditions have been met, scan from higher to lower wavelengths at 50 Å/minute with the recorder set at 4 inches/minute.

Theoretical Background

Imagine that two atoms, their energy determined by their electronic states, are brought together from infinite distance. As soon as they begin to interact, they will either repel each other, which means that the energy of the diatomic system is higher than the sum of their electronic energies, or they will attract each other, forming a bond as the distance between them is shortened. A bond is formed only if the energy level of the molecule in a particular electronic configuration lies below the sum of the electronic energies of the individual atoms. Given the same atomic states, different bonding and antibonding molecular states are formed. For each of these states the energy depends on the internuclear distance: it increases steadily for antibonding states and always possesses a minimum for bonding states. The internuclear distance for which the energy has a minimum is called the equilibrium bond length. The bond length and the shape of the potential curve, obtained when the energy E is plotted against internuclear distance r, are generally different for different states. The lowest molecular state is called the ground state and is given the symbol X, the next higher state is called A, the next B, and so on. This system is used to label all states, including those starting from higher atomic states. In addition, a similar nomenclature is used to describe the states in terms of quantum numbers. (A short review of that nomenclature is given in Appendix A. However, familiarity with it is not necessary for the present purpose.) Figure 6-2 shows potential curves that schematically represent the situation found in iodine. Potential curves, like that in Figure 6-2, can be calculated as the sum of potential and kinetic energy at each distance r, using valence bond or molecular orbital methods. Why, then, are they called potential curves?

At the beginning of this discussion two atoms in their electronic ground states were brought together until they formed a molecular bond. In turn, the energy of the molecular electronic state sank to a value characteristic of the distance, r_1, between the two nuclei. There is another much shorter distance, r_2, at which the molecule possesses the same electronic energy as at r_1. Since the total energy of the system must be conserved, another molecule or atom must exist to which the energy difference ΔE is transferred (see Figure 6-3). The energy difference, ΔE, is the difference between the molecular state at r_1 or r_2 and the sum of the two atomic states. The molecule is stable, but its bond length is either longer or shorter than the equilibrium bond length and it obviously possesses more energy than the minimal energy of the bonded species. The difference between the minimum and the occupied energy level must be stored as the vibrational energy of the two nuclei, and the two distances r_1 and r_2 represent the classical turning points for the vibrating nuclei. At r_1 and r_2 the kinetic energy and the momentum of the nuclei are zero and the total vibrational energy is stored in the form of potential energy. The total energy curve for the electronic state is equivalent to the potential energy curve for the oscillator and is therefore called a potential curve.

Unlike a classical oscillator, a molecular oscillator is quantized, and allowed vibrational levels can only be found at certain points within the potential curve. If the potential curve is parabolic, $U = \frac{1}{2}k \cdot x^2$ (where k is a constant, and $x = r - r_e$),

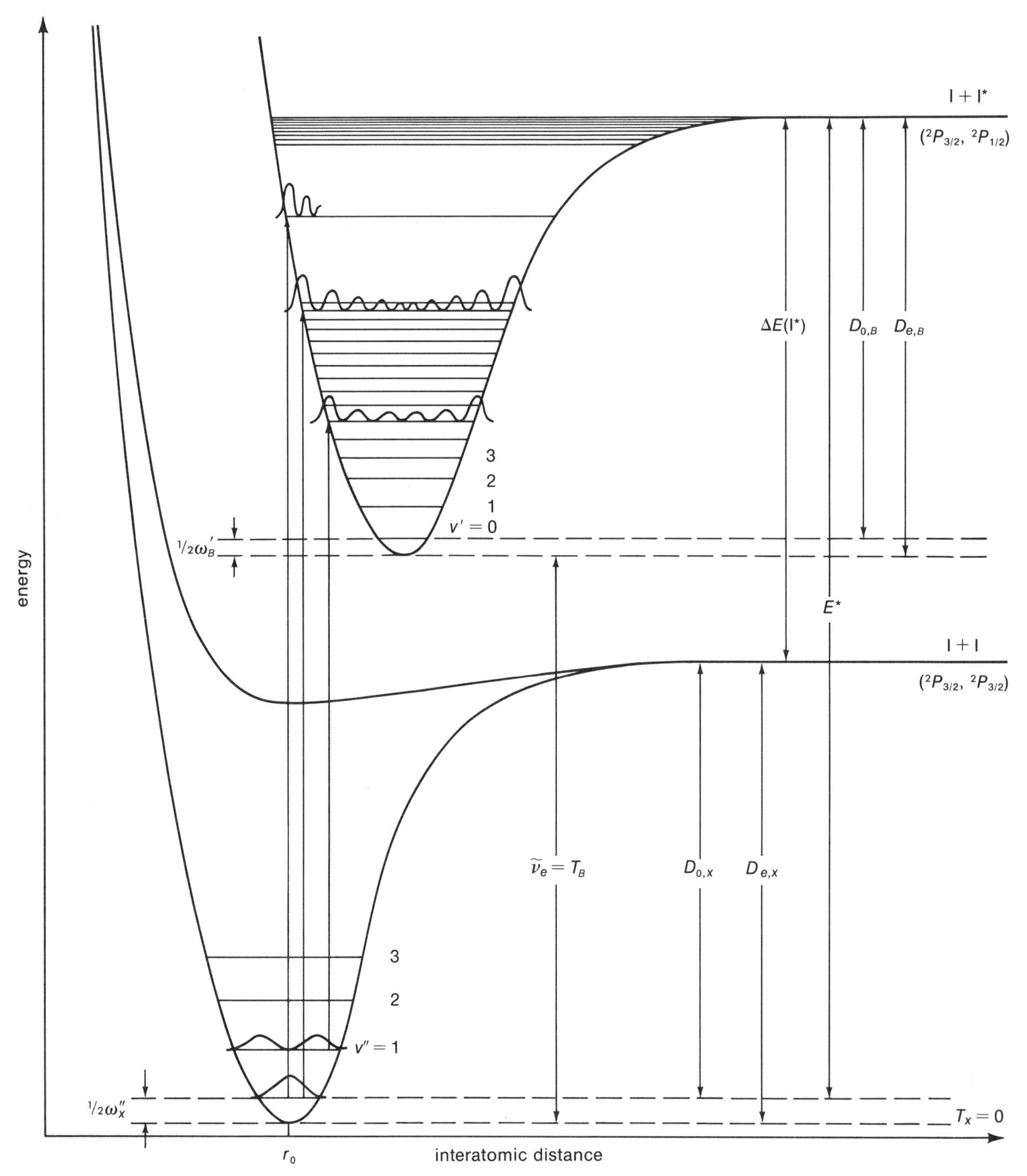

FIGURE 6-2
Representative potential curve for diatomic homonuclear halogens. [Adapted from Stafford (1962).]

the oscillator is harmonic. The solution of the Schrödinger equation for a harmonic oscillator provides the following expression for the energy, $\widetilde{G}(v)$, of allowed vibrational states:

$$\widetilde{G}(v) = (v + \tfrac{1}{2})\omega \qquad (1)$$

where $v = 0,1,2,3, \ldots$ is the vibrational quantum number and ω is the frequency of the oscillator in wave numbers. The symbol ~ indicates that the energy is expressed in wave numbers.

Equation (1) shows that even when $v = 0$ a residual vibrational energy $\frac{1}{2}\omega$ remains. Higher

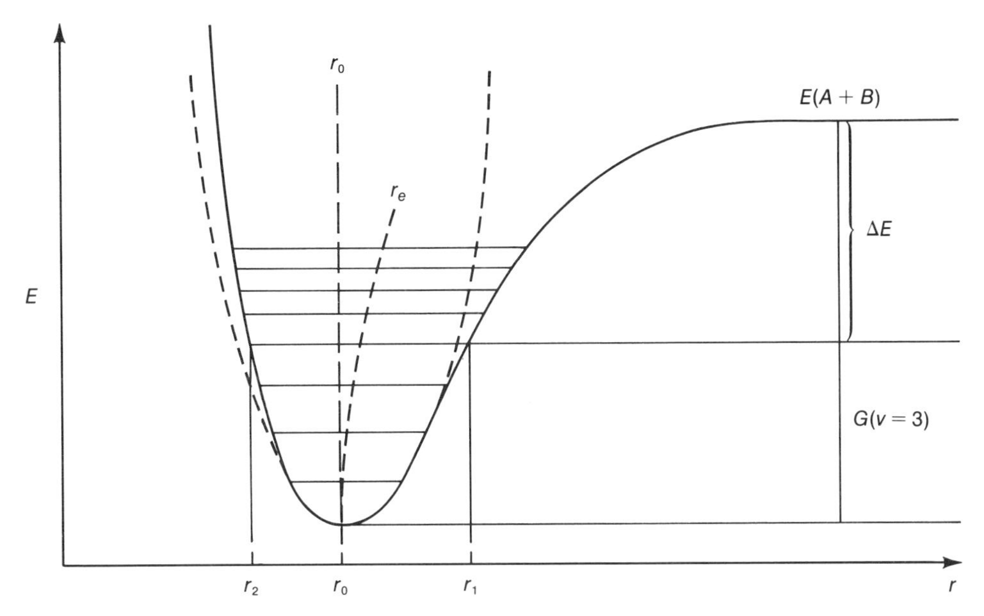

FIGURE 6-3
Arrangement of vibrational levels in an anharmonic potential, classical turning points for $v = 3$, and the dependence of r_e on vibrational level. A harmonic potential has been drawn for comparison (dotted line). Many rotational levels are found between two vibrational levels.

levels are equally spaced at intervals of ω. However, molecular potentials are never harmonic. A harmonic potential does not level off at any internuclear distance and therefore does not provide for dissociation. However, regions of low vibrational quantum numbers can be nearly harmonic. A convenient way to take care of the anharmonicity is to develop $G(v)$ in a power series

$$\widetilde{G}(v) = \omega_e(v + \tfrac{1}{2}) - x_e\omega_e(v + \tfrac{1}{2})^2 + y_e\omega_e(v + \tfrac{1}{2})^3 + \dots \tag{2}$$

in which x_e and y_e are constants that are usually quite small. The subscript e refers to the electronic state of the molecule since ω_e, x_e, and y_e will normally be different for different states.

Equation (2) shows that levels in an anharmonic potential are not equidistant; their separation decreases as v increases until, dissociation having occurred, the levels are virtually continuous. Only a finite number of levels can fit into an anharmonic potential.

In addition to vibrational energy, the molecule can also possess rotational energy. Quantization of energy $\widetilde{F}$ for a rigid rotor yields

$$\widetilde{F}(J) = \widetilde{B}_e J(J + 1) \tag{3}$$

where the quantum number of rotation is $J = 0,1,2,3, \dots$, and $\widetilde{B}_e$ is a constant depending on bond length r_e and masses m_A and m_B for atoms A and B, respectively (given here in wave numbers).

$$\widetilde{B}_e = \frac{h}{8\pi^2 \mu r_e^2 c} \tag{4}$$

with the reduced mass μ,

$$\mu = \frac{m_A m_B}{m_A + m_B}$$

and the light velocity c. $\widetilde{B}_e$ carries the subscript e because of the dependence on r_e, which may vary for electronic states. At high quantum numbers J, stretching of the bond may occur because of centrifugal forces. $\widetilde{F}(J)$ is then better represented by

$$\widetilde{F}(J) = \widetilde{B}_e J(J + 1) + \widetilde{D}J^2(J + 1)^2 \tag{5}$$

The approximation of a rigid rotor is not very good when applied to an anharmonic oscillator in a higher quantum state of vibration. The equilibrium distance r_e, defined as midway between turning points, then depends on v because of the asymmetry of the potential (see Figure 6-3). Therefore $\widetilde{B}_e$ should bear the additional suffix v to indicate the influence of the vibrational state on the rotational constant.

Because the rotational structure of bands cannot be resolved in this experiment, only equation (4) will be used as a good approximation.

Spectra and Selection Rules

In this study many fully resolved vibrational bands are superimposed on only one electronic transition. The rotational structure is only scarcely visible. The objective is to find out which vibrational transitions can occur. In addition, a brief study of rotational transitions will be made to explain the shape of the vibrational bands. (Some selection rules for transitions between different electronic states are given in Appendix A. However, it is not necessary to be familiar with them at this point.)

The Franck-Condon principle states that transitions are most likely to occur between classical turning points located at the same internuclear distance for upper and lower electronic states if high vibrational levels are involved. For vibrational ground states ($v = 0$), transitions arise from internuclear distances around r_e. Figure 6-2 illustrates these vertical transitions. This behavior can be explained in the classical approximation by assuming that electronic transitions are faster than nuclear motion. The molecule is thus in the same nuclear conformation and state of motion before and after the electronic transition. However, since the radiation field involved cannot change the kinetic energy or momentum of the nuclei, the nuclei must be in positions where the kinetic energy, as well as the momentum, is zero. (For high vibrational states, zero point energy may be disregarded.) Only the classical turning points fulfill this requirement. In all other positions a change in kinetic energy is unavoidable. For low vibrational levels an argument based on a classical analogy fails, and the quantum mechanical point of view must be considered. Transitions are predicted to occur from nuclear positions for which the nuclear distribution function possesses a maximum value in both states. These maxima are found around classical turning points for high vibrational states, at the equilibrium distance r_e for the vibrational ground state, and at intermediate positions for low levels.

The intensity of observed bands is not entirely governed by the Franck-Condon principle. In absorption the population of vibrational and rotational levels in the lower electronic state is also important.

The difference, then, in vibrational quantum numbers for upper and lower states can take any integer value. This is not true for rotational transitions. Instead, $\Delta J = \pm 1$. How does this affect the appearance of the spectrum?

It is spectroscopic practice to label the lower state $''$ and the upper state $'$. The total energy of the two states involved in a transition is written as

$$\widetilde{E}' - \widetilde{T}_e' + \widetilde{G}(v') + \widetilde{F}(J') \tag{6}$$

and

$$\widetilde{E}'' = \widetilde{T}_e'' + \widetilde{G}(v'') + \widetilde{F}(J'')$$

in accordance with the nomenclature explained previously. If all energy terms in equation (6) are written in wave numbers, the wave number of the transition $\widetilde{\nu}$ is obtained as the difference between the energies $\widetilde{E}'$ and $\widetilde{E}''$.

$$\widetilde{\nu} = \widetilde{E}' - \widetilde{E}'' = \widetilde{\nu}_e + \widetilde{G}(v') - \widetilde{G}(v'') + \widetilde{F}(J') - \widetilde{F}(J'') \tag{7}$$

where

$$\widetilde{\nu}_e \equiv \widetilde{T}_e' - \widetilde{T}_e''$$

If the lower state in the transition is the electronic ground state, then $\tilde{\nu}_e = \widetilde{T}_e'$, because $\widetilde{T}_e''$ is set equal to zero.

The contribution to the wave number of the transition arising from the difference in vibrational levels is frequently called $\widetilde{\nu}_v$. Equation (8) gives $\widetilde{\nu}_v$ in terms of the vibrational quantum numbers and anharmonicities of the two levels.

$$\begin{aligned}\widetilde{\nu}_v = \widetilde{G}(v') - \widetilde{G}(v'') &= (v' + \tfrac{1}{2})\omega_e' \\ &- (v' + \tfrac{1}{2})^2 x_e'\omega_e' + (v' + \tfrac{1}{2})^3 y_e'\omega_e \\ &- (v'' + \tfrac{1}{2})\omega_e'' + (v'' + \tfrac{1}{2})^2 x_e''\omega_e'' \\ &- (v'' + \tfrac{1}{2})^3 y_e''\omega_e''\end{aligned} \tag{8}$$

The vibrationless transition $\widetilde{\nu}_e$ (between levels with $v'' = 0$ and $v' = 0$) cannot be observed for I_2 because of the relative positions of the two potential curves (see Figure 6-2).

For a series of vibronic transitions starting from the same lower level a series of bands may be seen, each separated from the next by the difference in energy between the higher levels of two transitions. Other series starting from a different lower level will exhibit the same spacing of lines, but the whole series will be displaced by the difference between the lower levels of the two series. Figure 6-4 shows two such transitions and the resulting spectrum. Disregarding the rotation terms in equation (7), the wave number of each line in Figure 6-4B can be written in accordance with equation (7).

$$\widetilde{\nu} = \widetilde{\nu}_e + \widetilde{\nu}_v \tag{9}$$

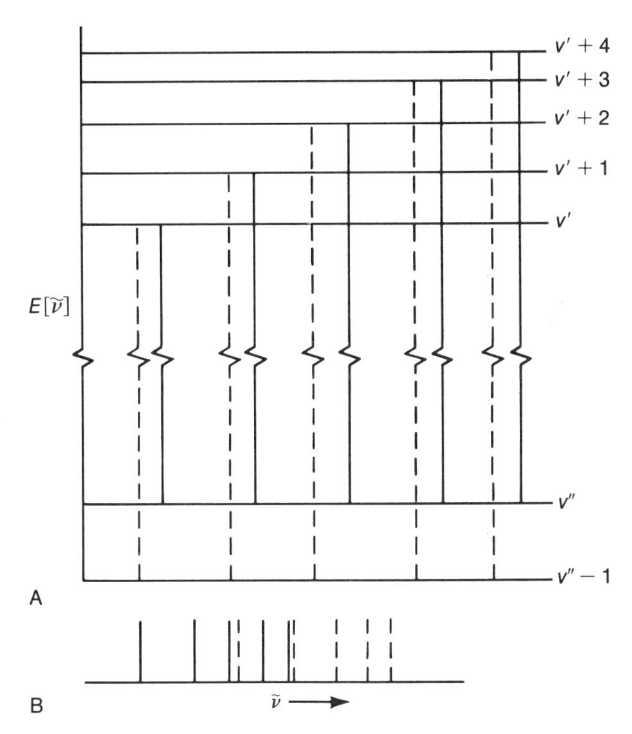

FIGURE 6-4
Two series of vibronic transitions and their resulting band structure.

The inclusion of rotation changes the band structure in Figure 6-4B. The third term in equation (7) reads:

$$\tilde{F}(J') - \tilde{F}(J'') = \tilde{B}_e'J'(J'+1) - \tilde{B}_e''J''(J''+1) = \tilde{\nu}_r \qquad (10)$$

according to equation (3).

Two different general equations can be derived for $\tilde{\nu}_r$, depending on whether $\Delta J = +1$ or $\Delta J = -1$. In the latter case the so-called *P* branch is observed in the spectrum. Wave numbers of rotational lines are derived from equations (10), (8), and (7) as:

$$\tilde{\nu} = \tilde{\nu}_e + \tilde{\nu}_v - (\tilde{B}_e' + \tilde{B}_e'')J'' + (\tilde{B}_e' - \tilde{B}_e'')(J'')^2 \qquad (11)$$

Wave numbers of rotational lines in the *R* branch arising from transitions with $\Delta J = +1$ are written as:

$$\tilde{\nu} = \tilde{\nu}_e + \tilde{\nu}_v + 2\tilde{B}_e' + (3\tilde{B}_e' - \tilde{B}_e'')(J'')^2 \qquad (12)$$

The rotational contributions in equations (11) and (12) are small, compared with the electronic and vibrational terms. Therefore, just one line may be selected from the spectrum shown in Figure 6-4B for examinations of the structure caused by rotational changes. In cases where the bond length is the same for the two electronic states involved, terms with $(J'')^2$ do not contribute, and the rotational structure will look like the vibration-rotation spectrum of a diatomic molecule (e.g., HCl or DCl). In I_2, where $r_e'' > r_e'$, the terms associated with $(J'')^2$ in equations (11) and (12) become negative. All lines in the *P* branch proceed (with increasing energy difference) toward regions of lower wave numbers.

For the *R* branch, equation (12) now includes one negative term. Lines in sequence of increasing J'' will hence be displayed in decreasing intervals toward regions of higher wave numbers until finally the magnitude of the negative term (due to the square of the quantum number) becomes large enough to prevent a further increase in wave number. The next rotational line will then be found at lower frequencies. Figure 6-5 shows the rotational structure of one vibronic line.

The location at which rotational lines turn around is called the band head, and the site of the rotationless ($J' = 0$, $J'' = 0$) transition is the band center. In cases where the rotational structure of the spectrum is not resolved, the band heads are substituted for the band centers in calculations involving different vibrational states. The error so introduced depends on the difference between B_e' and B_e'', which determines how much the band head and band center differ.

Information concerning the bond length can only be obtained from the resolved rotational structure. Nevertheless, the occurrence of bands with well-

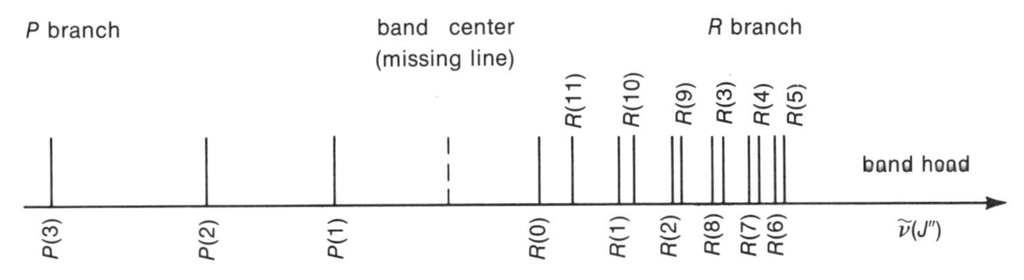

FIGURE 6-5
Rotational structure of one vibronic band for $B_e'' < B_e'$.

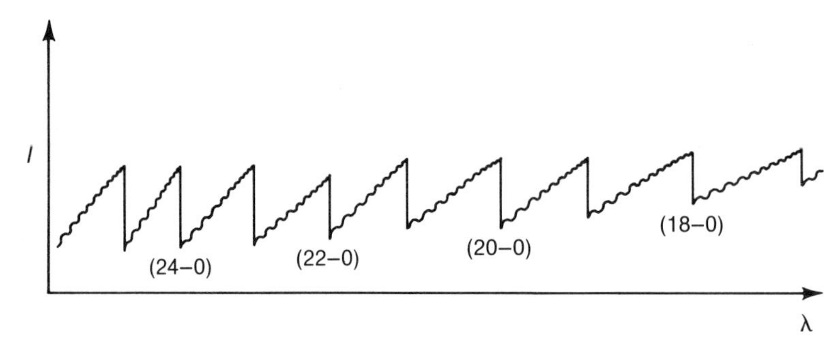

FIGURE 6-6
Schematic representation of an absorption spectrum for a diatomic molecule undergoing vibronic transitions. Rotational structure is unresolved, but shading to the red is observable.

developed band heads and intensity decreasing toward longer wavelengths (shading to the red) or shorter wavelengths (shading to the blue) shows at least that the bond lengths for upper and lower states must be different. It also indicates which state possesses the longer bond length. Figure 6-6 shows a vibronic spectrum with an unresolved rotational structure. Vibrational transitions are indicated in the form (v'–v'').

Determination of Vibrational Constants

In earlier discussions it was assumed that vibrational energy could always be expressed in some form of equation (2), in which higher terms are left out if they have been shown not to contribute significantly. Which higher terms to include must be decided on the basis of the observed spectrum. A very convenient method for the calculation of vibrational parameters has been developed by Birge and Sponer (1926).

The difference in wave number between two lines that start from (or go to) the same level is equal to the energy difference between the other two levels in the transition. A study of neighboring lines immediately reveals unequal spacing, an indication of the anharmonicity of the potential. For example, consider two lines that arise from transitions that start from the same lower level and go to higher levels one vibrational quantum apart. Their wave numbers can be written as

$$\tilde{\nu} = G(v') - G(v'') \tag{13}$$

and

$$\tilde{\nu} = G(v' + 1) - G(v'')$$

The wave number difference is

$$\Delta\tilde{\nu} = \tilde{G}(v') - \tilde{G}(v' + 1) \tag{14}$$

If all but the first anharmonicity term in equation (2) can be disregarded, then equation (14) reduces to

$$\Delta\tilde{\nu}_i = \omega_e' - 2x_e'\omega_e' - 2x_e'\omega_e'v_i' \tag{15}$$

where i labels the lower of the two states reached in the neighboring transitions.

Equation (15) shows $\Delta\tilde{\nu}_i$ to be a linear function of v_i', if all measured wave numbers fit equation (2) with the series broken off after the first term. A plot of $\Delta\tilde{\nu}_i$ against v_i' demonstrates how well observed data fit the assumed relationship. Deviations from a straight line indicate that terms of a higher order in equation (2) cannot be disregarded.

If lines with high vibrational quantum numbers v_i' have been measured, $\Delta\tilde{\nu}_i = 0$ can be extrapolated. The intercept with the v' axis yields the highest vibrational quantum number, v_∞', existing in this electronic configuration. At $\Delta\tilde{\nu}_i = 0$ the continuum, together with dissociation, has been reached.

Another useful quality of the Birge-Sponer plot is that whatever shape the curve may take, the area below it must be the sum of all energy differences between adjacent levels. Thus the area between $v' = 0$ and v_∞' must be equal to D_0' (see Figure 6-2). Similarly, the area below the curve from v_∞' to v_j' (see Figure 6-7) added to the wave number of the transition leading to the state v_j' must yield the energy difference between the dissociation limit of the upper electronic state and the vibrational level, v'', in the lower electronic state from which all transitions started. The highest vibrational level in the upper electronic state to which a transition was measured and included in the Birge-Sponer plot is v_j'.

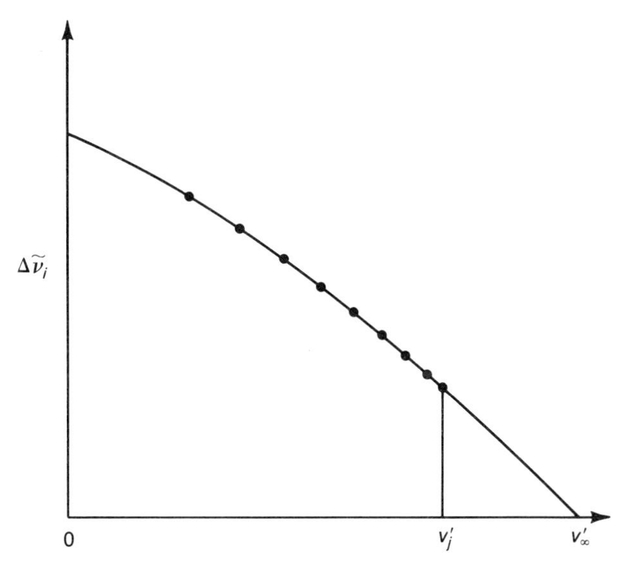

FIGURE 6-7
Birge-Sponer plot.

It may be necessary to include two anharmonicity terms in the expression for vibrational energy:

$$\widetilde{G}(v) = (v + \tfrac{1}{2})\omega_e - (v + \tfrac{1}{2})^2 x_e\omega_e + (v + \tfrac{1}{2})^3 y_e\omega_e \tag{16}$$

Under these circumstances equation (15) becomes

$$\Delta\widetilde{\nu}_i = \omega_e'(1 - 2x_e' + 3\tfrac{1}{4}y_e) - (2x_e'\omega_e' - 6y_e'\omega_e')v_i' + 3y_e'\omega_e' v_i'^2 \tag{17}$$

Whereas equation (15) can be made the subject of a linear least-squares fit, equation (17) must be treated by a quadratic least-squares fit to yield the best constants ω_e', $x_e'\omega_e'$, and $y_e'\omega_e'$.

In view of the great amount of work necessary in least-squares treatments of a large number of data points, this should only be tackled if a computer is available (a desk model will suffice). Otherwise, a graphic evaluation of data is more convenient. A program for a least-squares treatment by digital computer is included in Exercise 16. Whichever method of evaluation is chosen, one should bear in mind that potential curves can possess odd shapes. Thus, extrapolations contain great uncertainties. After all, curve fitting is only a means to express the form of the potential in the region where measurements have been performed.

Data

The data obtained with a Jarrel-Ash or comparable spectrograph yields the following information about the B state of I_2: the electronic energy T_B, the dissociation energies D_{0B} and D_{eB}, the vibrational frequency ω_B, and the first and second anharmonicity constants x_B and y_B. See Table 6-1 for comparison. Note that data have been fitted to a third-order equation.

TABLE 6-1
Spectroscopic constants for the upper state (B) of iodine.

$\omega_B = 125.237\ \text{cm}^{-1}$	$D_{eB} = 4391.0\ \text{cm}^{-1}$
$\omega_B x_B = 0.7016\ \text{cm}^{-1}$	$T_B = 15770.59\ \text{cm}^{-1}$
$\omega_B y_B = -0.00567\ \text{cm}^{-1}$	$r_{eB} = 3.027$ Å
$\omega_B z_B = 0.000032\ \text{cm}^{-1}$	$B_B = 0.028969\ \text{cm}^{-1}$

Source: Data from Klemperer (1965).

For the electronic ground state X, ω_X may be calculated, and with the help of a value for the energy difference between I and I* $[E(^2P_{3/2}) - E(^2P_{1/2}) = \Delta E(\text{I}^*)$ (see Figure 6-2)] taken from the literature, D_{0X} and D_{eX} can be obtained. For a comparison of results see Table 6-2.

TABLE 6-2
Spectroscopic constants for the electronic ground state (X) of iodine.

$\omega_X = 214.52\ \text{cm}^{-1}$	$D_{0X} = 12452.5\ \text{cm}^{-1}$
	$r_{eX} = 2.67$ Å
	$B_X = 0.03734\ \text{cm}^{-1}$

Source: Data from Verma (1960).

The first step in the calculation of the spectroscopic constants is to number the band heads correctly, with respect to the lower and upper states between which they occur, and to assign correct wavelengths to them. With the aid of the mercury wavelengths listed in Table 6-3, prepare a calibration curve ($\lambda_{\text{observed}} - \lambda_{\text{literature}}$) against $\lambda_{\text{literature}}$ for the observed first- and second-order

TABLE 6-3
Mercury wavelengths (in Å) for calibration.

5460.74 5790.66	The following lines can be used if observed in second order: 2536.5, 3025.62, 3023.47, 3021.50, and 3125.66.

mercury lines. Since approximately fifty band heads can be observed for transitions from $v_X = 0$ or 1, it is absolutely necessary to make corrections only when they are significant, and not as a matter of routine. The points to be considered are:

1. As long as only the differences in wave number of neighboring bands occur in calculations, systematic errors in wavelength measurements will almost cancel one another.
2. Because band heads are substituted for band centers, an error of about 1 cm^{-1} is introduced in all readings. (This estimate applies to I_2 in the observed spectral region. The true deviation, of course, changes with the wavelength.) Therefore, errors of less than 12 Å should not be corrected.

A guide for numbering band heads is given in Table 6-4. The band appearing in the region of the mercury line at 5460 Å is due to the transition ($v' = 25$–$v'' = 0$) and is framed by two band heads due to transitions from the first excited vibrational level of the electronic ground state (28–1) and (27–1).

TABLE 6-4
Characterization of band heads by vibrational quantum numbers.

v'	v''	Å	v'	v''	Å
17	0	5677.9			
20	0	5588.9			
21	0	5561.0			
22	0	5534.0	24	1	5547.6
23	0	5508.1	25	1	5522.6
24	0	5482.7	26	1	5498.2
25	0	5458.2	27	1	5474.6
26	0	5434.7	28	1	5452.4
27	0	5411.8			
28	0	5389.9			
29	0	5368.7			

Source: Data from Mecke (1923).

The next step is to convert wavelengths to wave numbers. Then by plotting $\Delta\widetilde{\nu}_i$ against v_i' [see equation (15)], determine ω_e', $x_e'\omega_e'$, and possibly $y_e'\omega_e'$. This work can be done by a computer by means of a linear and a quadratic least-squares data fit. In the first case it was assumed that vibrational energy can be expressed with the help of only one anharmonicity constant [see equation (15)]. In the latter case a second anharmonicity constant must be taken into account [see equation (17)]. The program, together with explanations and instructions on punching and arranging data cards, is furnished in Exercise 16. A review of the least-squares method is in Exercise 15. On the basis of the computer output and after plotting $\Delta\widetilde{\nu}_i$, as calculated by the computer, against v_i, decide whether vibrational levels of the upper electronic state are better represented by a function containing one or two anharmonicity constants. Use this function for further calculations.

Several parameters can be calculated from the Birge-Sponer plot for lines arising from $v'' = 0$. E^* (see Figure 6-2) is the sum of the area under the curve from the first experimental point to the point where the curve intercepts the v' axis plus the wave number of the transition of the lowest v' on the plot. The intercept of the curve with the v' axis also determines the highest possible vibrational state v_∞, since for this state $\Delta\widetilde{\nu} = 0$. The dissociation energy D_e' can thus be calculated as $\widetilde{D}_e' = (v_\infty + \frac{1}{2})\omega_e - (v_\infty + \frac{1}{2})x_e\omega_e + \ldots$

In halogen molecules the curve frequently bends away from the v axis for high v. In this case select the linearly extrapolated v' for calculation rather than the value extrapolated from the last observed points, since much more error is introduced by the parallel shifting of the curve than by disregarding the area under the curve that is cut off by linear extrapolation. Each case should be judged individually. If the observed Birge-Sponer curve deviates too much from a straight line, this method cannot be applied at all. Better values for D_{eB} are obtained by simply subtracting T_B from the energy of dissociation to I and I*, counted from the potential curve's minimum for the electronic ground state X, that is, $E^* + \frac{1}{2}\omega_X''$.

$$D_{eB} = E^* + \tfrac{1}{2}\omega_X'' - T_B''$$

To obtain ω_X'', simply subtract wave numbers of transitions from $v'' = 1$ to a common level v' from wave numbers of transitions originating from $v'' = 0$ to the same level v'. For example, consider $\widetilde{\nu}(28\text{–}0) - \widetilde{\nu}(28\text{–}1) = \omega_X''$. Independent measurements yield $\omega_X'' = 215$ cm^{-1} and $x_X\omega_X'' = 0.6$ cm^{-1} (see Herzberg 1950).

To obtain D_{0B}, subtract $\frac{1}{2}\omega_B'$ from D_{eB}.

$$D_{0B} = D_{eB} - \tfrac{1}{2}\omega_B'$$

Calculate T_B'' (which in this case is equal to $\widetilde{\nu}_e$ because T_X has been set equal to zero) from any observed transition originating from a $v'' = 0$ state.

The wave numbers of the transition can be written as:

$$\widetilde{\nu} = \widetilde{\nu}_e + \omega_B'(v' + \tfrac{1}{2}) - x_B'\omega_B'(v' + \tfrac{1}{2})^2 + \ . \ . \ .$$
$$- \omega_X'\tfrac{1}{2} + x_X'\omega_X'\tfrac{1}{4} - \ . \ . \ .$$

All parameters in this equation are known and T_B'' can be calculated simply as observed wave numbers minus calculated vibrational energy of the upper state plus vibrational energy of the lower state.

Obviously, the same method can be applied to any line originating from the lower level $v'' = 1$; however, calculation of the vibrational energy implies that a function that accurately describes the experimental behavior has been found. It is therefore advisable to select $v'' = 0$ and v' in a region where observed points are situated right on the Birge-Sponer straight line. The average of several calculations should be used.

The dissociation energy of the ground state, D_{eX}'' may now be calculated. Figure 6-2 shows that $\Delta E(\text{I}^*) + D_{eX}'' = T_B'' + D_{eB}' = E^* + \tfrac{1}{2}\omega_X''$, where $E^* = 76032 \text{ cm}^{-1}$ (Gaydon 1953). From this, D_{eX}'' can be calculated.

When observed through the spectrophotometer, the spectral region between 5000 Å and 5800 Å is sufficiently well resolved to allow the accurate measurement of band heads arising from the ground and the first excited levels of the electronic ground state. Above 5800 Å, a third band system originating from the second vibrational level of the electronic ground state begins to obscure the spectrum. It is difficult to assign bands in this region. [Refer to Mecke (1923) for the assignment of band heads.] The correct numbering of upper states is achieved by setting $v' = -(n_2 - 25)$ with n_2 referring to the nomenclature of Mecke. The value 25 has only recently been found to be correct as opposed to the previously used value, 26. [See Klemperer et al. (1965). This reference is a valuable source of information about spectroscopic constants of the upper electronic state.]

B. VIBRATION-ROTATION SPECTRUM OF DCl OR HCl

Procedure

Prepare gaseous DCl as described in Exercise 4C. Fill a 10-cm infrared gas cell with DCl with a pressure at least of 300 mm Hg on the vacuum line and record the spectrum on a Beckmann IR 5 spectrophotometer or comparable instrument. This spectrum shows the fundamental absorption with the rotational structure unresolved. Then fill a 10-cm Pyrex cell with DCl at about the same pressure and obtain the spectrum between 2.3 and 2.6 microns on a Cary 14 or similar instrument with sufficient resolution. The first overtone is observed in this region superimposed by the resolved rotational structure. When searching for higher overtones, extend the spectrum to lower wavelengths. Increase the pressure in the cell if absorptions are too weak.

If DCl has not been prepared, the experiment can be carried out with HCl, which is easily obtained by mixing a few drops of concentrated sulfuric acid and hydrochloric acid in the cell directly. If HCl is used, the first overtone is observed between 1.7 and 1.85 microns.

Theoretical Background

Much of the material covered in this section has been described in Section A of this exercise.

The simplest model suitable to describe a vibrating diatomic molecule is the harmonic oscillator, characterized by a parabolic potential

$$U = \tfrac{1}{2} \cdot k \cdot (r - r_e)^2 \qquad (1)$$

vibrating with the classical frequency

$$\nu = \frac{1}{2\pi}\left(\frac{k}{\mu}\right)^{1/2} \qquad (2)$$

where k, the force constant, can be related to the chemical bond strength, r_e is the equilibrium bond distance, and μ is the reduced mass; $\mu = m_A m_B/(m_A + m_B)$, with m_A and m_B being atomic masses.

Quantum mechanical treatment of the harmonic oscillator yields the following expression for energy eigenstates.

$$\widetilde{G}(v) = \frac{E}{hc} = (v + \tfrac{1}{2})\omega \qquad (3)$$

The energy of the level, called "term," is written in wave numbers, $1/\lambda = \omega = \nu/c$, where c is the velocity of light. The vibrational quantum number v can assume integer values starting with zero.

Transitions are allowed only between adjacent levels[1]

$$\Delta v = \pm 1 \tag{4}$$

$$\Delta\widetilde{G} = \widetilde{G}(v') - \widetilde{G}(v'') = \omega \tag{4a}$$

and since levels are equidistant, as equation (3) shows, only one frequency, namely ω, should be observed in the spectrum. This is obviously not true, since more than one absorption can be obtained in the spectrum of a diatomic molecule. The lowest observed frequency always shows the highest intensity and the other frequencies occur at wave numbers that are roughly given by $n \cdot \omega$, where n is an integer; the higher n is, the lower the intensity. This observation leads to the conclusion that transitions with $\Delta v \neq \pm 1$ are not impossible, but become less probable as $|\Delta v|$ increases.

The observed band of the lowest frequency (for which $\Delta v = 1$) is called the "fundamental." The higher frequencies are called "overtones." The first overtone possesses $\Delta v = 2$, the second, $\Delta v = 3$, and so forth. All transitions originate from the vibrational ground state. The higher the overtone, the more the observed frequency deviates from a simple multiple of ω. In general, the frequency is lower than $n\omega$. These effects are caused by the so-called anharmonicity of the potential, the deviation from the parabolic form (see Figure 6-3). The simplest analytical expression for a potential that shows the correct behavior of the resulting spectrum has been given by Morse:

$$U(r) = D_e\{1 - e^{-a(r-r_0)}\}^2 \tag{5}$$

in which D_e is the extrapolated dissociation energy (see Section A of this exercise), and a is a measure of the curvature of the potential.

The expression for the energy of vibrational states derived from a Morse potential is

$$\widetilde{G}(v) = (v + \tfrac{1}{2})\omega - (v + \tfrac{1}{2})^2 x\omega \tag{6}$$

with x being the first anharmonicity constant.

For the transitions between lower vibrational levels, the harmonic oscillator approximation will give nearly correct results, since the actual potential can always be nicely approximated by a harmonic potential in the vicinity of the bottom of a potential well (see Figure 6-3). But more than one anharmonicity term may have to be taken into account if higher transitions are involved. In many cases the assumption that the fundamental frequency in the spectrum yields ω directly is sufficient. The effect of isotopic substitution can thus be predicted by equation (2), since the potential, as well as k, remains unchanged. For two isotopes the ratio of fundamental frequencies should be approximately inversely proportional to the ratio of the square roots of the reduced masses.

$$\frac{\omega_1}{\omega_2} = \frac{\nu_1}{\nu_2} = \left(\frac{\mu_2}{\mu_1}\right)^{1/2} \tag{7}$$

While the molecule vibrates it also rotates. The simplest model that can be applied is the rigid rotor, which describes the molecule as two mass points of mass m_A and m_B separated by a rod of negligible mass and length r_0. Quantum mechanical treatment of the rigid rotor yields energy eigenstates

$$\frac{E}{hc} = \widetilde{F}(J) = \widetilde{B}J(J+1) \tag{8}$$

where the quantum number J can assume integer values only: $J = 0,1,2,3, \ldots$ Note the difference between the harmonic oscillator and the rigid rotor. For quantum number $J = 0$ the rotor possesses no energy, whereas the oscillator retains the zero point energy $\widetilde{G}(0) = \frac{1}{2}\omega$, which does not correspond to a classical vibration but can be explained by the uncertainty principle. The constant $\widetilde{B}$ is inversely proprtional to the reduced mass and the square of the bond length r_0.

$$\widetilde{B} = \frac{h}{8\pi^2\mu r_0^2 c} \tag{9}$$

The model of the rigid bond is not very good at high quantum numbers, J, since the centrifugal force tends to lengthen the bond through the pull on A and B. In this case equation (8) is better written as

$$\widetilde{F}(J) = \widetilde{B}J(J+1) + \widetilde{D}J^2(J+1)^2 \tag{10}$$

in which D is very small so that the second term contributes markedly only at high quantum numbers.

The fact that the molecule vibrates while it rotates does not make any difference if the vibration can be assumed to be harmonic. The time required for one rotation is much longer than for a vibra-

[1]The lower level is indicated by double prime, the higher by prime.

tion. The rotation thus effectively "sees" the molecule possessing the average bond length r_e.

Figure 6-3 shows quite clearly that the average bond length of an anharmonic oscillator is not the same for all vibrational levels. In general, the higher v is, the longer the bond. Thus the rotational constant $\widetilde{B}_v$ for a certain vibrational state v will be smaller than $\widetilde{B}$. The rotational constant $\widetilde{B}_v$ may be expressed as a function of $\widetilde{B}$ and the vibrational quantum number v.

$$\widetilde{B}_v = \widetilde{B} - \widetilde{\alpha}(v + \tfrac{1}{2}) \tag{11}$$

where $\widetilde{\alpha}$ is a constant.

The total energy of a molecule in a certain vibrational state v and rotational state J, taking anharmonicity and centrifugal stretching into account, can be expressed as follows:

$$\begin{aligned}\widetilde{G}(v) + \widetilde{F}(J) &= (v + \tfrac{1}{2})\omega - (v + \tfrac{1}{2})^2 x\omega \\ &\quad + [\widetilde{B} - \widetilde{\alpha}(v + \tfrac{1}{2})] \cdot J(J+1) \\ &\quad + \widetilde{D} \cdot J^2(J+1)^2\end{aligned} \tag{12}$$

The spacing of rotational levels is much closer than for vibrational levels, so that many rotational states can be situated between two vibrational levels.

The selection rule for rotational transitions is

$$\Delta J = \pm 1 \tag{13}$$

and applies to transitions within a certain vibrational state, as well as to transitions involving a change of vibrational and rotational levels. In the first case, +1 means absorption and −1, emission; but in the second case the loss or gain of one rotational quant can occur in absorption or emission. The infrared spectrum of a diatomic molecule thus consists of the fundamental transition and the overtones, each with the rotational structure superimposed.

The rotational structure shows two branches: the R branch, which extends from the "missing line" (the vibrational state changes, $\Delta v \neq 0$; the rotational state does not change, $\Delta J = 0$) to shorter wavelengths and is caused by transitions with $\Delta J = +1$, and the P branch, which proceeds to longer wavelengths and for which $\Delta J = -1$. Figure 6-8 shows the two branches schematically, that is, the position only and not the intensity of the lines. The lines in each branch are labelled by the rotational quantum number of the lower of the two states in the transition. Figure 6-9 shows the actual spectrum under high resolution.

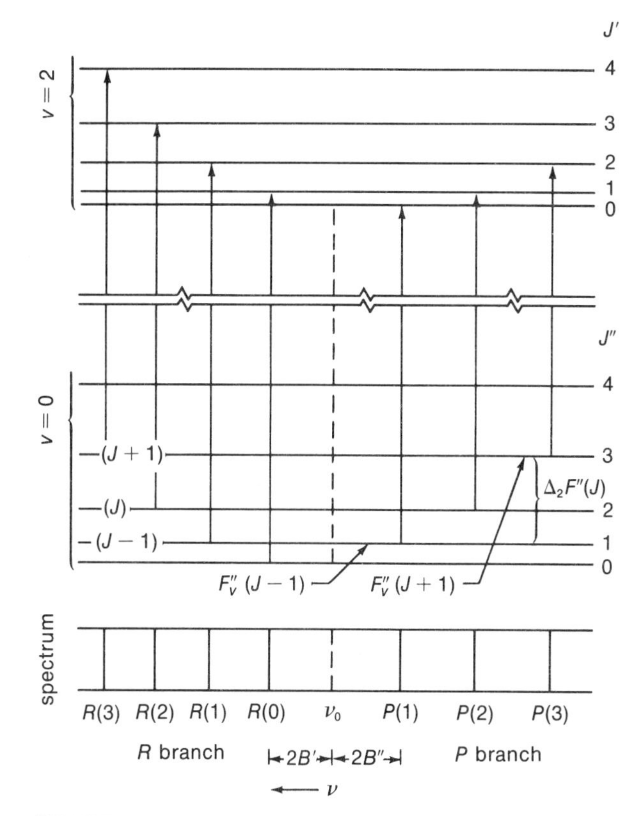

FIGURE 6-8
P and R branches and originating transitions (schematic). [From Stafford et al. (1963).]

Derivation of Molecular Parameters from the Spectrum

Bond Length

The two rotational constants $\widetilde{B}_0$ and $\widetilde{B}_2$ can be obtained from the rotational structure of the (2′–0″) band: $\widetilde{B}_0$ can be calculated from the difference between any two lines that end in the same upper state; $\widetilde{B}_2$ from any two lines that have a common lower level.

$$\begin{aligned}R(J-1) - P(J+1) &= \widetilde{F}''(J+1) \\ &= (J+1)(J+2)\widetilde{B}_0 \\ &\quad - (J-1)J\widetilde{B}_0 \\ &= 4\widetilde{B}_0(J + \tfrac{1}{2})\end{aligned} \tag{14}$$

Similarly,

$$R(J) - P(J) = 4\widetilde{B}_2(J + \tfrac{1}{2})$$

Since a number of line pairs can be used in this way, fairly accurate values for the rotation con-

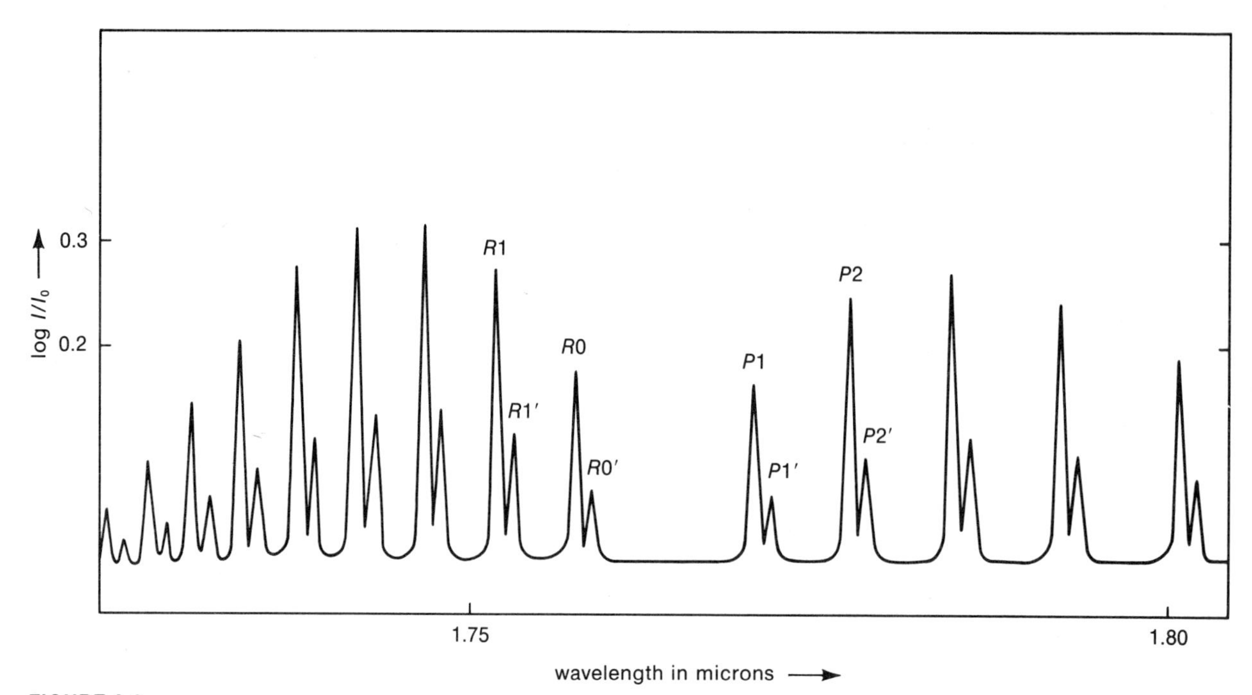

FIGURE 6-9
High-resolution spectrum of the first vibrational overtone of HCl. The lines in the branches are labelled prime for $H^{37}Cl$ and unprime for $H^{35}Cl$. [From Stafford et al. (1963).]

stants are obtainable. Systematic deviations in $\widetilde{B}$ with increasing J can frequently be traced to the centrifugal term in equation (12). In most cases the accuracy of data does not warrant the calculation of D from

$$R(J-1) - P(J+1) = 4\widetilde{B}_0(J+\tfrac{1}{2}) + 4\widetilde{D}(J+1)^3 \quad (15)$$

Isotopic Shift

By using equation (11), $\widetilde{B}$, as well as r_0, can be calculated from $\widetilde{B}_0$ and $\widetilde{B}_2$. The preceding calculations should actually be done for both isotopes Cl^{35} and Cl^{37}. By using the rotation constants for both levels and for each isotope, the frequency of the vibrational overtone can be obtained for both isotopic molecules, and the validity of equation (7) can be checked. It is also possible to do this with the fundamental frequency even though the rotational structure cannot be resolved. If desired, the two isotopic molecules HCl and DCl may be checked on low resolution instruments. Equation (7) may be expected to apply strictly at very low vibrational states, where the harmonic approximation is valid.

Even though the rotation constants are different for the isotopic molecules, r_0 should be the same in a consistent system. The same potential applies to both molecules, but the vibrational and rotational levels are situated differently according to the reduced mass. Use precise values of the isotopic masses in the calculation of the reduced masses. The conversion from wavelength to wave numbers must be done with accuracy. Carry wave numbers to five significant figures.

Anharmonicity

Once $\widetilde{B}_2$ and $\widetilde{B}_0$ are known, the frequency of the vibrational transition alone—that is, the first overtone—can be calculated very accurately from

$$R(0) - 2\widetilde{B}_2 = \widetilde{G}(v=2) - \widetilde{G}(v=0) = \widetilde{\nu}(2-0) \quad (16)$$

and

$$P(1) + 2\widetilde{B}_0 = \widetilde{G}(v=2) - \widetilde{G}(v=0) = \widetilde{\nu}(2-0)$$

where $\widetilde{\nu}$ = the wave number of the transition $(v' - v'')$.

Figure 6-8 may help to demonstrate that

$$R(0) - \widetilde{\nu}(2-0) = B_2 J(J+1)$$

where $J = 1$. Similar reasoning for $\tilde{\nu}(2-0) - P(1)$ follows directly.

Values of ω and the first anharmonicity constant $x\omega$ can be calculated from the fundamental vibrational frequency and the first overtone. These values cannot be expected to be very accurate because the fundamental frequency, which has been obtained from a low-resolution spectrum, is uncertain.

Since the spacing between adjacent levels is

$$\Delta\tilde{G}(v) = \omega - 2x\omega(v+1) \tag{17}$$

where v is the quantum number of the lower vibrational state [equation (17) can be easily derived from equation (6)],

$$\Delta\tilde{G}(0) = \tilde{\nu}(1-0) = \omega - 2x\omega \tag{18}$$

and

$$\Delta\tilde{G}(1) = \tilde{\nu}(2-0) - \tilde{\nu}(1-0) = \omega - 4x\omega \tag{19}$$

The first anharmonicity constant can be obtained from the first overtone minus twice the fundamental frequency:

$$\tilde{\nu}(2-0) - 2\tilde{\nu}(1-0) = -2x\omega \tag{20}$$

and ω can be calculated from

$$3\tilde{\nu}(1-0) - \tilde{\nu}(2-0) = \omega \tag{21}$$

Dissociation Energy

A rough estimate of the dissociation energy can be obtained under the assumption that all vibrational terms up to the dissociation limit are accurately described by the inclusion of just one anharmonicity term in equation (6). According to equation (17), ΔG then decreases linearly with v. However, Gaydon (1953) and Lewis and Randall (1961) maintain that this assumption is rarely fulfilled. Nevertheless, the estimate will yield the correct order of magnitude.

According to equation (17), the dissociation limit is reached when $\Delta G(v_{\text{lim}}) = 0$. Thus,

$$v_{\text{lim}} = \frac{\omega - 2x\omega}{2x\omega} \tag{22}$$

The dissociation energy D_e is equal to the vibrational energy of the limiting vibrational level:

$$D_e = (v_{\text{lim}} + \tfrac{1}{2})\omega - (v_{\text{lim}} + \tfrac{1}{2})^2 x\omega \tag{23}$$

From equations (22) and (23) we obtain

$$\tilde{D}_e = \frac{\omega^2}{4x\omega} - \frac{x^2\omega^2}{4x\omega} \tag{24}$$

The last term is negligible and equation (24) becomes:

$$\tilde{D}_e = \frac{\omega^2}{4x\omega}$$

and

$$\tilde{D}_o = \tilde{D}_e - \tfrac{1}{2}\omega = \frac{\omega^2}{4x\omega} - \tfrac{1}{2}\omega \tag{25}$$

for energies in cm^{-1}.

Construction of a Potential Curve

In constructing a Morse potential the constant a in equation (5) is expressed as (Gaydon, 1953)

$$a = \frac{2\omega}{4r_0(\tilde{B}\tilde{D}_e)^{1/2}} \tag{26}$$

All necessary parameters, ω, r_0, $\tilde{B}$, and $\tilde{D}_e$, have already been derived from spectroscopic data.

Although it is instructive to see how in the simplest case a potential curve can be constructed from spectroscopic data, bear in mind that predictions for higher vibrational terms are necessarily very imprecise, since only data from lower terms have been used.

If a Morse function is introduced into the Schrödinger equation as the potential for the nuclear motion, equation (6) is obtained.

Intensity of Absorption and Temperature

So far only the spectrum has been dealt with as a source of information about the energy of eigenstates for a molecule. Therefore only the measurement of wavelengths has been of concern. The measurement of band intensities provides information about the distribution of molecules over these eigenstates, and consequently about their temperature. The intensity of a line is a measure of how frequently the transition occurs. Absorption intensity measures the number of light quanta taken up by the system and depends on (1) the fraction of the total number of molecules in the light path that are in the lower state of the transition and (2) the probability with which the transition occurs. The latter quantity can be derived by quantum mechanical methods, the first by statistical mechanics. Both factors are considered in Beer's empirical law:

$$A = \log \frac{I_0}{I} = kn_j \qquad (27)$$

where A is the absorbance, I, the intensity at the end of the light path, and I_0, the original intensity.

According to the Beer's law, the absorbance is proportional to the number, n_j, of molecules in the light path that are in the state j from which absorption starts. The proportionality constant k contains, among other factors, the transition probability.

Usually, only the total number of molecules in the light path is known as the product of concentration, c, and volume, V, through which the light passes.

$$n = c \cdot V$$

if the concentration is measured as the number of molecules per unit volume. The volume, on the other hand, is determined by the cross section of the beam, s, and the length of the light path, b, as $V = s \cdot b$. The expression for the absorbance thus becomes

$$A = k \cdot n \frac{n_j}{n} = k \cdot s \cdot b \cdot c \frac{n_j}{n}$$

The fraction n_j/n of molecules in the state j is expressed by the Boltzmann distribution law as a function of temperature

$$\frac{n_j}{n} = \frac{g_j \exp(-\epsilon_j/kT)}{Q} \qquad (28)$$

where g_j is the statistical weight of the jth state, ϵ_j its energy, and Q the partition function.

$$Q = \sum_j g_j \exp(-\epsilon_j/kT)$$

In most experiments the absorbance, A, is considered to be a function of concentration, c, and path length, b. Since the temperature is usually constant, n_j/n is constant; and since s is an apparatus constant, both can be taken into the constant a to yield the familiar Lambert-Beer law:

$$A = a \cdot b \cdot c$$

with the absorption coefficient a. In this case we are interested in the absorption as a function of population of states; therefore b and c remain constant, too, and Beer's law is written as

$$A = \text{const} \frac{n_j}{n} \qquad (27a)$$

First, consider a vibrating molecule. The partition function for just one vibration is:

$$Q_{\text{vib}} = \sum_v g_v \exp(-\epsilon_v/kT) \qquad (29a)$$

where the vibrational energy is given in the harmonic approximation as [see equation (3)]

$$\epsilon_v = (v + \tfrac{1}{2})\omega hc = G(v) \qquad (30)$$

and $g_v = 1$ since there is only one vibration.

If equations (29a) and (30) are introduced into equation (28) the term due to zero point energy is cancelled. Thus in further discussion the state with $v = 0$ may be regarded as the state of lowest energy.

$$\epsilon_v = v\omega hc \qquad (31)$$

and equation (29) can be written as

$$Q_{\text{vib}} = \sum_v e^{-v\omega hc/kT} \qquad (29b)$$

$$= 1 + x + x^2 + x^3 + \ldots$$

$$= (1 - x)^{-1}$$

where $x = e^{-u} = e^{-\omega hc/kT}$.

The introduction of equations (29b) and (31) into equation (28) yields the fraction of molecules in the vibrational state v:

$$\frac{n_v}{n} = e^{-vu}(1 - e^{-u})$$

For molecules that possess large vibrational quanta ωhc, at room temperature $kT \ll \omega hc$, or $1 \gg e^{-u}$. Thus equation (28a) reduces to

$$\frac{n_v}{n} = e^{-vu}$$

For $v = 0$, $n_v/n = 1$ holds, which means that virtually all molecules are in the vibrational ground state. Thus only absorptions from the state $v = 0$ can be observed with any appreciable intensity. Check this result and the validity of the approximations using your experimental results.

The situation is completely different for rotation.

The energy of a rigid rotor in the Jth state is given by

$$\epsilon_j = \widetilde{B}hcJ(J+1) \quad (32)$$

[see equation (8)] but the state is $2J+1$ times degenerate in the absence of an external field and thus the rotational partition function can be expressed as

$$Q_{\text{rot}} = \sum_{J} (2J+1)\, e^{-J(J+1)hc\widetilde{B}/kT} \quad (29c)$$

The rotational quantum Bhc is much smaller than the vibrational quantum, which is indicated by the fact that many rotational levels fit between two vibrational levels. In most cases, therefore, the sum in equation 29c may be replaced by an integral.

$$Q_{\text{rot}} = \int_0^\infty (2J+1)\, e^{-J(J+1)\widetilde{B}hc/kT}\, dJ \quad (29d)$$

This integral can be evaluated after substituting the new variable $\xi \equiv J^2 + J$ and $d\xi = (2J+1)dJ$ to yield:

$$Q_{\text{rot}} = \frac{kT}{hc\widetilde{B}} \quad (29e)$$

The fraction of the molecules in the Jth state is then again expressed as in equation (28)

$$\frac{n_j}{n} = (2J+1)\frac{hc\widetilde{B}}{kT}\, e^{-J(J+1)hc\widetilde{B}/kT} \quad (28a)$$

Since the exponential part of equation (28a) decreases with J, whereas the preexponential factor increases, the function n_j/n goes through a maximum for a certain $J_{\text{max}} \neq 0$

$$J_{\text{max}} = \left(\frac{kT}{2hc\widetilde{B}}\right)^{1/2} - \frac{1}{2} \quad (33)$$

which shows that the maximum population moves to states of higher quantum number as the temperature increases.

Equations (27a) and (28a) may be combined to relate the intensity distribution in a vibration-rotation band to the temperature, except for the transition probabilities. These depend on the frequency of the absorbed light and on the statistical weight of the lower, as well as the upper, state of the transition. The frequencies at which the lines are observed differ only slightly, because the frequency is mainly determined by the vibrational transition. For the present purposes, the frequency may be considered to be equal for all lines. The average value of the statistical weight of the upper and lower states can be introduced into equation (28a), instead of the statistical weight of the lower state alone, to take care of the change in transition probability and the modified equation (28a) can be inserted into equation (27a).

$$A = \text{const}\,(J' + J'' + 1)\, e^{-J''(J''+1)hcB/kT} \quad (34)$$

(See Herzberg (1950) for the replacement of $2J''+1$ by $J' + J'' + 1$.) All constant values in the originating equations have been condensed into the constant in equation (34), among them the temperature T.

Conversion to the logarithmic form

$$\ln \frac{A}{J' + J'' + 1} = \frac{-J''(J''+1)hc\tilde{B}}{kT} + \ln \text{const} \quad (35)$$

shows that a plot of the left side versus $J''(J''+1)$ should yield a straight line, the slope of which is inversely proportional to the temperature. This method of temperature measurement is used to determine temperatures in flames, high-temperature arcs, the sun, and stars. The method should be employed carefully, however, because it requires equilibrium of the rotational states. To what extent does the temperature obtained from the plot agree with the measured temperature of the cell?

Spectroscopic Determination of Thermodynamic Values

Thermodynamic proprties can frequently be calculated more accurately from molecular parameters obtained through spectroscopy than by conventional methods. Statistical thermodynamics provides the necessary connection between thermodynamic functions and the partition function, which in turn can be calculated from spectroscopic data. Refer to standard texts on thermodynamics [e.g., Lewis and Randall (1961)] or physical chemistry [e.g., Daniels and Alberty (1966)] for the derivation of these relations. Table 6-5 contains thermodynamic functions expressed in terms of the partition function. The free energy function $(G - H_0)/T$ is particularly useful in calculations of equilibrium constants.

The partition function has been defined in equa-

TABLE 6-5
Thermodynamic functions in terms of the partition function.

(1*)	$E - E_0 = RT^2 \frac{d \ln Q}{dT} = R \frac{d \ln Q}{d(1/T)}$
(2*)	$c_p = \frac{d(E - E_0)}{dT} + R = R\left(\frac{1}{T^2} \frac{d^2 \ln Q}{d(1/T)^2} + 1\right)$
(3*)	$\frac{H - H_0}{T} = \frac{E - E_0}{T} + R = R\left(T \frac{d \ln Q}{dT} + 1\right)$
(4*)	$\frac{G - H_0}{T} = -R \ln Q + R \ln N$
(5*)	$S = \frac{H - H_0}{T} - \frac{G - H_0}{T} = R(1 - \ln N) + RT \frac{d \ln Q}{dT} + R \ln Q$

Note: All quantities are cited per mole and the index "0" is assigned to the value at absolute zero temperature. N = Avogadro's number; E = internal energy; H = enthalpy; S = entropy; c_p = heat capacity.

tion (29). The total partition function can be written as the product of the individual partition functions as long as the total energy of the system can be expressed as the sum of the translational, rotational, vibrational, and electronic energy:

$$E_{\text{total}} = E_{\text{trans}} + E_{\text{rot}} + E_{\text{vib}} + E_{\text{el}} \quad (36)$$

then

$$Q_{\text{total}} = Q_{\text{trans}} \cdot Q_{\text{rot}} \cdot Q_{\text{vib}} \cdot Q_{\text{el}} \quad (37)$$

This means that only weak interaction between the different modes of motion is allowed. In particular, only harmonic oscillators and rigid rotors will be dealt with.

The translational partition function can be evaluated if the spectrum of translational energies is known. The solution of the Schrödinger equation for a particle moving freely over the distance l in the x-direction yields:

$$E = \frac{n_t^2 h^2}{8\pi m l^2} \quad (38)$$

The quantum number n_t can assume integer values only. These energy eigenstates are situated so close to each other that they are practically continuous. The summation over n_t can therefore be replaced by an integration in equation (29) and the translational partition function for motion in one direction is derived as:

$$Q_{\text{trans}} = \frac{l(2\pi m k T)^{1/2}}{h} \quad (39)$$

For three dimensions and a volume of $l^3 = V$, the mole volume,

$$Q_{\text{trans}} = \frac{(2\pi m k T)^{3/2} \cdot V}{h^3} \quad (40)$$

or

$$Q_{\text{trans}} = \frac{(2\pi M R T)^{3/2}}{h^3 N^2} \cdot \frac{RT}{P} \quad (40a)$$

under the assumption that the behavior of the gas can be approximated by that of an ideal gas.

The rotational and vibrational partition functions have been derived in the discussion on "Intensity of Absorption and Temperature"; they are given in equations (29b) and (29e). While deriving equation (29e) an additional possibility for degeneracy in molecular rotation was neglected. This degeneracy depends on the number of equivalent positions that can be obtained by the rotation of the molecule, and is characterized by the symmetry number σ, by which the denominator in equation (29e) is multiplied. For homonuclear diatomic molecules, $\sigma = 2$; for heteronuclear diatomic molecules, $\sigma = 1$. The electronic partition function does not contribute if, as is usually the case, only the ground state is populated to any significant extent. Note that only the logarithm of the partition function occurs in expressions (1*) to (5*) of Table 6-5. The contributions to the logarithm of the partition function originating from translation, rotation, and vibration are additive.

From expression (4*) of Table 6-5, and equations (40a), (29b), and (29e), equation (41) is obtained for the free energy function.

$$-\frac{G - H_0}{T} \tag{41}$$
$$= \{4.574\ (\tfrac{3}{2}\log M + \tfrac{5}{2}\log T) - 7.282\}$$
$$+ \{4.574\ (\log T - \log \sigma - \log B) - 7.22\}$$
$$- \{4.574 \log(1 - e^{-u})\} \text{ cal/degree Kelvin}$$

The three bracketed terms represent the contribution of translational, rotational, and vibrational motion, respectively. The term $R \cdot \ln N$ in the translational contribution cancels.

The introduction of equations (40a), (29e), and (29b) into expression (5*) leads to

$$S = \{4.574\ (\tfrac{3}{2}\log M + \tfrac{5}{2}\log T) - 2.314\} \tag{42}$$
$$+ \{4.574\ (\log T - \log \sigma - \log B) + 1.264\}$$
$$- \{4.574\ [\log(1 - e^{-u}) - \frac{u}{e^u - 1}]\}$$

The contributions of each partition function are clearly shown. The term $-R \log N$ cancels one term in the translational contribution, whereas the additional R has been taken into rotational contribution. By the same procedure, from expression (2*) the following is obtained:

$$C_p = \tfrac{5}{2}R + R + R\,\frac{u^2 e^u}{(e^u - 1)^2} \tag{43}$$

For translation and rotation we thus arrive at the result predicted by the equipartition of energy theorem.

The functions $u/(e^u - 1)$, $\ln(1 - e^{-u})$, and $u^2e^u/(e^u - 1)^2$ are called "Einstein functions." Their values have been tabulated as a function of u (Sherman and Ewell 1942).

Calculations

1. Calculate the bond length of gaseous DCl or HCl.
2. Evaluate the isotopic shift and anharmonicity constant.
3. Obtain the dissociation energy.

The construction of a potential curve, the calculation of rotational temperatures, and the determination of thermodynamic values require additional study and are suitable for longer reports. They may be omitted from this experiment.

C. VIBRATION-ROTATION SPECTRUM OF ND_3

Normal Vibrations

To describe the motion of N atoms in space, $3N$ coordinates x_i, y_i, z_i are needed, where $i = 1,2,3, \ldots, N$, or $3N$ translational degrees of freedom. If these N atoms are combined to form a molecule, the three coordinates of the center of inertia are sufficient to describe the translational motion of the molecule. This means that three of the original $3N$ degrees of freedom are translational degrees of freedom. Three more belong to the rotational motion of the molecule around three mutually perpendicular axes. The remaining $3N - 6$ internal degrees of freedom belong to the motion of atoms relative to each other; they describe vibrational modes. Linear molecules are totally symmetrical around one axis, and therefore possess only 2 rotational degrees of freedom, or $3N - 5$ vibrational modes.

The vibrational motion of a molecule is very complex but can be described by a superposition of so-called normal modes. These are vibrations in which all atoms move with the same frequency and in phase but with different amplitudes and directions, as dictated by the condition that no rotational or translational motion may result. These vibrations are harmonic as long as the displacement of atoms from their equilibrium positions is slight. The number of normal modes is equal to the number of internal degrees of freedom. Not all normal modes have to have different frequencies; some may be degenerate. The number of degenerate modes depends on the symmetry of the molecule.

Nomenclature for Normal Modes

All vibrations that are symmetric with respect to a p-fold axis of rotation, C_p, are given the symbol A; all those that are antisymmetric carry the symbol B. Doubly degenerate vibrations are named E; triply degenerate ones, F; and so forth. Subscripts g or u refer to symmetry or antisymmetry with respect to a center of inversion. Subscripts + or − indicate symmetry or antisymmetry with respect to a C_2 axis. In a molecule with an axis C_p, $p > 2$, vibrations may be indexed $\parallel$ if the motion is mainly parallel to C_p and $\perp$ if the motion is mainly perpendicular to the axis. The relationships between types of vibration and the symmetry of a molecule are tabulated in any standard text on infrared spectroscopy (see References). Linear molecules have a different nomenclature. Still another distinction can be made for normal vibrations of molecules containing atoms of very different weights. In such cases, the amplitude of the lighter atoms is much greater than that of the heavier atoms for any given vibration. The heavier group can therefore be depicted as being virtually at rest, and vibrations can be categorized according to the motion of the lighter atoms—whether it is mainly parallel or perpendicular to the bond. The former, called bond-stretching vibrations, are given the symbol ν, and the latter, called bond-bending vibrations, are symbolized by δ.

Group Frequencies and the Assignment of Vibrations

The assignment of observed frequencies to normal modes or their combinations is complicated and frequently requires additional information. For molecules consisting of atoms having very different weights, the observation of group frequencies makes the assignment of observed frequencies less difficult. It has been found empirically that bond-stretching modes absorb in the region between 4000 cm^{-1} and 1000 cm^{-1}, whereas bond-bending modes appear at wave numbers smaller than 1500 cm^{-1}. In each case, however, bending and stretching modes are well separated. The frequency of the vibrations, as in diatomic molecules, depends on the force constant of the particular bond connecting the lighter atom to the heavier part of the molecule. Force constants seem to be nearly the same for similar bonds, which leads to the observation of group frequencies. A certain chemical group will absorb in roughly the same region of the spectrum independently of its molecular surrounding. For each such group there is one stretching and one bending frequency, which may not be entirely observed because of degeneracies. For ND_3 there are three ν and three δ modes. One pair in each class is found to be degenerate.

Vibration-Rotation States and Transitions

As mentioned previously, any vibration of a molecule can be described as a superposition of normal modes, which can be represented as harmonic oscillators capable of being in states of quantum numbers v, $v = 0,1,2,3, \ldots$, only. The total system is therefore unambiguously described by n vibrational quantum numbers v_i, $i = 1,2, \ldots, n$, where n is the number of normal modes. The vibrational energy of the molecule can be written as the sum of the energies of each harmonic oscillator representing a normal mode.

$$\widetilde{G}(v_1, v_2, \ldots, v_n) = \omega_1(v_1 + \tfrac{1}{2}) + \omega_2(v_2 + \tfrac{1}{2}) + \ldots + \omega_n(v_n + \tfrac{1}{2})$$

in cm^{-1}. Anharmonicities must be introduced if higher levels are involved, which complicates the description because of the appearance of cross terms. The general procedure is analogous to the way in which anharmonicity is handled in diatomic molecules. Further explanation is unnecessary for the present purpose.

Selection rules are the same as those for diatomic molecules. A restriction to harmonic vibrations allows $\Delta v = \pm 1$ only; for anharmonic vibrations, $\Delta v = \pm 1, \pm 2, \pm 3$ is allowed, but the probability of transitions decreases as Δv increases.

Transitions between the ground state and the first excited state of one normal mode are called "fundamentals." They appear as frequencies of the highest intensity in the absorption spectrum. The wave number of a fundamental is given by

$$\widetilde{\nu}_0 = \widetilde{G}(0,0, \ldots, 1,0) - \widetilde{G}(0,0, \ldots, 0,0) = \omega_0$$

in the absence of anharmonicities. It should be noted that normal modes of polyatomic molecules are frequently strongly anharmonic. For accurate work, corrections should be introduced even for transitions between the lowest levels.

The number of fundamental frequencies that can be observed in an infrared spectrum depends on the occurrence of degeneracies and on the fact that only transitions involving a change of dipole moment will give rise to light absorption or emission.

As in diatomic molecules, a number of rotational states will be found between vibrational states. Since the spacing of rotational levels and the appearance of the rotational fine structure of bands depends heavily on the structure of the molecule, this discussion will be restricted to symmetric tops, the class to which ND_3 belongs.

A molecule of C_{3v} symmetry has only two nondegenerate moments of inertia: the one in the direction of the C_{3v} axis (the figure axis) is called I_A; the two remaining moments, designated I_B, are in a plane perpendicular to the figure axis, which contains the center of inertia, and are degenerate because of the symmetry.

The allowed energy states of a rigid symmetric top are characterized by two quantum numbers J and K, where K stands for the absolute magnitude of the component in the figure axis of the vector J. Therefore K can possess $J + 1$ values.

$$K = J, J - 1, J - 2, \ldots, 0$$

The energy of allowed rotational states is then given by

$$\widetilde{F}(J,K) = \widetilde{B}J(J + 1) + (\widetilde{A} - \widetilde{B})K^2$$

in cm^{-1}, where

$$\widetilde{B} = \frac{h}{8\pi^2 c I_B}$$

and

$$\widetilde{A} = \frac{h}{8\pi^2 c I_A}$$

The following selection rules apply for vibration-rotation transitions. For $\parallel$ bands: $\Delta K = 0$, $\Delta J = 0, \pm 1$ (if $K \neq 0$) and $\Delta K = 0$, $\Delta J = \pm 1$ (if $K = 0$); for $\perp$ bands: $\Delta K = \pm 1$, $\Delta J = 0, \pm 1$. Since $\parallel$ bands are easier to analyze than $\perp$ bands, the discussion will be confined to $\parallel$ bands.

As in diatomic spectra, frequencies can be expressed in terms of the difference of vibrational and rotational energy of the states involved.

$$\begin{aligned}\tilde{\nu} = {} & \widetilde{G}(v_1', \ldots, v_n') - \widetilde{G}(v_1'', \ldots, v_n'') \\ & + \widetilde{B}_v' J'(J' + 1) \\ & + (\widetilde{A}_v' - \widetilde{B}_v')K^2 - \widetilde{B}_v'' J''(J'' + 1) \\ & - (\widetilde{A}_v'' - \widetilde{B}_v'')K^2\end{aligned}$$

For a vibrational fundamental, $\Delta v = 1$,

$$\begin{aligned}\tilde{\nu} = {} & \tilde{\nu}_0 + \widetilde{B}_v' J'(J' + 1) - \widetilde{B}_v'' J''(J'' + 1) \\ & + [(\widetilde{A}_v' - \widetilde{A}_v'') - (\widetilde{B}_v' - \widetilde{B}_v'')]K^2\end{aligned}$$

where $''$, as usual, indicates the lower state. The subscript v refers to the vibrational level.

For each permissible K ($K \leq J$), a sub-band will appear with a P, Q, and R branch, except for $K = 0$ where only P and R branches are allowed. As K increases, the number of lines contained in these sub-bands decreases, since $J < K$ is not allowed and transitions between large J have slight intensities. Figure 6-10 shows sub-bands and the total band for the vibrations of a symetric-top molecule (Herzberg 1945). The Q branch of any sub-band belonging to a particular K is defined as

$$\tilde{\nu}_{0,\mathrm{sub}} = \tilde{\nu}_0 + [(\widetilde{A}_v' - \widetilde{A}_v'') - (\widetilde{B}_v' - \widetilde{B}_v'')]K^2$$

Therefore $\tilde{\nu}_0$ lies on the short wavelength end of the Q branch. The general equations for lines in the P and the R branch are:

$$\tilde{\nu}_{P(J)} = \widetilde{B}'(J - 1)J - \widetilde{B}''J(J + 1) + \text{const}$$

$$\tilde{\nu}_{R(J)} = \widetilde{B}'(J + 1)(J + 2) - \widetilde{B}''J(J + 1) + \text{const}$$

and

$$\text{const} = \tilde{\nu}_0 + [(\widetilde{A}_v' - \widetilde{A}_v'') - (\widetilde{B}_v' - \widetilde{B}_v'')]K^2$$

which shows that the rotational constant for the upper vibrational level can be obtained from

$$\begin{aligned}\tilde{\nu}_{R(J)} - \tilde{\nu}_{P(J)} &= \widetilde{B}'[(J + 1)(J + 2) - J(J - 1)] \\ &= 2\widetilde{B}'(J + 1)\end{aligned}$$

The rotational constant for the lower vibratinal state, B'', can be calculated by using

$$\begin{aligned}\tilde{\nu}_{P(J+2)} = {} & \widetilde{B}'(J + 1)(J + 2) - \widetilde{B}''(J + 2)(J + 3) \\ & + \text{const}\end{aligned}$$

$$\begin{aligned}\tilde{\nu}_{R(J)} - \tilde{\nu}_{P(J+2)} &= \widetilde{B}''[(J + 2)(J + 3) - J(J + 1)] \\ &= 2\widetilde{B}''(2J + 3)\end{aligned}$$

From this equation the moment of inertia in a plane perpendicular to the figure axis can be calculated for the upper and lower vibrational levels.[2]

Inversion Doubling

For molecules having a pyramidal XY_3 structure there are, in fact, two equivalent positions for the X atom—one above and one below the plane of the three Y atoms. A plot of potential energy as

[2]The equations for $\widetilde{B}'$ and $\widetilde{B}''$ apply to the total band even though their derivation holds strictly for sub-bands only.

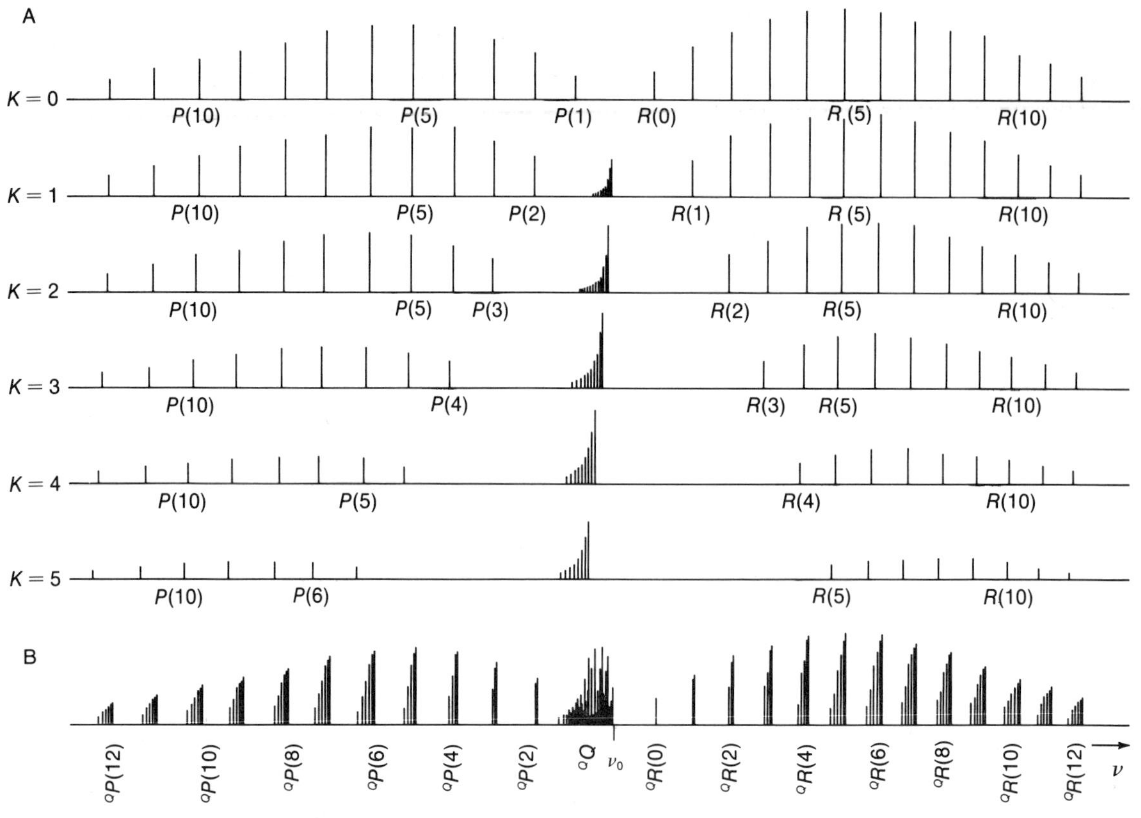

FIGURE 6-10
Sub-bands of (A) a parallel (||) band are (B) superimposed to yield the total band. Only a small difference between A'-B' and A''-B'' has been assumed. The structure of the P, Q, and R branches, due to different J, can frequently be resolved, whereas the finer structure due to different K cannot be seen. [From Herzberg (1945).]

a function of distance r of the X atom from the Y_3 plane is shown in Figure 6-11.

For two equivalent, but unconnected, harmonic oscillators allowed energy levels would be obtained as indicated by the broken lines in Figure 6-11. A classical particle with an energy level that is lower than the potential barrier would never leave one well and appear in the other; however, a quantum mechanical system can tunnel through the potential barrier. It must therefore be described by a stationary wave function that reflects the property of the system found on both sides of the barrier. This can be done simply by a superposition of the wave functions for the two individual wells. In a symmetric double-well potential there are two possibilities for superposition—subtraction and addition. One of the two new eigenfunctions will therefore be symmetric; the other one antisymmetric with respect to $r = 0$. These correspond to two energy levels, one lower and one higher than the original level. These levels are indicated by solid lines in Figure 6-11. The symmetric eigenfunctions are always associated with the lower sublevel and the antisymmetric with the higher sublevel. The sublevels are indexed + and − in Figure 6-11. Transitions are allowed only between levels of different symmetry ($+ \rightarrow -$; $- \leftarrow +$). The difference in observed infrared frequencies due to inversion splittings is thus the sum of the splittings of the upper and the lower levels.

Inspection of the normal modes of ND_3 as seen in Figure 6-12 reveals that none of them represents a pure motion of the D_3 plane perpendicular to the N atom. Normal vibrations ν_1 and ν_2 very nearly depict such a motion, particularly ν_2. Splitting can therefore be best observed in the absorption of these modes.

Vibration ν_2 is the motion of a particle of mass μ (reduced mass) in the potential shown in Figure 6-11.

$$\mu = \frac{m_D}{1 + (3m_D/m_N)}$$

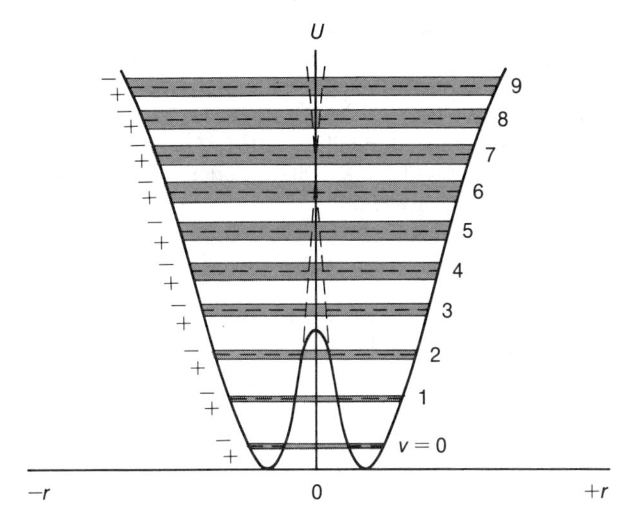

FIGURE 6-11
Potential energy as a function of the distance r of X from the Y_3 plane. [Adapted from Herzberg (1945).]

The probability of tunneling decreases as μ increases. The substitution of H for D thus leads to greater splitting for equivalent vibrational levels. Levels of heavier isotopes are much deeper in the potential well. This can be easily shown by setting $E = h\nu(v + \frac{1}{2})$, where $\nu = \omega \cdot c$, and keeping in mind that $\nu \propto (1/\mu)^{1/2}$. Splitting for equivalent levels will therefore be smaller for the molecule containing the heavier isotope. (See Figure 6-11 for an increase in level splitting).

Instruments

Record the spectrum of ND_3 on a Beckmann I.R. 9 spectrophotometer or an instrument of similar resolution. The assistant in charge of the experiment will handle the instrument. Observe the spectrum between 400 and 3000 cm^{-1}. Use a cell with KBr windows, containing ND_3 at a pressure of 100 mm Hg. In addition, record the region between 600 and 900 cm^{-1} on a fourfold extended chart.

If the Beckmann I.R. 9 is not available (it is commonly used for research), record the spectra of ND_3 and NH_3 using a Beckmann I.R. 5 spectrophotometer or an instrument of similar resolution. The resolution of this instrument is not high enough to observe inversion splitting in the spectrum of ND_3, but it is sufficient to do so in NH_3.

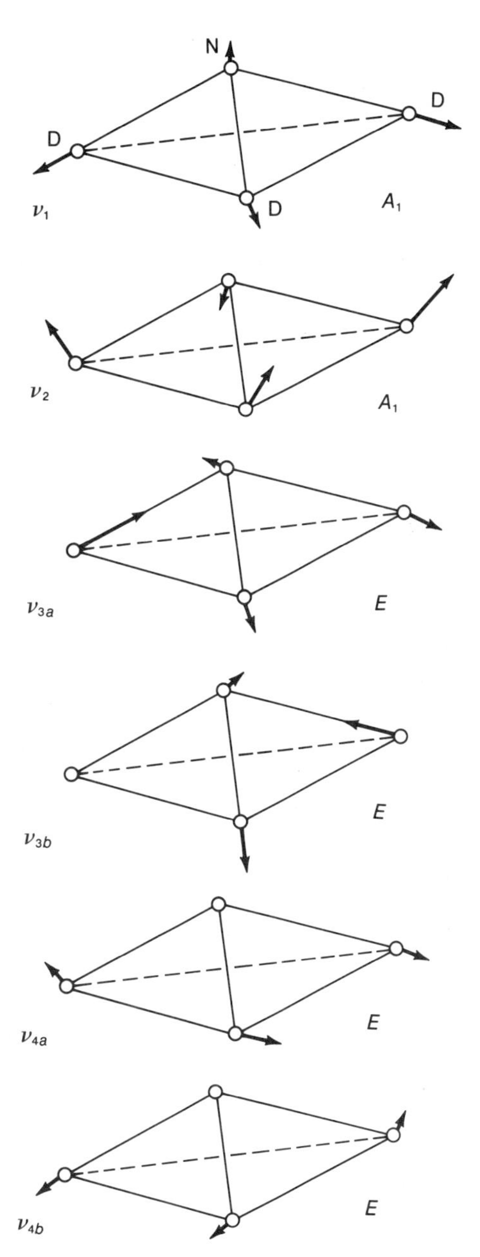

FIGURE 6-12
Normal vibrations of the ND_3 molecule. The length of the arrows indicates the magnitude of displacement during vibration. [Adapted from Herzberg (1945).]

Data

1. Assign as many of the bands you observe as possible. Indicate the stretching or bending mode using the appropriate symbols.
2. Prove that ND_3 is a symmetric-top molecule with evidence obtained from the spectrum.

3. Calculate I_B and, given the angle $\beta = 67°58'$ between r_{ND} and the height of the pyramid, h, calculate r_{ND}.

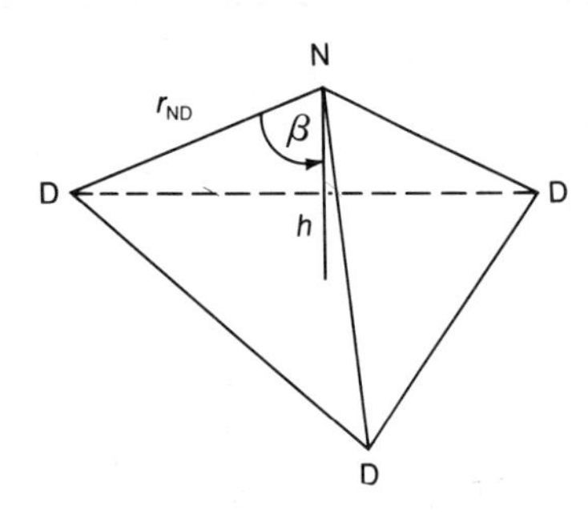

$$I_B = \frac{3m_y r^2(xy)}{2[1 + (3m_y/m_x)]}\left[2 - \left(1 - \frac{3m_y}{m_x}\right)\sin^2\beta\right]$$

$$r(xy) = r_{ND}$$

Calculate I_B from $B_{v=0}$, which can be obtained by using P and R lines that have the lower level in common. If this is not possible because the P branch cannot be observed well enough, use the relation that the distance between two neighboring lines is $2B$.

4. Estimate the sum of the splitting of two adjacent vibrational levels from the observed inversion splitting. If both the ND_3 and the NH_3 spectra have been registered, $r_{ND} = r_{NH}$ and β can be calculated from I_{BNH_3} and I_{BND_3}.

REFERENCES

Barrow, G. M. 1962. *Introduction to Molecular Spectroscopy.* New York: McGraw-Hill.

Bellamy, L. J. 1953. *Infrared Spectra of Complex Molecules.* 2d ed. New York: Wiley.

Birge, R. T., and Sponer, H. 1926. *Phys. Rev.* **28**, 259.

Daniels, F., and Alberty, R. A. 1966. *Physical Chemistry.* 3d ed. New York: Wiley.

Davis, J. C., Jr. 1965. *Advanced Physical Chemistry: Molecules, Structure, and Spectra.* New York: Ronald Press.

Gaydon, A. G. 1953. *Dissociation Energies and Spectra of Diatomic Molecules.* London: Chapman and Hall.

Herzberg, G. 1945. *Molecular Spectra and Molecular Structure: II. Infrared and Raman Spectra of Polyatomic Molecules.* Princeton, New Jersey: Van Nostrand.

———. 1950. *Molecular Spectra and Molecular Structure: I. Spectra of Diatomic Molecules.* 2d ed. Princeton, New Jersey: Van Nostrand.

Klemperer, W. 1965. *J. Chem. Phys.* **42**, 25.

Lewis, G. N., and Randall, N. 1961. *Thermodynamics.* 2d ed. Revised by K. Pitzer and L. Brewer. New York: McGraw-Hill.

Mecke, R. 1923. *Ann. Phys.* **71**, 104.

Moore, C. 1958. *Natl. Bur. Std. (U.S.), Circ. 467.* **3**, 106.

Sherman, J., and Ewell, R. B. 1942. *J. Phys. Chem.* **46**, 641.

Stafford, F. E. 1962. *J. Chem. Ed.* **39**, 626.

Stafford, F. E., Holt, C. W., and Paulson, L. G. 1963. *J. Chem. Ed.* **40**, 245.

Verma, R. D. 1960. *J. Chem. Phys.* **36**, 738.

EXERCISE 7

Spectra of Simple and Complex Atoms

A. INTRODUCTION TO SPECTRA OF ATOMS WITH MORE THAN ONE OPTICAL ELECTRON

Only the basic facts concerning the nomenclature of electronic terms in atoms or ions with several open-shell electrons will be considered in this exercise, together with a review of the selection rules. For further explanation see Davis (1965) or Kautzmann (1957). The intention here is to provide the general information necessary for the various experiments in this part of the manual. If you are unfamiliar with the nomenclature of "one electron" systems, refer to the discussion of theory in Exercise 5A.

Each eigenstate of a one-electron system is characterized by the quantum numbers listed in the left-hand column of Table 7-1. The corresponding quantum numbers for a system with N open-shell electrons are in the right-hand column. Russell-Saunders or L-S coupling is assumed, which means that the individual orbital angular momenta combine to yield the total orbital angular momentum. This also holds for the individual spins. J is then obtained by vectorial addition of the total orbital momentum and the total spin momentum, rather than by coupling of the individual resultant angular momenta of the electrons (j-j coupling).

The nomenclature for electronic terms is the same for one or more electron systems:

$$^{M}L_{J}$$

for example,

$$^{5}D_{0} \quad \text{(quintet } D \text{ zero)}$$

where M is the spin multiplicity, $M = 2S + 1$; L and J are defined in Table 7-1. In many cases the index J is left out because the splitting due to different J values is unimportant for the property under considerations.

Which terms can be found in a system possessing N open-shell electrons and which of them will be lowest in energy? The Pauli principle answers the first half of the question and will be taken up later. The answer to the second half is provided by the application of Hund's rules. Hund's first rule states that the lowest term must have the largest possible multiplicity. The second rule states that among levels of the same multiplicity the level with the largest orbital momentum is lowest in

TABLE 7-1
Quantum numbers for one- and *N*-electron systems.

One-electron quantum numbers	*N*-electron quantum numbers
Principal quantum number $n = 1, 2, 3, \ldots$	$\vec{L} = \sum_{i=1}^{N} \vec{l_i}$
Orbital quantum number $l_i = 0, 1, \ldots, n-1$	is obtained by vectorial summation of individual orbital angular momenta.
Spin quantum number $s_i = 1/2$	$S = \sum_{i=1}^{N} s_i, \left(\sum_{i=1}^{N} s_i\right) - 1, \left(\sum_{i=1}^{N} s_i\right) - 2, \ldots, 0 \text{ or } 1/2^*$
Magnetic quantum number $m_{li} = l_i, l_i - 1, \ldots, 0, \ldots, -l_i$	$M_L = \sum_{i=1}^{N} m_{li} = L, L-1, \ldots, -L$
Inner quantum number $j = l_i + s_i, \ldots, \lvert l_i - s_i \rvert$	$J = L + S, L + S - 1, \ldots, \lvert L - S \rvert$

*0 for *N* even; 1/2 for *N* odd.

energy. The third rule states that among terms of the same multiplicity and orbital momentum the term with the lowest J value is lowest in energy, provided the shell is less than half-full. For shells more than half-filled, the highest J value characterizes the lowest level.

The orbital quantum number of a state is most conveniently found through the relation $M_L = L, L - 1, \ldots, -L$. All possible M_L values can be derived by using the Pauli principle, after it has been decided how spins have to be oriented to give the largest multiplicity. The process can best be shown by example.

To find the ground state of a carbon atom, the configuration $2p^2$ is used. Two quantum numbers n and l are then already equal for these two electrons. According to the Pauli principle, their spins or their magnetic quantum numbers must be different. To obtain the largest multiplicity, spins must be parallel, which means that m_l must be different for the two electrons. For both electrons $l = 1$ (p state) is allowed. Each m_l can thus assume the values 1, 0, −1, and combinations of m_l for the two electrons are $1 + 0$, $1 + (-1)$, $0 + (-1)$, which yields $M_L = 1, 0, -1$.

This leaves only one possibility for L: $L = 1$. The multiplicity is $M = 3$, since $S = 1$ and $M = 2S + 1$. For the three components of the ground state so obtained, $J = L + S$, $L + S - 1$, and $\lvert L - S \rvert$ can assume the values 2, 1, and 0. Of $^3P_2$, $^3P_1$, and $^3P_0$, the last is the lowest in energy according to Hund's rule. For more electrons and higher values of l, the situation cannot be analyzed so easily and the states are counted more conveniently by the distribution of electrons over available states of m_l.

The V^{3+} ion has a $3d^2$ configuration. A determination of possible states is demonstrated in Table 7-2.

The ground state of a V^{3+} ion is ^3F, the lowest component of which is $^3F_2$. The ^3P term is not the next excited one, as one might assume. A singlet term, ^1D, originating from a configuration in which the spins are paired, is lower in energy than the ^3P. This shows clearly that Hund's rules are valid only for selection of the ground state.

In addition to quantum numbers, terms are characterized by their "parity," the property of the wave function to be even or odd with respect to inversion through the nucleus. Even states are frequently designated by an index g (gerade), odd states by u (ungerade). Terms arising from con-

TABLE 7-2
Determination of the ground state of V^{3+} by distribution of the two d electrons over all available values of m_l.

	m_l					M_L	Resulting states
	2	1	0	−1	−2		
Occupation number	1	1				3	$^3F_4$, $^3F_3$, $^3F_2$
		1	1			1	
			1	1		−1	
				1	1	−3	
		1		1		0	
	1		1			2	
			1		1	−2	
	1				1	0	$^3P_2$, $^3P_1$, $^3P_0$
	1			1		1	
		1			1	−1	

Note: Each individual M_L is obtained by the summation of the two occupied m_l values, $M_L = m_{l1} + m_{l2}$. The resulting M_L values can be shown to constitute one set, $M_L = 3, 2, 1, 0, -1, -2, -3$, plus another set, $M_L = 1, 0, -1$, characteristic of an F state and a P state, respectively.

figurations with an odd number of electrons in one-electron orbitals are odd. All others are even.

Selection rules are simple. Laporte's rule prohibits transitions between terms of equal parity. Furthermore, only transitions between states of the same multiplicity are allowed: $\Delta S = 0$; $\Delta L = \pm 1$; $\Delta J = 0, \pm 1$. For heavy atoms and strong *j-j* coupling, S and L cease to be good quantum numbers. In these cases, $\Delta J = 0, \pm 1$ is the only selection rule in addition to Laporte's rule.

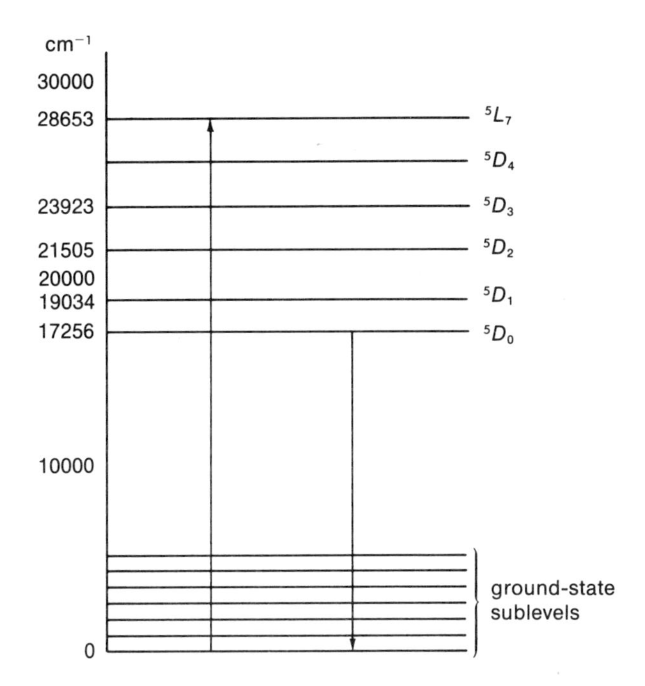

FIGURE 7-1
Energy level diagram of Eu^{3+}. Excitation at 3940 Å and fluorescence are included.

B. QUENCHING OF FLUORESCENCE IN Eu^{3+} SOLUTIONS: ENERGY TRANSFER PROCESSES IN SOLUTION

The transfer of intermolecular energy has become increasingly important in several areas of chemistry. Attempts have been made to use rare earth and transition metal ions in the production of liquid lasers [see Heller et al. (1967)]. Investigations of intermolecular energy transfer have indeed lead to the construction of working liquid lasers involving rare earth ions in which the solvent molecules are composed solely of relatively heavy atoms (e.g., selenium oxychloride) (Kropp and Windsor 1965). The close spacing of solvent vibrational levels makes radiationless transfer of the ion's excitation energy improbable. The resulting suppression of quenching allows the fluorescent level to become highly populated, making laser emission possible. In this experiment the intensity of Eu^{3+} fluorescence in H_2O solutions, in which hydrogen is progressively replaced by deuterium, will be measured to demonstrate the suppression of quenching.

Assign the ground state and its sublevels in Figure 7-1 (see Section A of this exercise for a description of the method to be used). Once this assignment has been made it should be apparent that transitions between any of the states cannot take place. Actually, the Eu^{3+} ion in either an H_2O or D_2O solution absorbs radiation in the region of 3900 Å, as observation of either ultraviolet spectrum shows. This phenomenon is caused by the electric field originating from the water dipoles in the ion's solvation shell. A slight mixing of the free-ion terms results—not enough to make the description of the spectrum in free ion terms impossible (which happens to transition metal ions), but enough to allow certain absorptions to occur. In addition, the labelling of heavy-ion terms based on *L-S* coupling is questionable.

The transition at 3940 Å brings the ion up into the $^5L_7$ state. The resulting fluorescence, however, originates from the $^5D_0$ level. The fluorescence wavelength is the same for both the H_2O and D_2O solutions, but intensities differ greatly.

A mechanism that possibly explains this phenomenon is caused by the electrostatic interaction of Eu^{3+} and the surrounding H_2O or D_2O molecules. A small coupling between europium f electrons and the H_2O (D_2O) dipoles in the electrostatic field makes the transfer of electronic energy from the europium to O-H or O-D bonds possible, where it is stored as vibrational energy. The vibrating dipole picks up energy from the electrostatic field by means of transitions, in which its dipole moment changes, that is, normal modes, overtones, or combination modes that are infrared active. The only requirement to be fulfilled is that the energy transferred in one step must consist of an integral number of electronic energy quanta of the ion, as well as an integral number of vibrational quanta of the dipole. How frequently the transfer occurs depends on the competition between radiative (emission) and nonradiative (transfer to solvent) energy loss of the central ion. The time required for the first process is about 10^{-7} seconds. The time needed for nonradiative energy loss depends on the "order" of the transition, that is,

TABLE 7-3
Infrared bands of liquid water.

Band type	H_2O	D_2O	HDO in D_2O	HDO in H_2O
X-O*X*′ bending	$\tilde{\nu} = 1645$	$\tilde{\nu} = 1215$	$\tilde{\nu} = 1447$	
	$a = 20.8$	$a = 16.1$	$a = 20^{*} \pm 5$	
	$\Delta\tilde{\nu}_{1/2} = 75$	$\Delta\tilde{\nu}_{1/2} = 60$	$\Delta\tilde{\nu}_{1/2} = 85^{*} \pm 5$	
Association*	$\tilde{\nu} = 2125$	$\tilde{\nu} = 1555$		
	$a = 3.23$	$a = 1.74$		
	$\Delta\tilde{\nu}_{1/2} = 580$	$\Delta\tilde{\nu}_{1/2} = 370$		
O-*X* stretching	$\tilde{\nu} = 3280$	$\tilde{\nu} = 2450$	$\tilde{\nu} = 3400$	$\tilde{\nu} = 2500$
	$a = 54.5$	$a = 55.2$	$a = 64 \pm 5$	$a = 42 \pm 5$
	$\tilde{\nu} = 3490$	$\tilde{\nu} = 2540$	$\Delta\tilde{\nu}_{1/2} = 255 \pm 5$	$\Delta\tilde{\nu}_{1/2} = 160 \pm 5$
	$a = 62.7$	$a = 59.8$		
	$\tilde{\nu} = 3920$	$\tilde{\nu} = 2900$		
	$a = 0.83$	$a = 0.598$		
	$\Delta\tilde{\nu}_{1/2}^{\dagger} = 400$	$\Delta\tilde{\nu}_{1/2}^{\dagger} = 330$		

Source: D. Eisenberg and W. Kauzmann, 1969, *The Structure and Properties of Water*. By permission of The Clarendon Press, Oxford.

Note: $\tilde{\nu}$ = frequency in cm^{-1}; a = extinction coefficient × 10^{-3} cm^2 mol^{-1}; $\Delta\tilde{\nu}_{1/2}$ = width of band in cm^{-1} at half-maximum intensity; *X* and *X*′ stand for H or D.

*The composition of the association band is not completely known. It may contain overtones of intermolecular modes as well as combinations of molecular and intermolecular modes.

†Halfwidth of entire band composed of two principal maxima and a shoulder.

the number of phonons transferred to the acceptor. The electronic energy quantum matched by a transition in a single normal mode ($\Delta v = 1$) results in a first-order transition or one-phonon process. If an overtone ($\Delta v = 2$) or a combination mode (e.g., $\nu_i - \nu_k$, where ν_i and ν_k are normal modes and $\Delta v_i = 1$, $\Delta v_k = -1$, and thus $\Delta v_{\nu_i - \nu_k} = 2$) is required, a second-order transition or two-phonon process is encountered. One-phonon processes usually occur fast enough to compete successfully with emission. The higher the order of a transition, the longer it takes to occur, thereby increasing the likelihood that the competing emission will take place instead. Table 7-3 contains a selection of vibrational modes in liquid H_2O and D_2O, together with their halfwidths. Figure 7-1 and Table 7-3 indicate that fluorescence from any of the sublevels of excited states, except $^5D_0$, can be completely quenched by a series of one- or two-phonon energy transfers to O-H or O-D bonds, respectively. In fact, no fluorescence can be found experimentally from any level except $^5D_0$. The slight mismatching of electronic level distances and normal frequencies is compensated by the large halfwidths of solvent bands due to hydrogen bonding.

Investigations beyond the scope of this experiment show that not all ions, excited by the incident radiation, reach the $^5D_0$ level, but for the present purpose it can be safely assumed that the population of the $^5D_0$ level is the same for both H_2O and D_2O solutions. Nevertheless, a great increase in fluorescence intensity is observed in going from H_2O to D_2O solutions. Thus H_2O is a much more effective quencher for Eu^{3+} fluorescence than D_2O, although D_2O solutions show some quenching. Whether the quenching in both solutions is due to nonradiative transfer of energy to O-H and O-D bonds will be determined in this experiment.

Formal Treatment

In the following derivation the quenching mechanism is assumed to be the same for either H_2O or D_2O solutions (nonradiative energy transfer to O-H or O-D vibrations), but the probability with which the energy transfer occurs is different because of a difference in phonon number for O-H or O-D bonds as acceptors. The elementary steps by which Eu^{3+} in the $^5D_0$ state loses its excitation energy are:

$$(Eu^{3+})^* \xrightarrow{k_f} Eu^{3+} \tag{1}$$

by fluorescence and

$$(Eu^{3+})^* + H_2O \xrightarrow{k'_H} Eu^{3+} + H_2O \tag{2}$$

and

$$(Eu^{3+})^* + D_2O \xrightarrow{k'_D} Eu^{3+} + D_2O \qquad (3)$$

by nonradiative energy transfer.

The rate laws for the deexcitation are:

$$R.R._{\cdot fl} = -\frac{d[(Eu^{3+})^*]}{dt} = k_f[(Eu^{3+})^*] \qquad (4)$$

by fluorescence alone and

$$R.R._{\cdot total} = -\frac{d[(Eu^{3+})^*]}{dt} \qquad (5)$$

$$= k_f[(Eu^{3+})^*] + k'_H[H_2O][(Eu^{3+})^*]$$

by fluorescence and nonradiative transfer in H_2O. An equation formally equivalent to equation (5) can be obtained for D_2O solutions.

If a certain fraction x_D of H_2O is replaced by D_2O, equation (5) becomes:

$$R.R._{\cdot total} = k_f[(Eu^{3+})^*] + \{(1 - x_D) \cdot k'_H[H_2O] \qquad (6)$$
$$+ x_D \cdot k'_D[D_2O]\}[(Eu^{3+})^*]$$

The concentration of H_2O and D_2O able to receive energy from the excited ion does not change during the time required for measuring since there is an excess of H_2O and D_2O over Eu^{3+} and vibrational energy is quickly dissipated. Concentrations may be combined with the respective constants. Equation (6) then reads:

$$R.R._{\cdot total} = \{k_f + k_H + x_D(k_D - k_H)\}[(Eu^{3+})^*] \quad (7)$$

with $k_D = k'_D[D_2O]$, and $k_H = k'_H[H_2O]$.

Since the system is in a steady state, the loss of excited europium ions in equations (7) or (8) must be balanced by absorption. The absorbance according to the Lambert-Beer law is:

$$A = -\ln\frac{I}{I_0} = abc \qquad (8)$$

with I_0 being the original intensity of the exciting radiation, I the intensity after passage through the solution, a the absorption coefficient of the Eu^{3+} ion in the H_2O or D_2O solution, b the optical path length, and c the concentration of Eu^{3+}.

That the absorption coefficient does not change when H_2O is exchanged for D_2O must be shown experimentally. Thus the intensity I_{ex} taken up by the system is:

$$I_{ex} = I_0 - I = I_0(1 - 10^{-2.304A}) \qquad (9)$$

according to the Lambert-Beer law.

For $A \leq 10^{-2}$, I_{ex} can be developed in a series:

$$I_{ex} = I_0\left(2.304A - \frac{(2.304A)^2}{2!} + \frac{(2.304A)^3}{3!} - \ldots\right)$$

The series is cut off after the first term. Therefore:

$$I_{ex} = I_0 \cdot 2.304 \cdot A \qquad (10)$$

Since only a certain fraction, fr, of the originally excited ions reaches the fluorescing $^5D_0$ state, the folowing equation may be written:

$$R.R._{\cdot total} = 2.304 \cdot \mathrm{fr} \cdot I_0 \cdot A = \mathrm{fr} \cdot I_{ex} \quad (11)$$

The observed fluorescence intensity I_{fl}, on the other hand, is proportional to the reaction rate because of fluorescence:

$$R.R._{\cdot fl} = \mathrm{fr}' \cdot I_{fl} \qquad (12)$$

See equation (4). The proportionality constant fr′ is determined experimentally by the observation of only a certain fraction of the total fluorescence. The fluorescence yield ϕ, the fraction of excited ions that lose their excitation by means of fluorescence, can now be defined in terms of reaction rates.

$$\phi_{H,D} = \frac{\mathrm{fr}' I_{fl}}{\mathrm{fr} I_{ex}} = \frac{k_f}{k_f + k_H + x_D(k_D - k_H)} \qquad (13)$$

Equation (13) is written when both H_2O and D_2O are present in solution.

Since the two constants fr′ and fr are unaffected by the exchange of H_2O for D_2O with otherwise unaltered experimental conditions, they cancel out of equations of relative fluorescence yields. As standard values either the fluorescence yield of H_2O or of D_2O solutions may be chosen. Using the fluorescence yield of H_2O solutions:

$$\frac{\phi_H}{\phi_{H,D}} = 1 + \left(\frac{k_D - k_H}{k_f + k_H}\right) \cdot x_D = \frac{(I_{fl})_{H_2O}}{(I_{fl})_{H,D}} \qquad (14)$$

This linear relationship between the ratio of measured fluorescence intensities and the fraction of D_2O in solution must be experimentally verifiable, if fluorescence quenching is actually due to energy transfer from excited europium ions to O-H or O-D bonds of the solvent. The fraction x_D of D_2O in solution must actually be interpreted as the fraction of O-D bonds in the solvation shell of

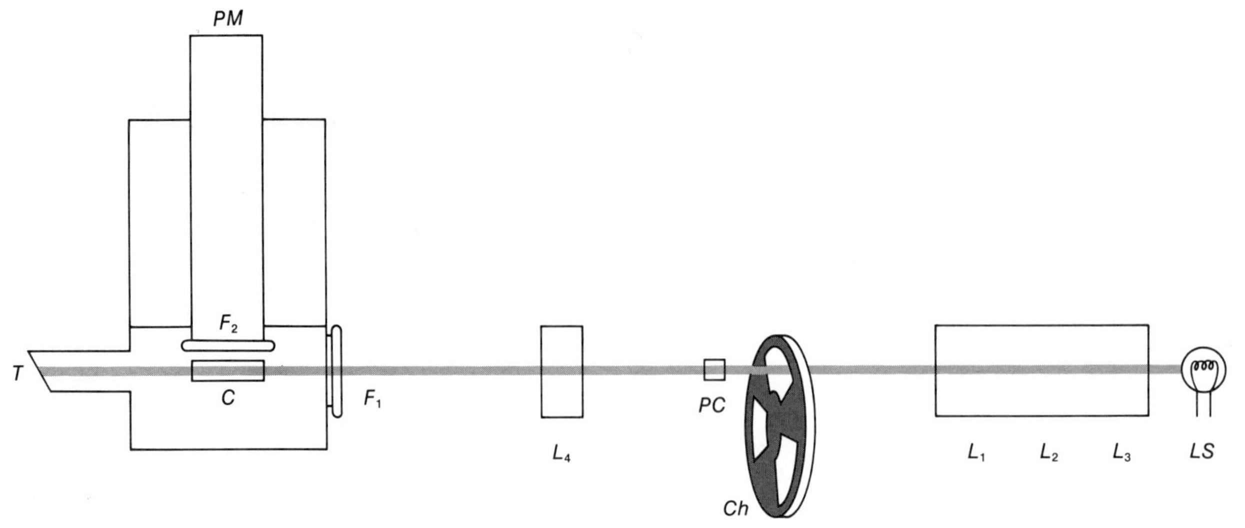

FIGURE 7-2
Optical arrangement viewed from above. *LS*, tungsten filament lamp, 25 W, 6.5 V; L_1, condenser system, T.D.C. matched tridor projection lens, f = 3.5 cm; L_2, plano convex lens, f = 9.2 cm; L_3, plano convex lens, f = 19.0 cm; *Ch*, chopper; *PC*, photocell; L_4, biconcave lens, f = 4.0 cm; F_1, entrance filter, Corning 7-59; *C*, cell; F_2, exit filter, Corning 3-67; *T*, light trap; *PM*, photomultiplier, R.C.A. 6217.

the Eu^{3+} ion, since H and D atoms are quickly interchanged in solution, according to:

$$H_2O + D_2O \rightarrow 2HDO$$

Apparatus

Figure 7-2 shows the optical arrangement of the fluorimeter to be constructed for the experiment that follows. The light from the tungsten filament lamp is guided by the lens system through the chopper and entrance filter into the cell and the light trap behind it. Exact focussing is not necessary since selection of the exciting wavelength is done by a filter that transmits 80% of the incident intensity at 3940 Å, 65% at 4180 Å, and 10% at 4650 Å. The fluorescence observed at right angles to the exciting beam passes a filter of 10% transparency at 4650 Å, 50% at 5520 Å, and 90% at 5900 Å. The exciting radiation is thus kept from entering the photomultiplier. The condenser ensures the capture of a wide solid angle of radiation emitted by the light source. The remaining lenses guide the beam through the chopper, incident filter, and cell into the light trap, thereby keeping intensity losses on the way at a minimum. The inside of the cell compartment and the light trap is blackened to avoid the reflection of exciting radiation into the photomultiplier. The top of the cell compartment is removable to permit the easy exchange of cells; a cell holder keeps them in a fixed position opposite the photomultiplier. As soon as the top is lifted, a safety switch shuts off the voltage from the power supply for the photomultiplier. A large part of the fluorescent light, emitted at random to all sides, is lost, since it is observed from only one side. The fraction of observed fluorescence can be enlarged by covering all sides with a reflecting material, except the one in front of the photomultiplier and those through which the incident beam passes. [See Heller (1967).]

Since the light source is not very strong and the signal from the H_2O solution is weak, it is necessary to improve the ratio of signal to noise. This is done by connecting a lock-in-amplifier to the chopper in the light path. The lock-in-amplifier effectively subtracts the dark current or noise from the photomultiplier (measured at a time when the beam does not pass through the cell) from the total output of the photomultiplier (the fluorescence signal plus noise). Figure 7-3 shows the electronic arrangement.

Light from the stabilized source passes the chopper and falls on a small photocell, which indicates the shutter position to the lock-in amplifier. Whenever the chopper opens the light path, a signal from the photocell leads the output of the photomultiplier into sample-and-hold amplifier *A*. If the chopper blocks the light path no photocell signal is obtained and the photomultiplier output is diverted into sample-and-hold amplifier *B*. Both electrical paths are exactly equivalent. Amplifica-

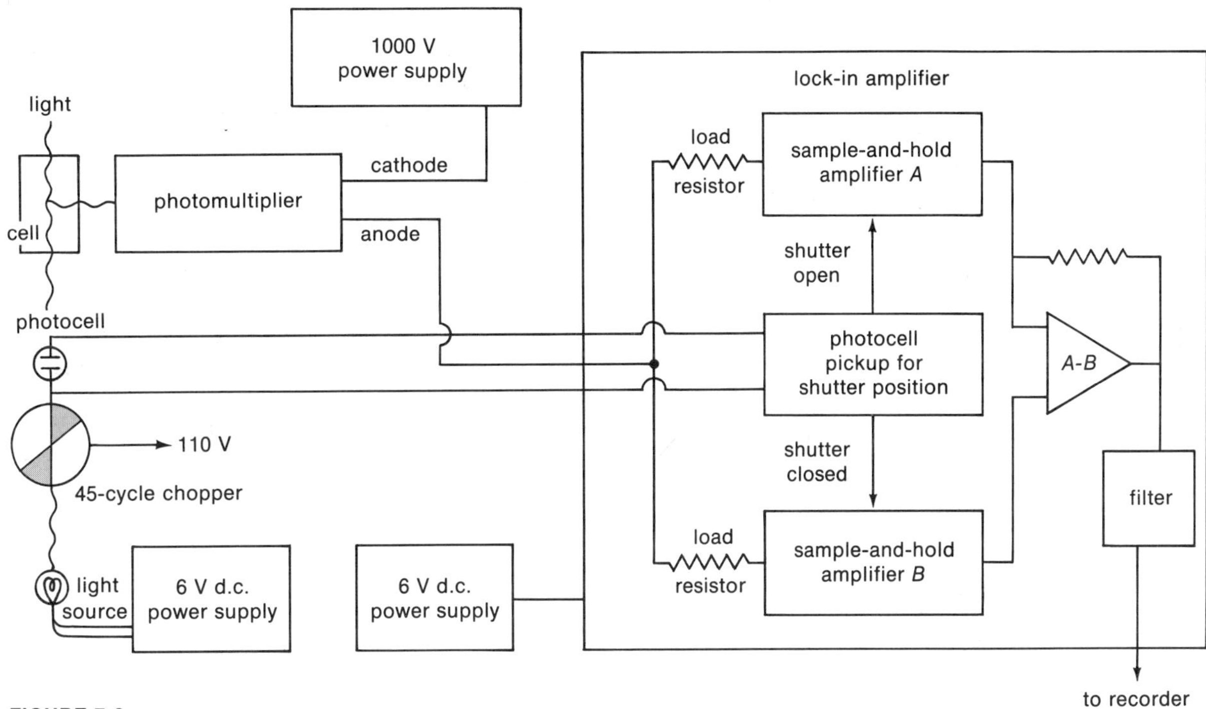

FIGURE 7-3
Electronic arrangement of lock-in-amplifier for fluorimeter.

tion is performed by operational amplifiers. A demodulator eliminates the phase shift between the output *A* and *B* and both are fed into the last operational amplifier, indicated by *A-B* in Figure 7-3. (For the way in which operational amplifiers work see Exercise 12.) A signal proportional to the difference between photomultiplier output due to fluorescence plus noise and output due to noise alone is received from *A-B* and read from an oscillograph, recorder, or galvanometer.

Procedure

Obtain five 20-mm × 40-mm cells, each filled to a suitable height with a 0.1 *M* solution of $EuCl_3$ in either 100% H_2O, 75% H_2O + 25% D_2O, 50% D_2O, 25% H_2O + 75% D_2O, or 100% D_2O. (Since the materials are expensive and the results depend heavily on accurate proportions of D_2O to H_2O, cells that are already filled and sealed are provided by the instructor. Atmospheric water contaminates solutions containing a high percentage of D_2O.)

Obtain the absorption spectrum of 0.1 *M* $EuCl_3$ in H_2O and 0.1 *M* $EuCl_3$ in D_2O in the ultraviolet and visible regions on an appropriate spectrophotometer (e.g., a Beckmann DU or DC, or a Cary 14). If the cells do not fit into the available apparatus, obtain these spectra from the instructor. Measure the fluorescence intensity of each of the five europium chloride solutions. Check whether your data fit equation (14) and discuss your result. It may be advantageous to make the D_2O solution the reference and to modify equation (14) accordingly, in case the signal from the fluorescence of the H_2O solution is too weak to be read with accuracy. Keep in mind that small traces of H_2O in the D_2O solution change the observed fluorescence intensity appreciably.

Questions

1. How many phonons have to be transferred to an H_2O molecule and how many to a D_2O molecule to cause a radiationless transition of Eu^{3+} from the $^5D_0$ state to a suitable sublevel of the ground state of the ion? Which modes of the H_2O or D_2O molecule are most likely to receive the transferred energy?
2. Does the presence of still other kinetically first-order transfer processes render the finding of a linear relationship similar to equation (14) impossible? To obtain an answer, include in the derivation of equation (14) an additional

term caused by transfer following the rate law

$$-\frac{d[(Eu^{3+})^*]}{dt} = k_x \cdot [(Eu^{3+})^*]$$

where $k_x = k'_x[M]$ is possible, and the concentration of acceptor M remains constant throughout the measurement.

C. SPECTROSCOPY OF TRANSITION METAL COMPLEXES BASED ON CRYSTAL FIELD THEORY

Many puzzling phenomena concerning the magnetic and structural properties of transition metal complexes are explained by crystal field theory. An advantage of crystal field theory is that it permits the simple interpretation of transition metal spectroscopy. Crystal field theory advanced from the original assumption that the interaction between the central atom and the ligands is purely electrostatic to include the covalent bonding between them (frequently called ligand field theory). By way of explanation of the optical phenomena encountered in the experiments that follow, this discussion will be restricted to the results of crystal field theory. An introduction to crystal field theory may be found in Orgel (1960) and Dunn, McClure, and Pearson (1965). For detailed instructions on the methods and results of ligand field theory refer to Griffith (1961).

The colors displayed by transition metal complexes present a problem. These metal ions must possess closely spaced electronic levels between which transitions are permitted. These levels do not agree with the free-ion terms for which the first transitions occur in the ultraviolet region of the spectrum. Furthermore, all transitions between terms originating from d electrons are prohibited by the selection rules for electric dipole radiation; namely, $\Delta S = 0$, $\Delta L = \pm 1$, and Laporte's rule (since all terms originating from the same kind and number of electrons must possess identical parity). The V^{3+} ion serves as an example (see Section A of this exercise). The ground state of this d^2 configuration is ^3F; the next excited state is ^1D, followed by ^3P, ^1G, and ^1S. Transitions between these terms are clearly prohibited. Nevertheless, for $[V(H_2O)_6]^6$ two bands are found in the visible spectral range. The origin of these absorptions must be sought in the splitting of free-ion terms due to the electrostatic field of the ligands. Transitions between the resulting terms are permitted.

It has been known for a long time that electronic terms split under the influence of external electric fields (the Stark effect). A similar splitting can be expected if the ion is exposed to the field originating from neighboring ions or dipoles. This field, the crystal field, has a definite symmetry according to the positions of the neighbors. Crystal fields of octahedral symmetry will be under consideration here, since these are encountered most frequently. The way in which terms split is easily explained. In an ion of d^1 configuration (e.g., Ti^{3+}), the single d electron can occupy any of the five degenerate orbitals d_{xy}, d_{xz}, d_{yz}, d_{z^2}, and $d_{x^2-y^2}$. (Consult textbooks on physical chemistry for the derivation and characteristics of these orbitals.) An electron in any of the first three orbitals has its maximum density between the x, y, and z axes. If it occupies one of the other two, the maximum density is found around the z axis or in the x-y plane. Six negatively charged particles are allowed to approach the central ion along the $\pm x$, $\pm y$, and $\pm z$ directions in octahedral symmetry. It then definitely requires more energy to put the single d electron in either the d_{z^2} or the $d_{x^2-y^2}$ orbital than in any of the other orbitals. In other words, the energy in three of the five d orbitals is lowered, in two it is raised. Figure 7-4A shows this level splitting. The magnitude of the splitting depends on the strength of the crystal field. Figure 7-4B shows the splitting of the ^2D ground state of a d^1 configuration as a function of crystal field strength for octahedral symmetry. The nomenclature of the resulting crystal field terms is derived from group theory and refers to the symmetry properties of the terms. For a basic working knowledge it is not necessary to know the origin of the designations. The subject is treated in *The Theory of Transition Metal Ions* (Griffith 1961). Figure 7-4A illustrates that the product of level degeneracy times upward shift is equal to the product of the shift of lower levels times their degeneracy. This is a general phenomenon called "preservation of the center of gravity."

Only one selection rule in addition to Laporte's rule holds for transitions between crystal field terms namely $\Delta S = 0$. The transition $^2T_{2g} \rightarrow {}^2E_g$ would thus be allowed[1] were it not for Laporte's rule (Figure 7-4B). Laporte's rule, on the other hand, can be partly satisfied by a slight contribution of p or f orbitals, which partly alters the even

[1]In Figure 7-4A the transition $^2T_{2g} \rightarrow {}^2E_g$ corresponds to the removal of the electron from an e_g orbital to one of the t_{2g} orbitals.

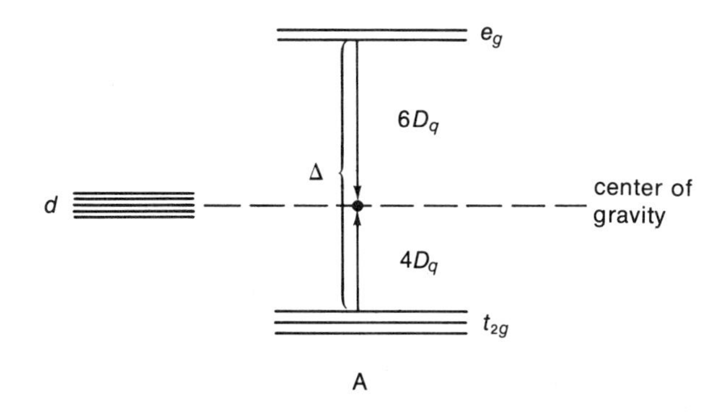

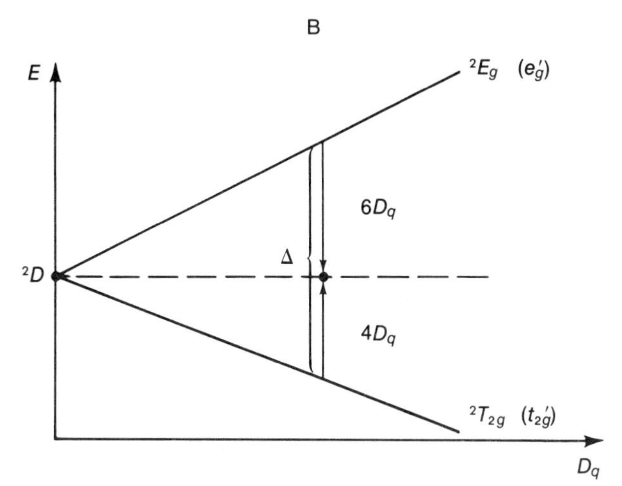

FIGURE 7-4
A. Configuration diagram for octahedral splitting. The magnitude of 10 D_q depends on crystal field strength.
B. Term diagram for octahedral crystal field splitting of a d^1 system. Different central atoms and ligands are found on different places of the D_q scale.

parity of the wave functions. This is the case for all geometric configurations that do not contain a center of symmetry in which the central ion is situated. In other cases, vibrations that are not totally symmetric remove the center of symmetry through distortions of the complex and thus serve to circumvent Laporte's rule. The distortions are too small to have much influence on the term arrangement, which can be successfully based on the undistorted symmetry (for instance, octahedral). From this discussion one might expect one band in the visible spectrum of Ti^{3+} in a complex. The position of this band depends on the nature of the ligands that surround the central ion and the wavelength absorbed is an indication of the ligand field strength expressed in D_q.

$$h\nu(^2T_{2g} \rightarrow {}^2E_g) = 10D_q = \Delta$$

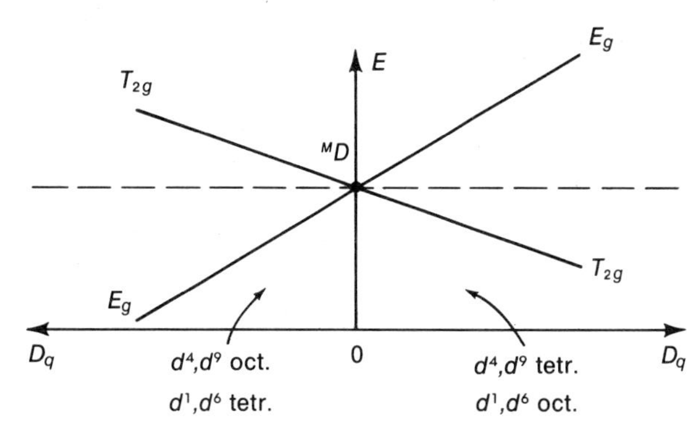

FIGURE 7-5
Term splitting for d^1, d^6, d^4, and d^9 in octahedral and tetrahedral crystal fields. The magnitude of splitting is, of course, different for each ion, ligand, and symmetry.

In fact, a "spectrochemical series" of ligands has been assembled based on the observation of many splittings by different ligands for different central ions. In the series, splitting increases from the left to the right:

$$I^- < Br^- < Cl^- < SCN^- < F^- \sim OH^- \sim NO_2^-$$
$$< \text{oxalate} < H_2O < NH_3 < \text{ethylenediamine}$$
$$< SO_3^{2-} < \text{phenanthrolene} < CN^-$$

(The series as written here is not complete.)

There are d^r configurations in additions to d^1 for which only one crystal field transition is observed. Among these is d^6. The d^6 configuration splits into e_g and t_{2g} levels as did the d^1. The argument is the same for one additional electron placed into an already half-filled shell as for a single electron placed in an empty shell. The ^5D ground state of a d^6 configuration is thus split into a low $^5T_{2g}$ term and a high ^5E_g term. A transition is possible between these crystal field terms. The d^9 systems such as $Cu^{2+}(^2D)$ display the same behavior, but here one electron is missing from an otherwise filled shell. A missing electron is considered to be a positive hole and thus the system is referred to as a "one-hole system" instead of a one-electron system. The argument used for d^1 will, in this case, yield inverse results because of the positive charge of the hole. The five d levels split into two lower e_g levels and three upper t_{2g} levels. Thus ^2E_g becomes the new ground state and $^2T_{2g}$ the first excited state. Analogous to d^6, $d^4(^5D)$ is a configuration with one hole in an otherwise halffilled shell, and the same type of splitting is expected for d^4 as for d^9. Figure 7-5 is a schematic diagram for d^1, d^6, d^9, and d^4 in an octahedral crystal field.

Figure 7-5 shows term splitting in tetrahedral surroundings as well. In a tetrahedral crystal field, ligands approach the electron orbitals from between the axes. The energy of the d_{xy}, d_{xz}, and d_{yz} orbitals rises compared with the remaining two. The splitting pattern is thus inverted with respect to that for octahedral symmetry. However, the magnitude of splitting is much less for given ions or ligands in a tetrahedral arrangement. This is not indicated in Figure 7-5. Diagrams such as Figure 7-5 are called Orgel diagrams.

Of the d^r configurations remaining (d^2, d^3, d^5, d^7, and d^8), d^2 and d^8 form a pair, as do d^7 and d^3. These pairs can be discussed together. For d^2 and d^8, two electrons or two holes, respectively, are distributed over the two kinds of orbitals. For d^7, two electrons are distributed over the already half-filled octahedral crystal field levels and for d^3, two holes are distributed. What happens in the term diagrams is not immediately obvious; therefore the d^2 system must be examined more closely. From the brief discussion of the free V^{3+} ion it is known that the ground state is a ^3F term. This term was derived in Section A of this exercise through the distribution of two electrons with parallel spins over five degenerate orbitals characterized only by different values of m_l. In switching to a more convenient set of five equivalent orbitals (d_{xy}, d_{xz}, . . .), the degeneracy has been found to be lifted in an octahedral or tetrahedral crystal field. By distributing two electrons over this new set of levels, three different configurations are arrived at: t_{2g}^2, $t_{2g}^1 e_g^1$, and e_g^2. These configurations must be of different energies because energy is spent in order to lift one electron from a t_{2g} level into an e_g level. Three different terms arise from the ground state free-ion term. The multiplicity of these three terms is the same as that for the parent term since spins remain parallel. Free-ion terms always split into the same crystal field terms. The crystal field terms to be expected from various free-ion terms are given in Table 7-4.

TABLE 7-4
Free-ion terms and resulting crystal-field terms.

Free-ion terms	Crystal-field terms				
	A_{1g}	A_{2g}	E_g	T_{1g}	T_{2g}
S	1				
P				1	
D			1		1
F		1		1	1
G	1		1	1	1

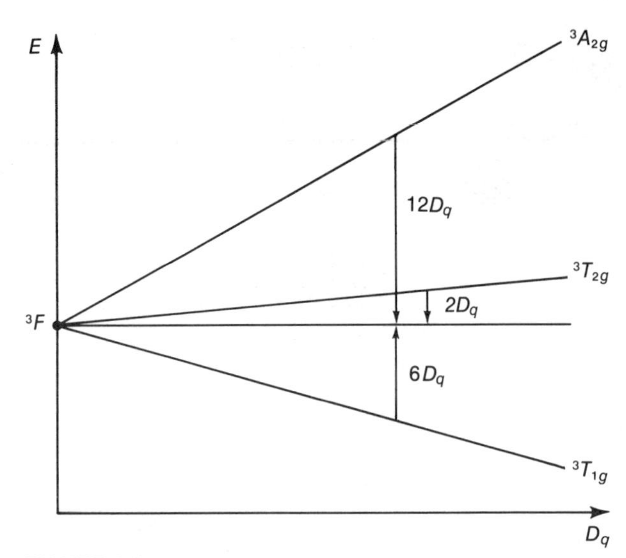

FIGURE 7-6
Octahedral splitting of the ^3F ground state of a d^2 configuration.

The crystal field splitting of the ^3F term of a d^2 configuration is shown in Figure 7-6. The A_{2g} term must be single and the T_{2g} and T_{1g} terms must be triply degenerate to conserve the center of gravity.

So far, higher free-ion terms have not been taken into account. From the discussion of the V^{3+} ion, we know that a ^1D term (paired spins) and a ^3P term (parallel spins) are the lowest excited terms. Of these the ^3P state, which has the same multiplicity as the ground state, shall be considered. As shown in Table 7-4, a P term is not split by the octahedral crystal field but converted into a T_{1g} term. The ^3P term gives rise to another $^3T_{1g}$ term. Two terms of like symmetry are said to interact with each other if their energies are not too dissimilar. In particular, they are not allowed to cross. Figure 7-7 shows the influence of interaction on the two $^3T_{1g}$ terms. Parent terms are indicated in parentheses. The $^3T_{1g}(P)$ term is also triply degenerate. At infinitely high crystal field strength the two terms $^3T_{1g}(P)$ and $^3T_{2g}(F)$ have to arrive at the same energy since they belong to the same configuration, namely $t_{2g}^1 e_g^1$. In fact, the two terms are of different energies and thus two absorptions are possible: $^3T_{1g}(F) \rightarrow {}^3T_{2g}$ and $^3T_{1g}(F) \rightarrow {}^3T_{1g}(P)$. For $V^{3+}(H_2O)_6$, these transitions are situated at 590 mμ and 400 mμ, respectively. The transition $^3T_{1g}(F) \rightarrow {}^3A_{2g}$ lies in the ultraviolet region where various other types of bands occur, which makes assignment difficult. The absorption of the last transition should also be very weak, since it is

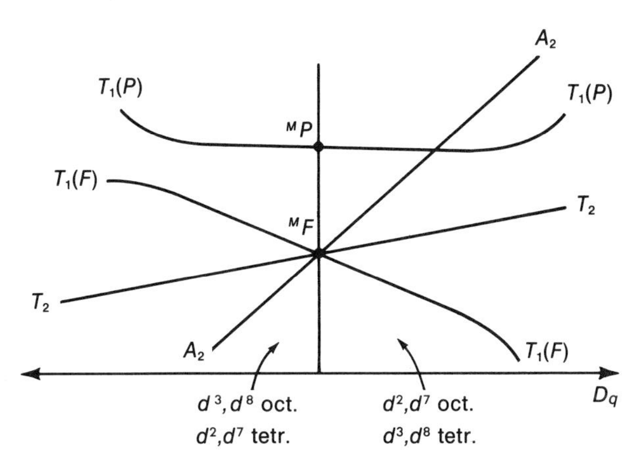

FIGURE 7-7
Octahedral and tetrahedral splittings of F and P terms for d^2, d^3, d^7, and d^8 configurations.

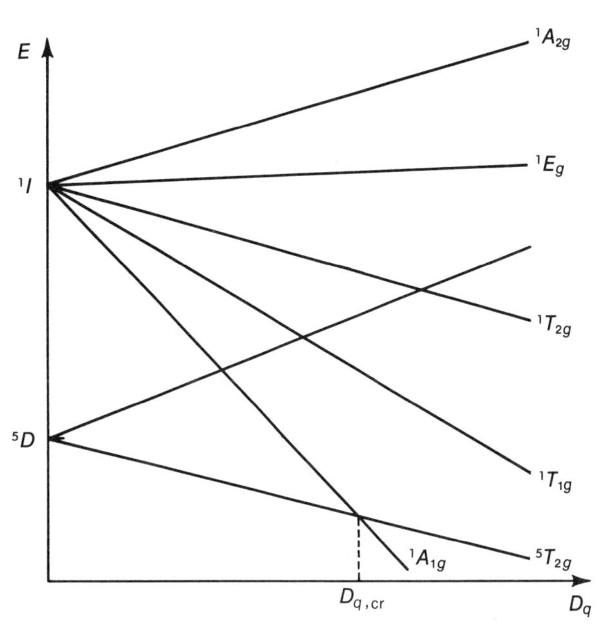

FIGURE 7-8
Crystal field levels originating from the lowest quintet and the lowest singlet free-ion terms of a d^6 configuration. The splittings are not drawn to scale.

a two-electron transition. The two former transitions occur between the same configurations but different terms. The magnitude of the crystal field splitting, $8D_q$, can only be obtained from the transition $^3T_{1g}(F) \rightarrow {}^3T_{2g}$.

Notice that the index g has been left out in Figure 7-7. Since both tetrahedral and octahedral crystal field terms are given in this diagram, the designation g cannot be used. It should be evident from the discussion of configurations lacking a center of symmetry that tetrahedral terms cannot be even (gerade).

The ^1D and other excited free-ion terms of the V^{3+} ion will not be given further consideration here because they are of little importance in absorption spectroscopy. However, this does not apply to some other d-electron configurations, notably d^4, d^5, d^6, and d^7. Here excited free-ion terms of a multiplicity different from that of the ground state may be exceedingly important. A closer examinations of the d^6 configuration is of interest in this regard. Figure 7-8 shows the splitting of the ^5D ground state and a ^1I excited state in the crystal field. At some critical field strength $D_{q,\mathrm{cr}}$ the lowest crystal field terms, one originating from ^5D, the other from ^1I, cross. From this point on, the ground state of the complexed ion is a singlet term. It is immediately evident that spin pairing occurs to the right of $D_{q,\mathrm{cr}}$. The reason for this change is simple. At low crystal field strengths the splitting of terms is small compared with electron-electron repulsion. It is favorable with regard to energy to place three electrons into the t_{2g} subshell, then two electrons into the e_g subshell (all of them with parallel spins), and, finally, the remaining electron into the t_{2g} subshell with antiparallel spin. The configuration $t_{2g}^4e_g^2$ is then reached. Splitting of the levels increases as the crystal field strength increases, and at $D_{q,\mathrm{cr}}$ the energy difference between the t_{2g} and the e_g subshells is larger than the electron-electron repulsion. Less energy is required to put all electrons with paired spins into the t_{2g} subshell than to put some in the e_g shell, which remains empty. The configuration t_{2g}^6 is thus obtained. The spectral implications of these configurations is evident. To the left of the "crossover point" ($D_{q,\mathrm{cr}}$) a spectrum of quintet-quintet transitions is observed; to the right one of singlet-singlet transitions appears. The number and position of bands is characteristically changed. In low-spin cobalt complexes only transitions to $^1T_{1g}$ and $^1T_{2g}$ are observed. The remaining transitions are hidden by other bands. Similarly, high-field and low-field complexes are possible for d^4, d^5, and d^7 configurations. Figure 7-9 illustrates these configurations schematically. The configuration pertinent for a particular compound can be decided with the help of magnetic measurements (see Exercise 9). In practice, ions are found to be either nearly always high-spin or nearly always low-spin in their different complexes.

The interpretation of a spectrum is frequently accomplished with the help of "Tanabe-Sugano

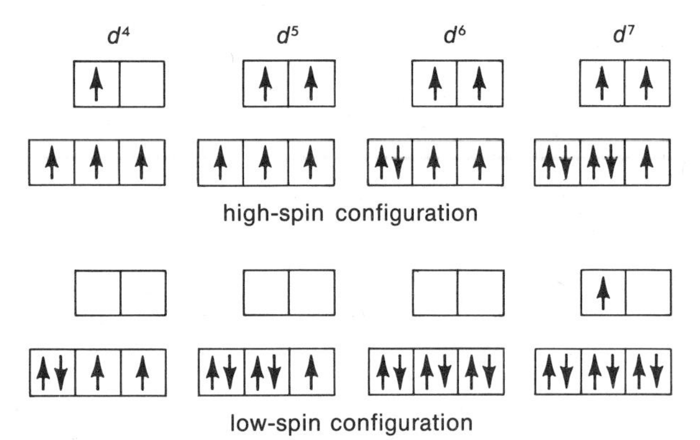

FIGURE 7-9
High- and low-spin configurations for four to seven *d* electrons.

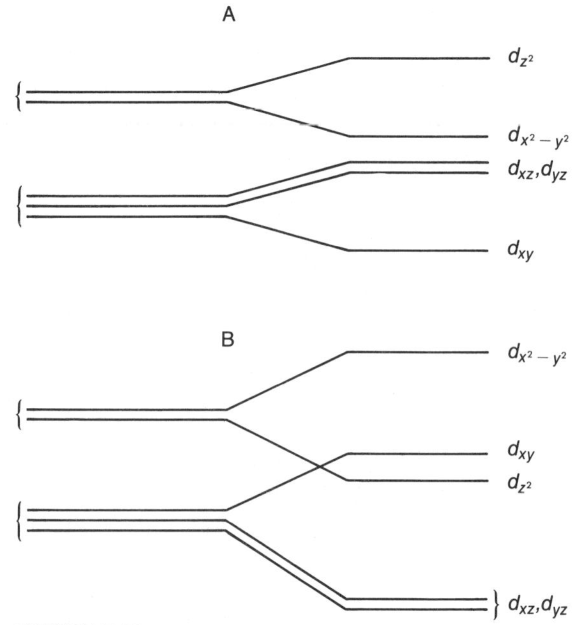

FIGURE 7-10
Splitting of octahedral levels for (A) shortening and (B) lengthening the ligand distance on the *z* axis.

diagrams," which are based on calculations that include many terms. They are similar to Orgel diagrams, except that the ground state is always drawn as a horizontal line; therefore all terms change slope at the crossover point. The energy E and crystal field strength Δ are usually divided by the so-called "Racah parameter," which is related to the term separations in the free ions. This division makes the diagram valid for different ions of the same d-electron configuration. A number of Tanabe-Sugano diagrams are presented by Cotton and Wilkinson (1966).

The d^5 configuration has been excluded from this discussion since the ground state ^6S, a half-filled d shell, does not split in any crystal field. All other terms are derived from excited free-ion terms of altered multiplicity. The very weak color of such complexes stems from forbidden transitions between crystal field terms of different multiplicities—so-called "intercombination bands."

Many octahedral complexes are actually distorted and thus of lower symmetry, a fact that may have considerable consequences for the crystal field spectrum of the complex. Obviously, if two of the six ligands differ from the other four, the ligand field itself cannot be of perfect octahedral symmetry. The more the ligands are separated in the spectrochemical series, the more pronounced the effect. For configurations such as d^1, d^4, and d^9, the "Jahn-Teller effect" lowers the structural symmetry. The Jahn-Teller theorem may be formulated as follows: a molecule of certain structural symmetry, predicted to possess an orbitally degenerate ground state, is actually found to have distorted symmetry, so that the degeneracy of the ground state is removed. What does this imply for octahedral complexes? The two simplest ways in which to distort an octahedron are either to lengthen or to shorten the ligand distance on the z axis. In both cases the degenerate t_{2g} and e_g orbitals do not remain equivalent. Figure 7-10 shows the level splittings for these cases.

In shortening the ligand distance more energy is required to bring an electron into an orbital with maximum density on or near the z axis; thus the energy of the d_z^2 and d_{xz}, d_{yz} orbitals is raised. The energy of the other orbitals must then be lowered to preserve the center of gravity. Lengthening the ligand distance produces the opposite result. This result can also be obtained by regarding a tetragonally distorted octahedron as somewhere between the regular octahedral and a square-planar arrangement. This situation is the result of removing the two ligands from the octahedron along the z axis (see Figure 7-11). In most cases, one axis is elongated. The total energy of a d^1 system is lowered for a distorted complex, compared with that of a regular octahedral arrangement, by exactly the amount by which the lowest t_{2g} orbital was lowered. For a d^4 or d^9 system, the decrease in energy is equal to the lowering of the lowest e_g orbital. To convince yourself of this, place the required number of electrons in orbitals of a regular octahedral or a distorted complex. By the same reasoning, a Jahn-Teller effect should be expected for low-spin d^7 complexes.

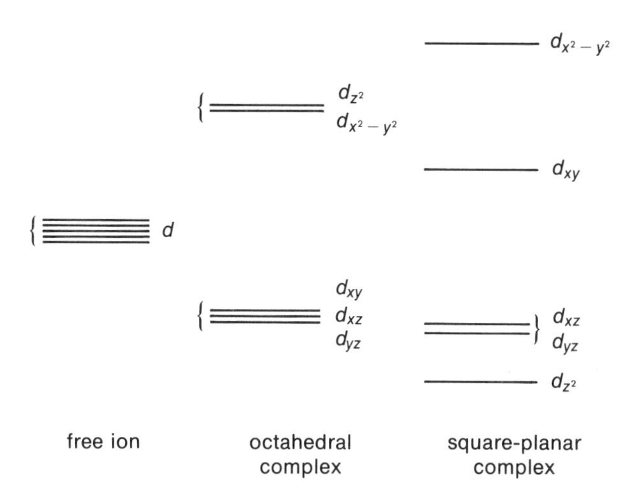

FIGURE 7-11
Configuration diagrams for octahedral and square-planar splitting.

The magnitude of Jahn-Teller splittings is of the order of a few hundred wave numbers. The splittings are therefore seldom resolved, since bands in solution are already broadened through solvent interaction and vibrations. The splitting is indicated mainly by unusually broad and asymmetric absorption bands. There is much evidence of bond lengthening in one axis for complexes of Cu^{2+} and Cr^{2+} (d^9, d^4, respectively), as demonstrated in Exercises 3B and 12C. If level splitting is caused by asymmetry in the ligand field itself, as in $[Co(en)_2Cl_2]^+$, much larger separations of the split levels are to be expected. The ground state of this complex, as well as that of $[Co(en)_3]^{3+}$, is $^1A_{1g}$. This term is not split by the distortion of the ligand field. It is easy to see why this is so by putting the six electrons into the appropriate orbitals. No energy is saved by arranging them in "distorted orbitals" as compared with "regular orbitals." For the excited state $^1T_{1g}$, which corresponds to the configuration $t_{2g}^5e_g^1$, level splitting must have an effect. Two new terms arise because an electron in an upper or lower e_g level produces two situations that differ in energy (see Figure 7-12 for details).

There are changes in intensity in absorption bands due to the distortion of the ligand field. The removal of a center of symmetry always leads to stronger absorption, since the restrictions of Laporte's rule are less applicable (see, for example, cis and trans $[Co(en)_2Cl_2]^+$).

Besides crystal field and intercombination bands the spectrum of a complex usually contains bands due to other sources such as absorption of the ligands (ligand bands). Ligand bands can be determined by comparison of the free ligand spectrum with spectra of different complexes containing the same ligands. Also, there are "charge-transfer bands" due to the transfer of an electron between orbitals of the central atom and the ligands. Charge-transfer bands are characterized by their high extinction coefficients. They are found in the short wave region of the visible range or in the ultraviolet range of the spectrum. One of the substances prepared in Exercise 1B displays such a charge-transfer band. The charge transfer can be verified by exposing a concentrated solution of $[Co(NH_3)_5Cl]^{2+}$ to ultraviolet light (see Exercise 11C) for a few hours or overnight. After this amount of time the pH of the solution will have changed from neutral to slightly basic and some coagulated brown $Co(OH)_2$ can be observed. The reaction is:

$$Co^{3+}(NH_3)_5Cl^- + h\nu \longrightarrow Co^{2+}(NH_3)_5Cl$$

$$Co^{2+}(NH_3)_5Cl + H_2O \longrightarrow Co(OH)_2 + 2NH_4^+ + 3NH_3 + \tfrac{1}{2}Cl_2$$

Charge-transfer bands are most likely to be observed in complexes with halogen or pseudohalogen ligands: (F^-, Cl^-, Br^-, I^-, CN^-, SCN^-, N_3^-, ONO^-, and NO_2^-). They are observed in many other ligand systems and are frequently used in analytical chemistry (spectrophotometry) because of their high extinction coefficients. The spectrophotometric determination of iron with *ortho*-phenanthroline (Exercise 13B) makes use of a charge-transfer band of the iron *ortho*-phenanthroline complex.

Procedure

Record the spectra of several or all of the following solutions in the visible range:

1. 0.1 *M* $TiCl_3$ in H_2O
2. 0.01 *M* $Cu(NO_3)_2$ in H_2O
3. 0.01 *M* $Cu(NO_3)_2$ in 0.1 *M* ethylenediamine, or use a portion of the titrated solution in Exercise 12C
4. 0.01 *M* $Cu(NO_3)_2$ in 12 *M* ethylenediamine
5. 0.2 *M* $Ni(NO_3)_2$ in H_2O[2]
6. 0.2 *M* $Ni(NO_3)_2$ in 1 *M* ethylenediamine
7. 0.01 *M* $[Co(en)_3]Cl_3$ in H_2O (see Exercise 1A)

[2]If possible, record the spectra of numbers 5 and 6 between 3000 Å and 12000 Å.

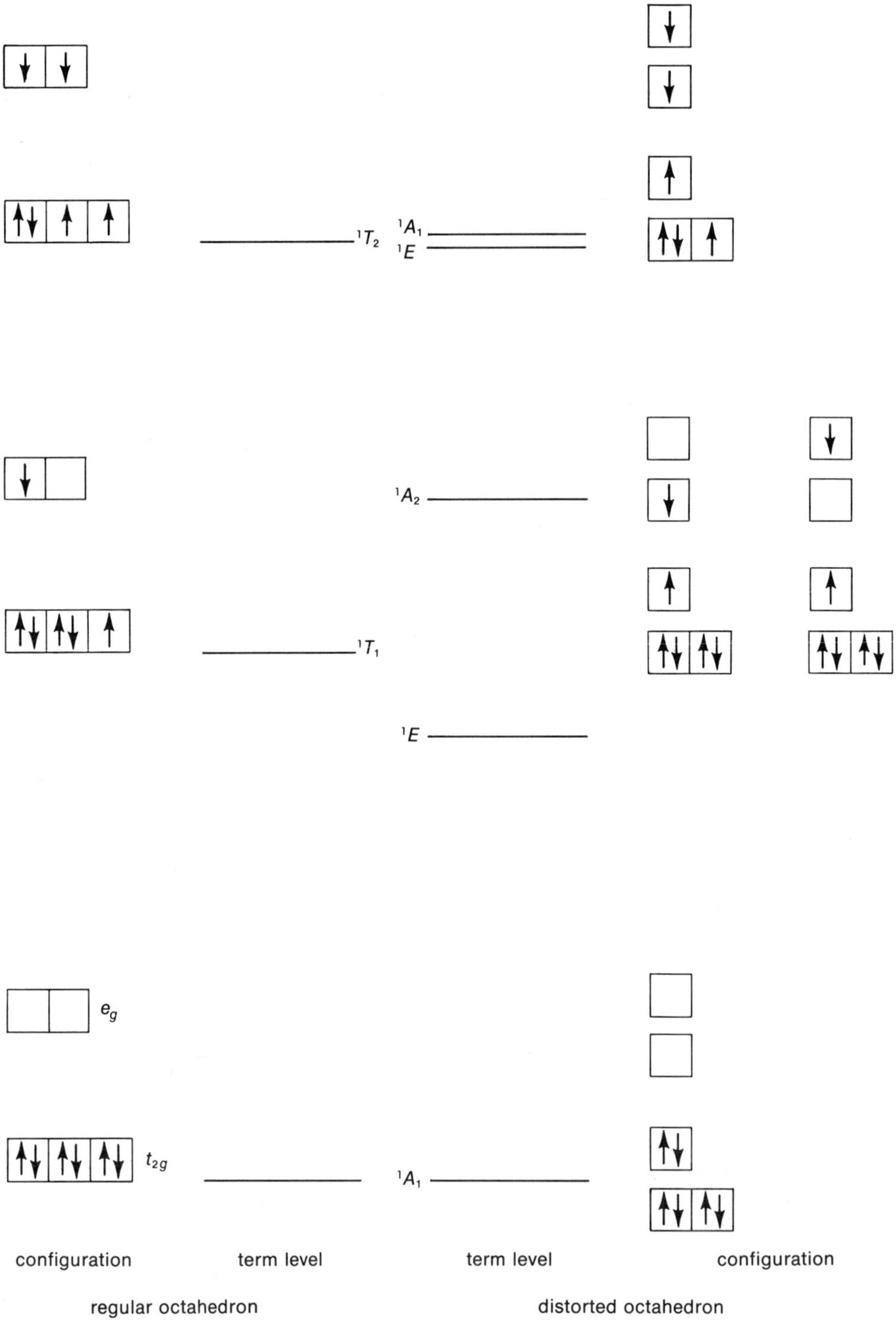

$^1T_2$ $^1A_1$ ^1E

$^1A_2$

$^1T_1$

^1E

e_g

t_{2g} $^1A_1$

configuration term level term level configuration

regular octahedron distorted octahedron

FIGURE 7-12
The d^6 electron configurations and term levels in a tetragonally distorted ligand field. Low-spin situation. Note the preservation of the center of gravity in term levels and configuration levels.

8. 0.01 *M* *cis*-$[Co(en)_2Cl_2]Cl$ in H_2O (see Exercise 1C)
9. 0.01 *M* *trans*-$[Co(en)_2Cl_2]Cl$ in H_2O (see Exercise 1C)
10. 0.01 *M* $[Co(NH_3)_5Cl]Cl_2$ in H_2O (see Exercise 1B) in the visible and ultraviolet ranges (dilute for the latter)
11. 0.01 *M* $Co(NH_3)_5H_2O(NO_3)_2$ in H_2O (see Exercise 1D) in the visible and ultraviolet ranges

Define the complex to which the spectrum is due. Label transitions and prepare a schematic sketch of the term diagram. Calculate D_q in cm^{-1}. Indicate cases in which a splitting or broadening of bands due to reduced symmetry of the ligand field or the Jahn-Teller effect is observed. Compare D_q values for identical ligands but a different central ion. What is the influence of the central-ion charge? Compare D_q values for different ligands and identical central atoms.

The spectra of solutions 7, 8, 9, and 10 were taken immediately following their preparation in the exercises indicated and retained for comparative purposes. Use these spectra to identify *cis*- and *trans*-$[Co(en)_2Cl_2]Cl$. Compare the spectra of solutions 10 and 11 for evidence of the charge-transfer band in solution 10.

Additional Work

Record a pure intercombination spectrum from a solution of $MnCl_2$ in H_2O. All absorptions are due to sextet-quartet transitions. They can be identified by using a Tanabe-Sugano diagram [see Cotton and Wilkinson (1966)].

REFERENCES

Cotton, F. A., and Wilkinson, G. 1966. *Advanced Inorganic Chemistry*. New York: Wiley.

Davis, J. C., Jr. 1965. *Advanced Physical Chemistry: Molecules, Structure, and Spectra*. New York: Ronald Press.

Dunn, T. M., McClure, D. S., and Pearson, R. G. 1965. *Some Aspects of Crystal Field Theory*. New York: Harper & Row.

Eisenberg, D., and Kauzmann, W. 1969. *The Structure and Properties of Water*. Oxford: Oxford University Press, p. 229.

Griffith, J. S. 1961. *The Theory of Transition Metal Ions*. Cambridge: Cambridge University Press.

Heller, A. 1967. *Phys. Today*. **20**, 35.

Kauzmann, W. 1957. *Quantum Chemistry: An Introduction*. New York: Academic Press.

Kropp, J. L., and Windsor, M. W. 1965. *J. Chem. Phys.* **42**, 1599.

Orgel, L. E. 1960. *An Introduction to Transition Metal Chemistry*. London: Methuen.

EXERCISE 8

Optical Activity

A. OPTICAL ROTATION, OPTICAL ROTATORY DISPERSION, AND CIRCULAR DICHROISM

Optically active substances possess the power to rotate plane-polarized light. This property can be found for matter in any state of aggregation. It is always connected with the possibility of arranging a group of molecules (or atoms in a molecule) in two forms that are identical, except that one form is the nonsuperimposable mirror image of the other. The two forms rotate the plane of polarized light to exactly the same extent, but in opposite directions.

The wavelength dependence of optical rotation, the so-called "optical rotatory dispersion" (O.R.D.), and the related phenomenon of "circular dichroism" (C.D.), which leads to the generation of elliptically polarized light, are increasingly employed to determine the structure of organic and inorganic compounds, to determine the helicity in polymer and biochemistry, and to assign electronic transitions to observed absorption spectra. In view of the growing importance of this field, a discussion of the origin of the phenomena before proceeding to their measurement and application is appropriate.

Optical Rotation

Light can be described as electromagnetic waves in which the electric and magnetic field vectors are perpendicular both to each other and to the propagation direction of the beam. If the electric field vector E stays in one plane throughout one wavelength, the light is plane-polarized. The magnetic field vector H of plane-polarized light is also confined to one plane, perpendicular to the plane of the electric vector. It is therefore sufficient to describe a light wave by the behavior of the electric vector alone. Figure 8-1 is a schematic representation of plane- and circularly polarized light.

In circularly polarized light the magnetic and electric field vectors rotate around the direction of propagation in such a way that at comparable points, separated by one wavelength, the same state is always restored. Plane-polarized light can be obtained by the superposition of right and left

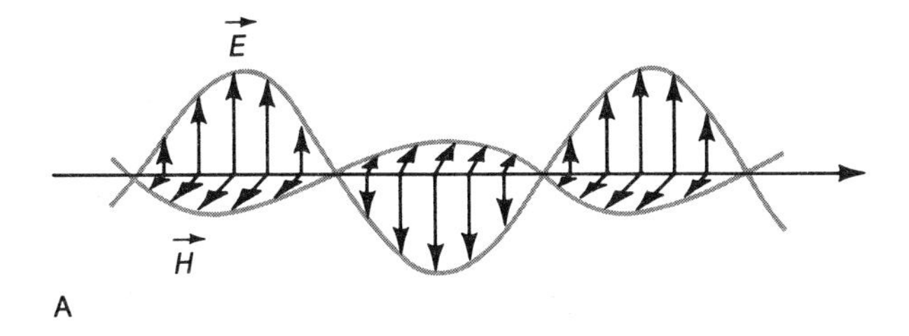

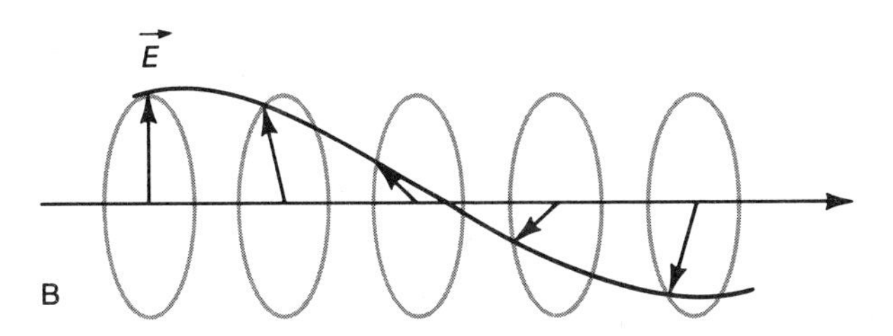

FIGURE 8-1
A. Plane-polarized light. B. Circularly polarized light.

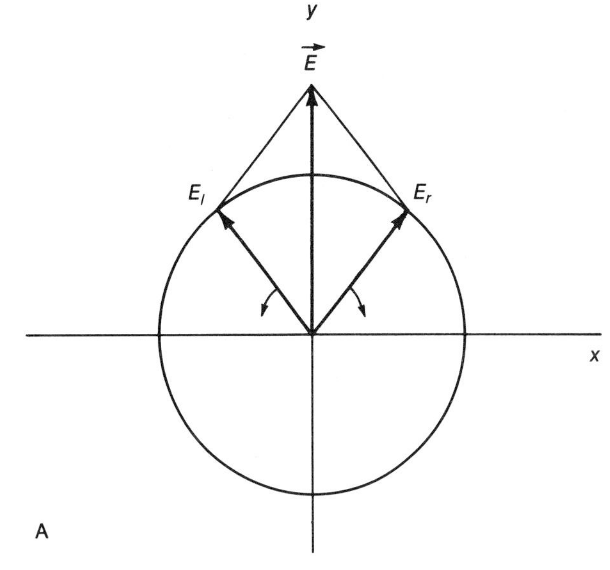

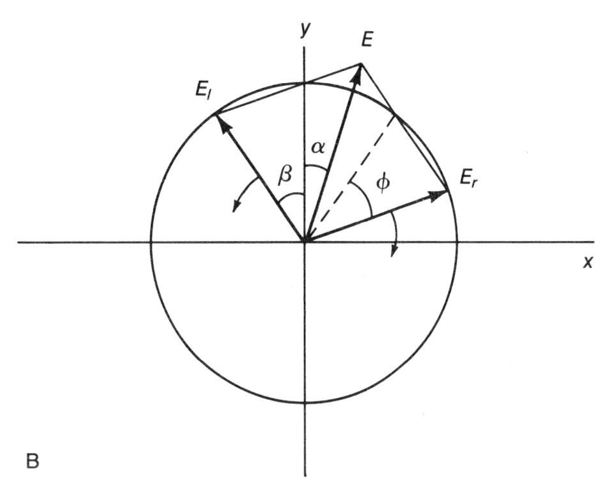

FIGURE 8-2
A. Construction of plane-polarized light from left and right circularly polarized light. The direction of propagation is perpendicular to the plane of this page. Only the electric vector is shown. B. Plane of polarization tilted by an angle α as compared with (A). This is due to a phase difference ϕ between left and right polarized components.

polarized light of identical frequency and amplitude. The plane in which the electric field vector of the resulting wave is found depends on the phase difference between right and left polarized light (see Figure 8-2).

In an optically active medium the refractive index of left and right polarized light is different, $n_l \neq n_r$. This means that the right and left circularly polarized components of the incident beam travel at different velocities. A phase difference results, and upon recombination of the two components a rotation of the resulting plane-polarized beam is observed. The angle of rotation α is directly related to the difference in refractive indices n_r and n_l.

The phase difference ϕ due to a difference in the refractive index can be calculated if the angular velocity $\omega = 2\pi\nu$ and the times t_l and t_r, which each component needs to traverse the sample length l, are known. The phase difference in radians is given by:

$$\phi = 2\pi\nu t_l - 2\pi\nu t_r$$

With $t = l/v$ the following is obtained:

$$\phi = \frac{2\pi\nu l}{v_l} - \frac{2\pi\nu l}{v_r} \qquad (1)$$

where v_r and v_l are the light velocities for the right and left polarized light components, respectively, in the medium.

Since the refractive index is the ratio of light velocity in vacuo to that in the medium, $v = c/n$, and $c = \nu \cdot \lambda$, equation (1) can be modified to yield:

$$\phi = \frac{2\pi l}{\lambda}(n_l - n_r) \qquad (2)$$

The angle α, by which the plane of the polarized light is tilted, amounts to $\frac{1}{2}\phi$ (see Figure 8-2B). Since the resultant of two vectors of equal magnitude always divides the angle between the two in half, $\beta + \alpha = \beta - \alpha + \phi$. β is the angle between E_l and the resultant in Figure 8-2A or E_l and the

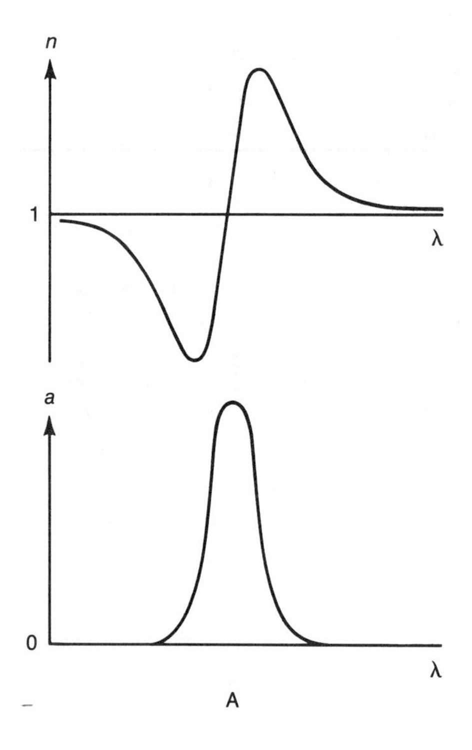

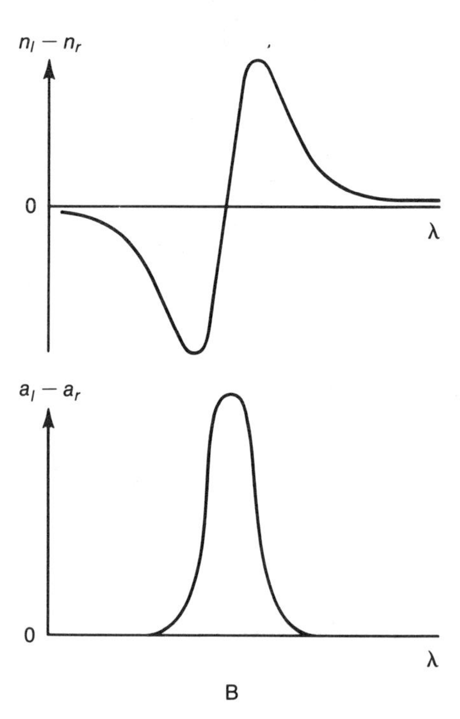

FIGURE 8-3
Behavior of (A) refractive index n and absorption coefficient a around region of absorption and (B) the difference in refractive indices of right and left polarized light and the difference in absorption coefficient in the neighborhood of an optically active band.

y axis in Figure 8-2B. The angle of rotation can thus be expressed in degrees as:

$$\alpha^\circ = \frac{180}{\lambda} l(n_l - n_r) \tag{3}$$

where $(n_l - n_r)$ depends on the nature of the optically active compound and its concentration in solution, on temperature, and on the wavelength of the incident light.

The angle of rotation is expressed empirically as follows:

$$\alpha^\circ = [\alpha]_\lambda^T \cdot l \cdot c \tag{4}$$

where the sample length l is measured in decimeters and the concentration c in grams per cubic centimeter. Comparison with equation (3) shows that the difference in the refractive index is assumed to be proportional to the concentration of the solution. The specific rotation $[\alpha]_\lambda^T$ is then independent of concentration. Temperature control and constancy of the wavelength measurement are necessary for large concentration ranges and precise work.

Optical Rotatory Dispersion

The wavelength dependence of α is directly related to the wavelength dependence of the difference in refractive indices of right and left circularly polarized light. Just as the refractive index of a normal substance increases slightly with wavelength until the region of an absorption band is reached, where anomalous dispersion is observed, so the angle of rotation changes only slightly until an "optically active absorption" is reached and "anomalous behavior" begins. Around an optically active band the absorption coefficients for left and right polarized light become different. Figure 8-3 shows anomalous dispersion for an absorption band observed with unpolarized light and for an optically active band observed with polarized light.

Not all absorption bands of an optically active medium display anomalous rotatory dispersion. It is necessary that the transition causing the absorption involves the center of asymmetry. From the discussion of the behavior of rotatory dispersion it should be obvious that the sign of rotation at one wavelength is no indication of the absolute configuration (right- or left-handedness) of the molecules causing the rotation. The sign varies for different wavelengths.

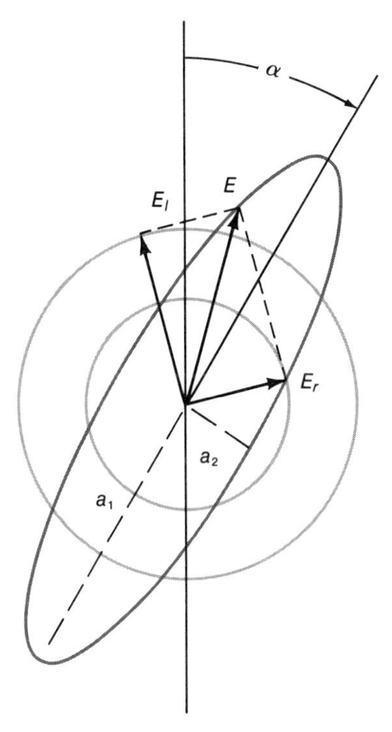

FIGURE 8-4
Creation of elliptically polarized light from right and left circularly polarized light due to the difference in absorption coefficients for the two components. The major axis of the ellipse is tilted by an angle α because of the difference in refractive indices of the two components.

Circular Dichroism

In an optically active absorption band the phenomenon of different absorption coefficients for left and right polarized light is called circular dichroism. It gives rise to an elliptical polarization of the formerly plane-polarized light, which can be verified by proving that the tip of the resulting electric field vector in Figure 8-4 describes an ellipse. The angle between the major axis of this ellipse and the vertical coordinate corresponds to the angle of rotation produced by the difference in refractive indices.

The ellipticity of the emerging light is measured by the ratio of the semiminor axis a_2 to the semimajor axis a_1 and is expressed in the form of the angle θ (see Figure 8-4).

$$\tan\theta = \frac{a_2}{a_1} \tag{5}$$

Light intensity I is proportional to the square of the magnitude of the electric vector $\vec{E}$: $E^2 = \text{const}\ I$. Since left and right polarized light were originally of the same intensity I_0, the intensities of the emerging components can be expressed by their transmittances T in the optically active medium.

$$I_r = I_0 T_r \tag{6}$$

and

$$I_l = I_0 T_l$$

Therefore,

$$|E_l| = \text{const}'\ |E_0| \cdot \sqrt{T_l} \tag{7}$$

and

$$|E_r| = \text{const}'\ |E_0| \cdot \sqrt{T_r}$$

The quantities a_1 and a_2 are the sum and the difference, respectively, of E_l and E_r. From equation (7) the following is obtained:

$$a_1 = \text{const}'\ |E_0|\ (\sqrt{T_r} + \sqrt{T_l}) \tag{8}$$

$$a_2 = \text{const}'\ |E_0|\ (\sqrt{T_r} - \sqrt{T_l})$$

or from equation (5):

$$\tan\theta = \frac{\sqrt{T_r} - \sqrt{T_l}}{\sqrt{T_r} + \sqrt{T_l}} \tag{9}$$

The numerator and denominator are multiplied by the denominator:

$$\tan\theta = \frac{T_r - T_l}{T_r + T_l + 2\sqrt{T_r T_l}} \tag{10}$$

Since θ is always small and $T_r \approx T_l$, equation (10) can be approximated as follows:

$$\theta = \frac{1}{4}\frac{\Delta T}{T_{\text{av}}} \tag{11}$$

in which $\Delta T = T_r - T_l$ and $T_{\text{av}} = (T_r + T_l)/2$.

The absorbance A can now be expressed by the Lambert-Beer law:

$$A = a \cdot b \cdot c = -\log T \tag{12}$$

in which $T = 10^{-A}$. Since the differences in transmittance are always small, $\Delta T = -10^{-A} \cdot \ln 10 \cdot \Delta A$; T is the average transmittance since A is obtained from a normal absorption spectrum, $\Delta A = (a_r - a_l) b \cdot c$. Thus equation (11) yields:

$$\theta = \frac{2.304}{4}\,\Delta A = \frac{2.304}{4}\,(a_l - a_r) \cdot b \cdot c \tag{13}$$

in radians, or

$$\theta^\circ = \frac{\theta 360}{2} = 33.0(a_l - a_r) \cdot b \cdot c$$

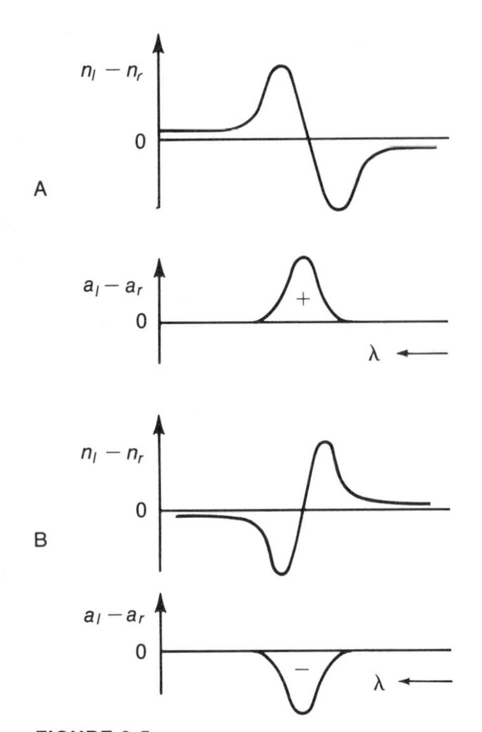

FIGURE 8-5
Cotton effects: (A) positive and (B) negative. Both effects are depicted for a single electronic transition.

in degrees. A frequently reported value is the molecular ellipticity defined as:

$$[\theta^\circ]_\lambda = \frac{\theta^\circ}{10} \frac{M}{b \cdot c} \tag{14}$$

where M is the molecular weight of the optically active substance, c its concentration in grams per cubic centimeter, and b the sample length in centimeters. The dimension of $[\theta]_\lambda$ is then degrees cm^2/decimole. The relation between equations (13) and (14) is easily seen.

Applications

The measurement of optical rotation at one or several wavelengths in regions far from optically active bands is useful for analytical purposes. In most cases $[\alpha]_\lambda^T$ is sufficiently constant to allow the calculation of concentration in solution from equation (4). However, if precise values are to be obtained, it is necessary to check the concentration dependence of the specific rotation.

In the absence of reactions that remove the optical activity, the angle of rotation of a mixture of optically active substances can be regarded as the sum of the individual angles:

$$\alpha = \left(\sum_i c_i\right)^{-1} \sum_i [\alpha_i]_\lambda^T \cdot c_i \tag{15}$$

A simple example is that of a racemic mixture in which both optically active forms are present in equal amounts. No optical activity is observed since the specific rotation is of equal magnitude but of opposite sign for the enantiomers.

For the purpose of structure determination it is necessary to study the wavelength dependence of the optical rotation. Anomalous optical rotatory dispersion and circular dichroism around an optically active band are called the Cotton effect. For any single optically active transition O.R.D. and C.D. curves are strictly related in sign and magnitude. A positive difference $(a_l - a_r)$ is always associated with an increase of $(n_l - n_r)$ from the long-wave region and is called a "positive Cotton effect." A negative Cotton effect displays a negative difference $(a_l - a_r)$ and a decrease of $(n_l - n_r)$ from regions of long wavelength. Figure 8-5 shows positive and negative Cotton effects. The trough-to-peak distance in the O.R.D. spectrum is proportional to the maximum of the C.D. curve.

Since O.R.D. and C.D. curves are so intimately related, in principle more information cannot be gained from one type of curve than from the other, and measurement of one or the other is sufficient. Experimentally, it is easier to obtain O.R.D. than C.D. curves, but since the C.D. does not change sign in a band arising from one transition it is very helpful in resolving the structure of overlapping band systems. Figure 8-6 is a schematic presentation of a band system composed of two transitions of different Cotton effects.

Classical theory of optical rotation (Kauzmann 1957) already indicates that the scattering of light by a right-handed helical molecule leads to a positive (dextro) rotation, whereas under the same circumstances left-handed helices produce a negative (levo) rotation. However, to date more complicated theories have failed to predict reliably the sign of the Cotton effect in small molecules. Empirical methods based on the comparison of the Cotton effect for the same type of transition in different transition metal complexes are discussed in recent literature [see Ballard, McCaffery, and Mason (1962)]. Section B of this exercise deals with the standard compound available $D^*(+)[Co(en)_3]^{3+}$ for which the absolute configuration is known from x-ray investigations.

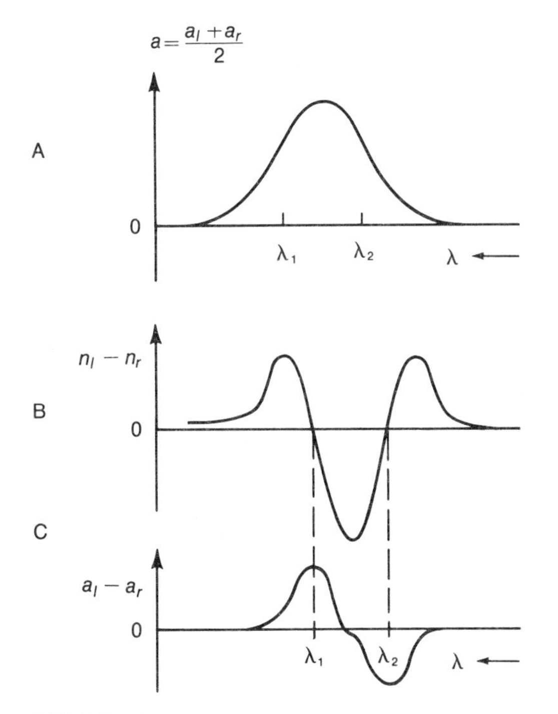

FIGURE 8-6
The (A) absorption spectrum, (B) O.R.D. curve, and (C) C.D. for an absorption band due to two electronic transitions of different Cotton effects.

Apparatus

Polarimeter

The principal elements in a polarimeter are Nicol prisms or similar devices, which convert normal light into plane-polarized light. They are fabricated from birefringent materials such as calcite in which light that is plane-polarized with respect to the major optical axis travels faster than the perpendicularly polarized component. The crystal therefore exhibits different refractive indices for parallelly or perpendicularly polarized light. The prism is cut and cemented together again with a material that possesses, as nearly as possible, the refractive index of one light component. This component can then pass through the cement in the calcite without restraint. The other component is reflected by the cement because of the large difference in refractive index and adsorbed in a black layer on the side of the prism.

Since the orientation of the prism defines the plane of polarization, rotation of the prism at a certain angle around the direction of propagation causes a revolution of the plane of emerging polarized light at the same angle. Two Nicol prisms are "parallel" if the planes of the emerging beams are parallel. Two parallel Nicol prisms on either side of a cell filled with an optically active solution constitute a simple polarimeter. The first Nicol prism is called the polarizer, the second, the analyzer. The optically active substance in the cell tilts the plane of light that has left the polarizer. Since only the parallel component of light leaving the cell can pass the analyzer, a dimming of the light is observed behind the analyzer. The angle of rotation due to the substance in the cell can then be measured as the angle at which the analyzer must be turned to restore maximum brightness. The human eye is much more capable of comparing the brightness of two pictures observed at the same time than of determining the absolute brightness of a single picture. "Half-shade" instruments operate on this principle (see Figure 8-7).

Rotate the auxiliary Nicol prism behind the polarizer at a small angle with respect to the polarizer. Without any optically active material in the cell, by turning the analyzer an angle can be found for which as much light passes through the polarizer plus the half-shade prism (auxiliary Nicol prism) as through the part of the polarizer not obstructed by the half-shade prism. In this case the two fields observed through the eyepiece are of equal brightness as shown in Figure 8-7. Note the angle of rotation. Upon insertion of the optically active solution, rotate the analyzer to obtain equal brightness again. The angle of rotation for the

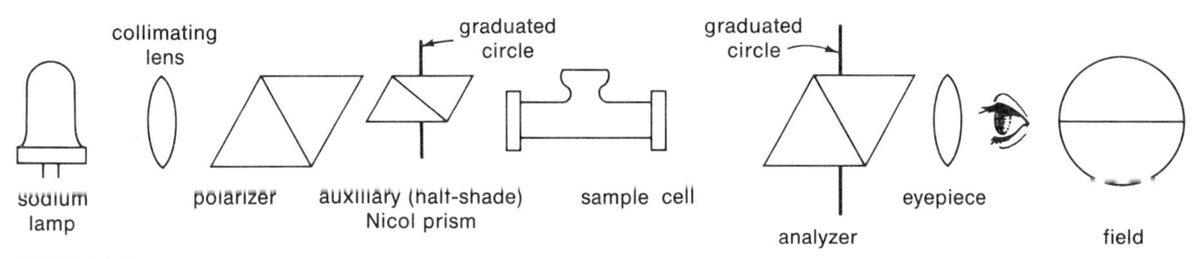

FIGURE 8-7
Schematic diagram of a half-shade polarimeter.

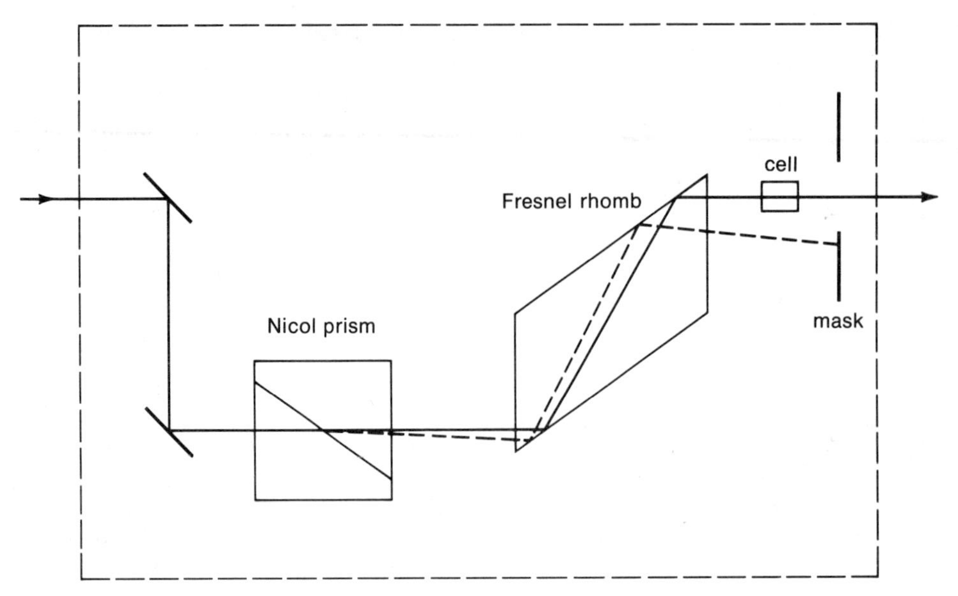

FIGURE 8-8
Creation of circularly polarized light from plane-polarized light by means of a Fresnel rhomb. Optical arrangement as used in the C.D. accessory of the Cary 14 spectrophotometer. [By permission of Varian Instrument Division/Cary Products.]

substance is simply the difference between the two readings. Rotation to the right is positive and to the left, negative, according to convention and procedure. Precision polarimeters permit the adjustment of the position of the auxiliary Nicol prism, so that illumination of the two half-fields is exactly uniform. This adjustment must be altered for different light intensities, wavelengths, or solutions of widely different optical densities.

Spectropolarimeter

Any polarimeter can be transformed to a spectropolarimeter in principle by the addition of a monochromator and the substitution of a white light source for the line source. The human eye as a radiation detector is replaced by photosensitive electronic equipment. Care must be taken that collimated light enters the polarimeter system from the monochromator. Slits and Nicols must be arranged to accommodate the partial polarization of light due to reflections in the monochromator system. Automatic recording spectropolarimeters are commercially available.

Circular Dichroism Apparatus

Three kinds of devices are used to produce circularly polarized light from plane-polarized radiation: quarter-wave plates, Fresnel rhombs, and Pockel cells. Quarter-wave plates are cut from birefringent material in such lengths that the two components emerge with a phase difference of 90° (a quarter-wavelength). Circularly polarized light is obtained upon the recombination of the two perpendicularly plane-polarized components. This device is strongly dependent on wavelength, that is, each plate works only within a very limited wavelength range and is thus scarcely used. In Fresnel rhombs the difference in the refractive index for right and left circularly polarized light in some materials is used. The rhomb is cut so that two internal reflections of the entering beam separate the two components sufficiently to blend out one of them (see Figure 8-8). Matched arrangements of a Nicol prism and a Fresnel rhomb can be placed in the sample and the reference path of a double-beam spectrophotometer so that in one path the absorption of right polarized light is observed; the absorption of left polarized light is observed in the other. The instrument records the difference of absorbances for right and left circularly polarized light, which is proportional to the molecular ellipticity [see equations (13) and (14)]. The separation of left and right circularly polarized light is slightly dependent on wavelength because of the wavelength dependence in refractive indices.

Pockel's electrooptical modulator consists of

a thin plate of potassium dideuterium phosphate (KDP), or a similar piezoelectric material, cut perpendicular to the single optical axis. A high electric potential difference is applied across both surfaces, which makes the crystal birefringent. The difference in refractive indices for the two mutually perpendicular plane-polarized light components depends on the applied voltage. It is adjusted in such way that a phase difference of 90° results for the two components; thus at the plate end circularly polarized light is obtained. The applied voltage can be programmed to follow the change in wavelength, when scanning the spectrum, so that the 90° phase difference is always maintained. The direction of polarization (right or left) depends on the sign of the applied voltage. Inversion of the voltage means inversion of polarization. A Pockel-type electrooptical polarizer is employed in the Cary model 60 C.D. accessory.

Procedure

Analysis of *Dextro-* and *Levo*-tris-(ethylenediamine)cobalt(III) Iodide

Fill the polarimeter cell with water and measure the position of the analyzer for identical brightness in the two fields of a half-shade polarimeter. Adjust the position of the auxiliary Nicol prism if possible and necessary. Be careful to fill the cell completely without the inclusion of air bubbles. Unscrew the plane windows if necessary to clean the cell. Do not press the windows too tightly on the cell. Induced strain in the glass causes polarization of transmitted light. Repeat the procedure again using a solution of 0.1 g of the *dextro*-iodide in 25 ml of water, followed by a solution of 0.1 g of the *levo*-iodide in 25 ml of water. Rinse the cell twice with a small amount of the solution before filling it for measurement. Do not adjust the auxiliary Nicol prism between the measurements using water and the measurement using the solutions.

Calculate the percentage of the resolved dextro or levo complex in your product.

$$\% \text{ optically active compound} = \frac{\alpha}{b \cdot c[\alpha]_\lambda^T} \times 100$$

where α is the measured angle. Derive this equation from equation (15) for optically inactive contamination or the unresolved racemate present. The specific rotation of tris(ethylenediamine)-cobalt(III) iodide is $[90°]_{Na_D}^{20}$. Base the calculation of yield for the preparation of optical isomers on the purity of the compound. The total content of tris(en)Co(III) in each sample can be obtained from the absorption spectrum. The molar extinction coefficient is $a = 84$ liter/mole cm at 469 mμ.

Measurement of O.R.D. and C.D.

If a spectropolarimeter is available, measure the O.R.D. curve of *levo-* and *dextro*-tris(en)Co(III) iodide between 400 mμ and 600 mμ. Compare it with the absorption spectrum in the same wavelength region. Compare it with the C.D. curve also, if an instrument for measurement is available. Report your observations concerning the Cotton effect of each band and optical isomer. Is the band centered around 4700 Å due to a single electronic transition?

B. OPTICAL ACTIVITY AND ABSOLUTE CONFIGURATIONS OF COMPLEXES

Optical isomers can be identified by their absolute configurations and by their signs of optical rotation at a certain wavelength. The sign of rotation at the Na_D line is denoted by (+) for dextrorotatory compounds and with (−) for levorotatory compounds. If the rotation has been measured at a wavelength λ different from Na_D, a subscript to the sign [for example, $(+)_\lambda$] specifies the wavelength λ.

The symbols D* and L* or Λ and Δ designate the absolute configuration. Λ and Δ pertain to the appearance of the complex when viewed along a C_3 axis. D* and L* designate the chirality observed along a C_2 axis. Any threefold chelated complex such as $Co(en)_3^{3+}$ in a Λ configuration has the form of a left-handed screw or, better, a three-bladed, left-handed propeller around the C_3 axis (Figure 8-9). Nevertheless, a look along the C_2 axis of the same complex suggests a right-handed, two-bladed propeller (Figure 8-10). Thus the configuration may as well be called D*. The asterisks have been affixed to D and L to avoid confusion with the old nomenclature in which *D*, *d*, or (+) indicated positive optical rotation at the Na_D line. Unfortunately, a total agreement on nomenclature has not yet been reached (Mason 1963). The complete information about a substance is given in the following sequence: first, the sign for the absolute

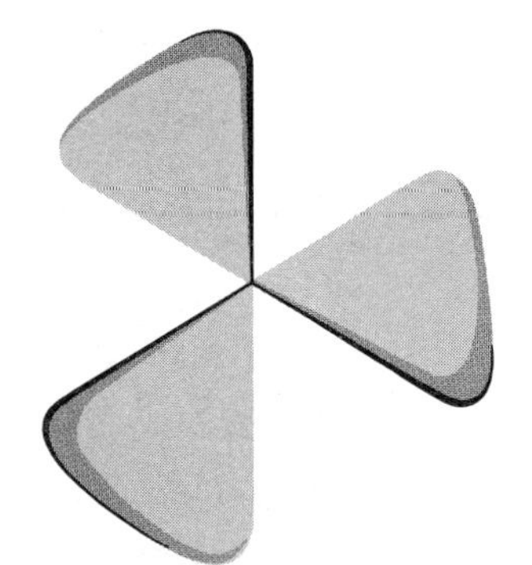

FIGURE 8-9
View along a C_3 axis.

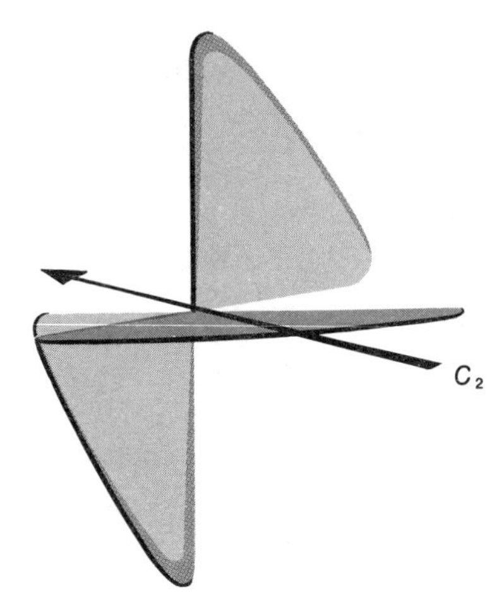

FIGURE 8-10
View along a C_2 axis.

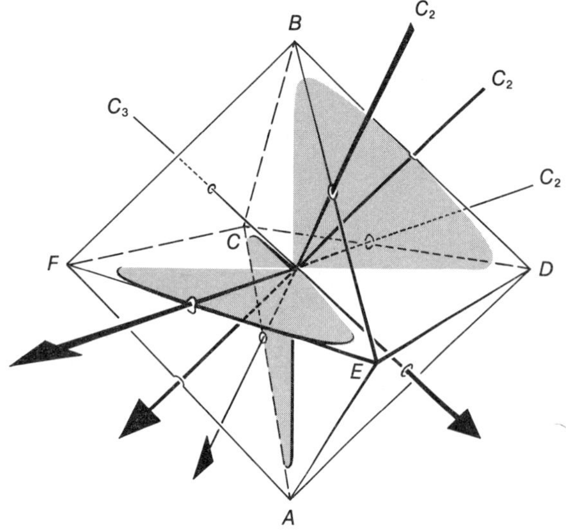

FIGURE 8-11
The C_3 axis passes through triangles *FBC* and *EDA*. The C_2 axes go through edges *B–D*, *A–C*, and *E–F*, respectively.

configuration, followed by the sign of optical rotation at the Na_D line, and finally, the name of the compound.

The sign of the Cotton effect and its magnitude must somehow be related to the nature of the electronic transition and to the chirality of the absorbing system. The Cotton effect is caused by the preferred absorption of right or left circularly polarized light. Which of the two polarization states is chosen by the molecules depends on the direction in which the electrical charge is displaced within the helical path because of geometric configuration of the molecule during the optical transition. Unfortunately, it is not yet possible to calculate the sign and magnitude of the Cotton effects for transitions in a system as complicated as a complexed ion and avoid ambiguity. But in a few cases evidence from different types of experimental investigation, together with a knowledge of the kind of transition causing the optically active band, enables us to relate absolute configuration to the sign of the Cotton effect. The first and most well-known of these examples is the subject of this experiment.

In Exercise 7C the spectrum of $[Co(en)_3]Cl_3$ was measured and the observed bands were assigned to transitions betweeen crystal field terms. In Section A of this exercise the C.D. spectra of (+)- and (−)-$[Co(en)_3]Cl_3$ were obtained and the observation was made that the band assigned to $^1A_{1g} \rightarrow {}^1T_{1g}$ in Exercise 7C is actually composed of two transitions caused by the very slight splitting of $^1T_{1g}$ (see Figure 7-12) due to the decrease in symmetry encountered during the change from perfect octahedral symmetry to a helical form. The $^1T_{2g}$ state (Figure 7-12) is also slightly split into a ^1E and a $^1A_1$ component, but the transition $^1A_1 \rightarrow {}^1A_1(T_2)$ is not possible. Ligand field theory classifies the two types of transitions that still occur in the spectrum as magnetic dipole allowed. Their transition moments are stated with respect to the C_3 axis of the molecule:

$$^1A_1 \rightarrow {}^1A_2 \text{ transition moment} \parallel \text{to } C_3$$

$$^1A_1 \rightarrow {}^1E \text{ transition moment} \perp \text{to } C_3$$

In a classical picture of the absorption process the direction of the transition moment may be identified as the direction in which the magnetic dipole causing the transition fluctuates.

The C.D. spectrum is due to three transitions (in the visible range); two possess a transition moment parallel to C_3, one perpendicular to C_3.

A look at the geometric configuration of our complex ion (see Figures 8-9 and 8-10) indicates that the helicity of the system in the C_3 direction is different from that which is perpendicular to C_3 (i.e., in the C_2 direction); if left in C_3 then right in C_2. Thus we expect two bands to show the same Cotton effect, whereas the third is of the opposite sign. The problem of how to relate the sign of the Cotton effect to the direction of the transition moment and the helicity of the system in this direction still remains.

Fortunately, the absolute configuration of one of the optical isomers of $[Co(en)_3]^{3+}$ has been obtained by means of x-ray analysis. This isomer has the configuration D* or Λ. It is contained in the double salt $2[Co(en)_2]Cl_3 \cdot NaCl \cdot 6H_2O$ in which the C_3 axis of the complex is parallel to the optical axis in single crystals of the salt. If circularly polarized light is passed through the crystal along the optical axis, the Cotton effect of the long-wave transition (493 mμ) is found to be positive. The connection between the sign of the Cotton effect and the helicity of the system has been made via the direction of the transition moment. In a C.D. curve of the salt solution where all orientations of the complexed ion to the direction of radiation occur, we expect to find the Cotton effects of the remaining two transitions (in the visible region) in accordance with the first transition.

Procedure

Record the C.D. curve of the two optically isomeric forms of $Co(en)_3Cl_3$ in aqueous solution. (The O.R.D. curve is too complicated because of small band splittings to yield interpretable results.) See Section A of this exercise for experimental background.

Determine which of the two optical isomers prepared in Exercise 1A has the D* and which the L* configuration. Explain the basis of this determination.

Note: The importance of relating the Cotton effect to the absolute configuration in $[Co(en)_3]^{3+}$ lies in the fact that this compound can now serve as a standard for the configuration determination of other d^3 and d^6 dihedral complexes. Ballard, McCaffery, and Mason (1962) have suggested "that dihedral d^3 and d^6 complexes have the same absolute configuration as the D*-$[Co(en)_3]^{3+}$ ion if the spin-allowed transition of lowest energy has an E component with a positive rotatory power."

REFERENCES

Ballard, R. E., McCaffery, A. J., and Mason, S. F. 1962. *Proc. Chem. Soc.* October, 331.

Djerassi, C. 1960. *Optical Rotatory Dispersion.* New York: McGraw-Hill.

Kauzmann, W. 1957. *Quantum Chemistry.* New York: Academic Press.

Lowry, T. M. 1964. *Optical Rotatory Power.* New York: Dover.

Mason, S. F. 1963. *Quart. Rev.* **27**, 57.

Velluz, L., Legrand, M., and Grosjean, M. 1965. *Optical Circular Dichroism.* New York: Academic Press.

Willard, H. H., Merrit, L. I., Jr., and Dean, J. A. 1965. *Instrumental Methods of Analysis.* New York: Van Nostrand.

EXERCISE 9

Magnetism

A. MEASUREMENT OF MAGNETIC SUSCEPTIBILITY

Any substance placed in a magnetic field becomes magnetized. That is, the total field inside a substance, B, differs from the external field, H, by the amount of induced magnetization. The intensity of induced magnetization is symbolized by the letter I. B is expressed in equation (1):

$$B = H + 4\pi I \tag{1}$$

The quantities B, H, and I are measured in units of gauss or oersted: 1 gauss = 1 oersted = $(\text{erg cm}^{-3})^{1/2}$. The intensity of induced magnetization is the magnetic moment created by the field inside the substance.

$$I = \frac{M^{(m)} \cdot n}{V} = M^{(m)} \cdot N \tag{2}$$

where $M^{(m)}$ is the average atomic magnetic moment parallel to the field and $N = n/V$ is the number of atoms per unit volume.

The intensity I is frequently found to be proportional to the external field and is thus expressed as:

$$I = \chi^{(V)} \cdot H \tag{3}$$

where $\chi^{(V)}$ is the magnetic volume susceptibility. With the aid of equation (3) equation (1) may be written as:

$$B = (1 + 4\pi\chi^{(V)}) \cdot H \tag{4}$$

Clearly, $\chi^{(V)}$ characterizes the magnetic behavior of the substance. Equation (5) shows its relation to the average magnetic moment:

$$\chi^{(V)} = \frac{M^{(m)} \cdot N}{H} \tag{5}$$

Thus $\chi^{(V)}$ can be calculated if $M^{(m)}$ can be derived from our knowledge of the microscopic properties of matter. Before attempting to do so, however, we shall briefly examine observed types of magnetic behavior.

Types of Magnetic Behavior

χ Is Independent of H

Here the susceptibility is determined by the properties of single atoms or molecules. There is no magnetic interaction between the units contributing to the magnetic moment.

Diamagnetism. Substances with $\chi < 0$ are diamagnetic. When a substance is placed in a magnetic field, the field inside the substance is weaker than the surrounding vacuum. Substances of this kind are repelled by an inhomogeneous field and are pushed into regions of less field strength. (For an application see the section of this exercise that deals with the Gouy balance.) Diamagnetism is a universal property of matter since it originates from the orbital motion of electrons. The magnetic susceptibility χ is very small, $\chi \sim -10^{-6}$, and is easily masked by additional magnetic behavior of a different kind and larger susceptibility.

Paramagnetism. The spins of unpaired electrons, together with contributions from the orbital angular momentum, give rise to behavior resembling that of small magnetic dipoles, which can be aligned by an applied field. This increases the field strength in the substance compared with the surrounding field, $\chi > 0$, and the substance is drawn into the region of greatest field strength when placed in an inhomogeneous field. Since thermal motion counteracts the alignment, χ will decrease as the temperature increases. Substances of this kind are paramagnetic. Paramagnetic susceptibilities are about one hundred times greater than diamagnetic susceptibilities, $\chi \sim 10^{-4}$.

χ Depends on H

This behavior is due to magnetic dipoles arising from spins, which interact in a specific way with each other.

Ferromagnetism. A specific interaction of unpaired electrons on different atoms causes the parallel alignment of spins in regions of the lattice called domains. An applied field has the tendency to align the domains. The degree of alignment depends on the field strength, so χ becomes field dependent. Hysteresis amd permanent magnetism are observed; χ is quite large, $\chi \sim 10^{+2}$, and becomes strongly dependent on temperature only above the Curie temperature at which domains cease to exist. Normal paramagnetism is observed at temperatures above the Curie temperature.

Ferrimagnetism. Domains of oriented spins are formed in a substance possessing two interpenetrating lattices containing atoms with unequal numbers of unpaired electrons. Spins in each sublattice are parallel, but spins in one sublattice are antiparallel to those in the other sublattice. The magnetic behavior is the same as that of ferromagnetism, but χ is much smaller, since only the difference of spins in the sublattices contributes, and atoms are more widely separated and interaction therefore weaker: $\chi \sim 10^{-3}$. Above the Curie temperature domains do not exist and the substance becomes paramagnetic in accord with the Curie-Weiss law.

Antiferromagnetism. The situation is the same as in ferrimagnetism, except that atoms in the two sublattices possess the same number of unpaired electrons. Spins in domains therefore cancel each other; χ becomes very small, $\chi \sim 10^{-5}$–10^{-7}, and increases slowly until the Neel temperature is reached, at which point the order in domains is destroyed and the substance becomes paramagnetic.

There are other types of magnetic behavior, which will not be discussed here. A rough theoretical outline of diamagnetic and paramagnetic behavior is presented in the following sections. For a more detailed discussion of the magnetic behavior of transition metal complexes see Section B of this exercise.

Origin of Diamagnetism

The application of a homogeneous magnetic field to an atom causes a precession of all otherwise unaltered electron orbits around the direction of the field, according to the Larmor theorem. This imposed precession possesses the frequency

$$\omega_L = -\frac{e}{2m_e c} \cdot H \tag{6}$$

where e and m_e are the electronic charge and mass, respectively, and c is the velocity of light. It can be considered to represent an electric current around the field direction that generates a magnetic field opposed to the direction of the applied field.

The magnetic moment produced by a current j, which corresponds to one electron in the Larmor precession, is:

$$M^{(m)} = \frac{j}{c} A \tag{7}$$

where A is the area enclosed by the current loop and the current itself is:

$$j = \frac{e\,\omega_L}{2\pi} = e\,\omega_L \tag{8}$$

The field is applied in the z direction and A is thus given by the projection of the resulting precession orbit of each electron onto the x-y plane. The introduction of equation (8) into equation (7) then yields:

$$M^{(m)} = \frac{e\,\omega_L}{2\pi c}\,(\overline{x_i^2} + \overline{y_i^2}) \tag{9}$$

where x_i and y_i are the projection coordinates of the ith electron. For several electrons the coordinates are expressed in the form of the average precessional radii r_i. For spherical symmetry the following relations hold:

$$\overline{x^2} = \overline{y^2} = \overline{z^2}$$

thus

$$\frac{\overline{x^2} + \overline{y^2}}{\overline{x^2} + \overline{y^2} + \overline{z^2}} = \frac{\overline{x^2} + \overline{y^2}}{\overline{r^2}} = \frac{2}{3}$$

and equation (9) becomes

$$M^{(m)} = \frac{e\,\omega_L}{3\pi c}\sum_i \overline{r_i^2} = -\frac{e^2 \cdot H}{6\pi m_e c^2}\sum_i \overline{r_i^2} \tag{10}$$

after the introduction of equation (6).

The insertion of equation (10) into equation (5) leads to equation (11) for the diamagnetic volume susceptibility.

$$\chi^{(V)} = -\frac{e^2 \cdot n}{6\pi m_e c^2 V}\sum_i \overline{r_i^2} \tag{11}$$

The diamagnetic susceptibility is negative.

In addition to $\chi^{(V)}$, the gram susceptibility $\chi^{(g)}$, and the mole susceptibility $\chi^{(M)}$, are frequently used. The expression $\chi^{(g)}$ is the susceptibility per gram and is obtained from $\chi^{(V)}$ by dividing by the density: $\rho = m/V$.

$$\chi^{(g)} = \frac{\chi^{(V)}}{\rho} \tag{12}$$

$$= -\frac{e^2 \cdot n}{6\pi m_e c^2 m}\sum_i \overline{r_i^2} \qquad [\mathrm{cm^3 g^{-1}}]$$

The expression $\chi^{(M)}$ is obtained from $\chi^{(g)}$ by multiplying by the molecular weight M.

$$\chi^{(M)} = \chi^{(g)} \cdot M = \chi^{(g)} \cdot \frac{m}{n} \cdot N_L \tag{13}$$

$$= -\frac{e^2 \cdot N_L}{6\pi m_e c^2}\sum_i \overline{r_i^2}$$

$$= -2.83 \cdot 10^{10}\sum_i \overline{r_i^2} \qquad [\mathrm{cm^3}]$$

Since $\overline{r_i^2}$ is of the order of 10^{-16} cm, the correct order of magnitude (about 10^{-6}) is arrived at for diamagnetic susceptibilities.

Since all molecules have orbiting electrons, all substances possess a diamagnetic susceptibility component. However, other factors, in particular unpaired spins, may give rise to more pronounced paramagnetic effects that overshadow the diamagnetism. In experiments on paramagnetism, the measured results can be corrected for diamagnetism. The methods of correction are discussed in a later section of this exercise.

Paramagnetism

Paramagnetism results from the spin and orbital angular momentum of unpaired electrons. For the purpose of simplification, an atom with only one electron in an open shell shall be considered first. Furthermore, the electron shall not be in an s state. From a simple classical point of view the electron is expected to revolve around the nucleus in a circular orbit of radius r with angular velocity ω. The classical angular momentum is thus $m\omega r^2$. On the other hand, quantum mechanics yields an angular momentum $\sqrt{l(l+1)} \cdot \hbar$ for an electron with orbital quantum number l. The quantization of the circular orbit may be introduced by:

$$m_e \omega r^2 = \sqrt{l(l+1)} \cdot \hbar \tag{14}$$

In equations (7) and (8) the magnetic moment resulting from the orbital motion of the electron was expressed as

$$M^{(m)} = \frac{e \cdot \omega}{c \cdot 2\pi} \cdot A \tag{15}$$

By replacing A with $A = 2\pi r^2$ and introducing ωr^2 from equation (14) the magnetic moment of an electron in orbit l is obtained as

$$M_l^{(m)} = \frac{e}{2 \cdot m_e \cdot c}\,\hbar\sqrt{l(l+1)} \tag{16}$$

$$= \mu_B \,.\, n_{\text{eff}} = \mu_{\text{eff}}$$

where n_{eff} is the "effective number of Bohr magnetons," and μ_{eff} is the effective magnetic moment. The expression μ_B is the Bohr magneton.

$$\mu_B = \frac{e \cdot \hbar}{2 \cdot m_e \cdot c} \tag{17}$$

$$= 0.927 \cdot 10^{20} \qquad [\text{erg/gauss}^{-1}]$$

The spin of the electron also exhibits the nature of an angular momentum giving rise to a magnetic moment. The spin, however, does not allow for the construction of a classical model and a derivation analogous to the one in equation (16) becomes impossible. In contrast to (16), quantum mechanics yields:

$$M_s^{(m)} = 2\sqrt{s(s+1)} \cdot \mu_B = \mu_B n_{\text{eff}} = \mu_{\text{eff}} \tag{18}$$

where $s = 1/2$ for one electron. The factor 2 on the right-hand side is called g, the Landé splitting factor. Thus $g = 1$ holds for magnetism due to orbital motion, and $g = 2$ for magnetism resulting from spin.

Actually, the state of an atom with respect to its total angular momentum is represented by the quantum number J. For atoms with several open-shell electrons and L-S coupling, J is the sum of the resulting orbital quantum number L and the resulting spin quantum number S. The state of the atom is thus characterized by L, S, and the values that J can assume, namely, $J = L + S$, $L + S - 1, \ldots, L - S$. States of different J are not degenerate. In fact, the splitting λ between them increases with the number of electrons the atom possesses and for paramagnetic atoms or ions $\lambda \gg kT$ holds. The splitting is so large that only the lowest of the multiplet states is populated. Thus the calculation of the magnetic moment can be based on the properties of the lowest multiplet component only. (For the selection of the lowest state see Exercise 7A). The atom is assumed to be in state J. Then its total angular momentum is $\sqrt{J(J+1)} \cdot \hbar$, from which an expression may be deduced for the magnetic moment analogous to equations (16) and (18). However, the object is to calculate the average atomic moment in the field direction and that can be done simply.

Apply the magnetic field in the z direction to find the component of the vector J in the z direction, m_J, which provides the angular momentum in the field direction $m_J \cdot \hbar$. Any atom in state J can assume an m_J value from the set of allowed quantum numbers $m_J = J, J-1, \ldots, -(J-1), -J$.

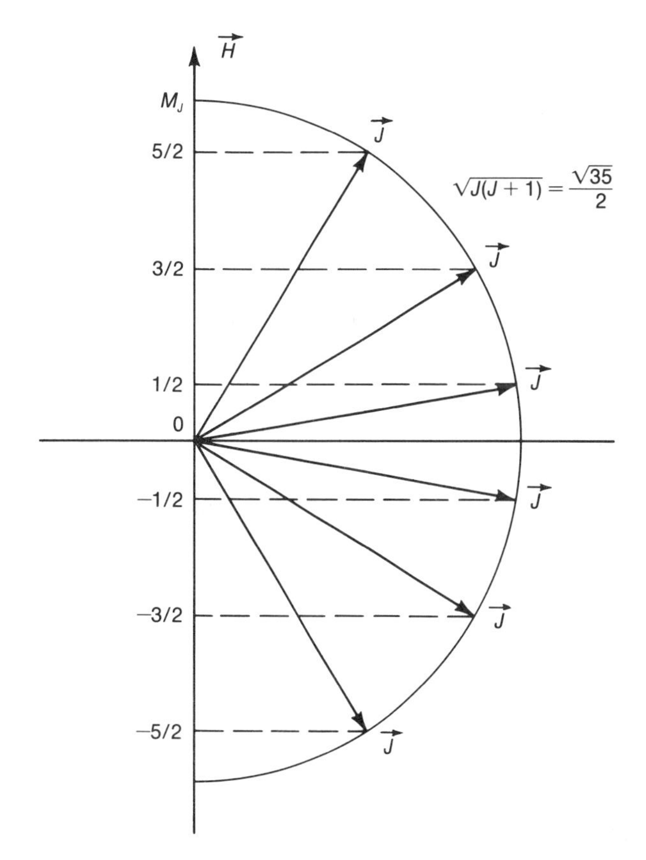

FIGURE 9-1
Allowed orientations of the angular momentum vector $\vec{J}$ and values of M_J in a magnetic field $\vec{H}$, where $J = 5/2$. The M_J values are given by $J, J-1, \ldots, -(J-1), -J$. Note that there are $2J+1$ (in this case six) values of M_J.

The distribution of atoms over possible m_J values in a probe consisting of many atoms is determined by the fact that in a magnetic field the degeneracy of states with different m_J values is raised. The energy difference between the split state m_J and the original state is:

$$\Delta E = -\mu_B \cdot g \cdot H \cdot m_J = \beta H m_J \tag{19}$$

where g has been derived by Landé as

$$g = 1 + \frac{J(J+1) + S(S+1) + L(L+1)}{2J(J+1)} \tag{20}$$

The state $m_J = 0$ (if any) remains at the same level of energy upon the application of a magnetic field because it does not possess an angular momentum component in the H direction (see Figures 9-1 and 9-2).

The average magnetic moment in the field direc-

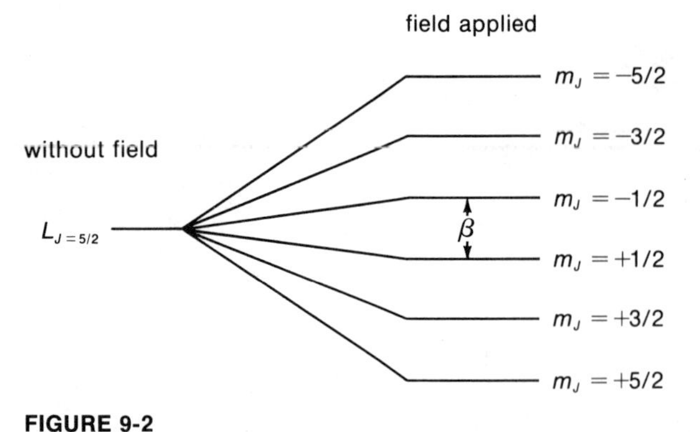

FIGURE 9-2
Splitting of a degenerate term L with inner quantum number $J = 5/2$ into states of different m_J upon the application of a magnetic field H. The energy difference between a split state and the original degenerate state is $\Delta E = -\mu_B \cdot g \cdot H \cdot m_J = \beta H m_J$. The splitting factor β states the energy difference between any two neighboring split states. The phenomenon is called the first-order Zeeman effect.

tion can now be calculated via the determination of an average value for m_J, namely, $\overline{m_J}$.

$$\overline{m_J} = \frac{\sum_{p=-J}^{+J} n_p m_J}{\sum_{p=-J}^{+J} n_p} \tag{21}$$

where n_p is the number of atoms in state $m_J = p$. The average magnetic moment in the field direction is then

$$M^{(m)} = g \cdot \mu_B \cdot \overline{m_J} \tag{22}$$

If a Boltzmann distribution of atoms over the resulting $2J + 1$ energy levels is assumed, the following is obtained for the number of atoms in state p:

$$n_p = \frac{n}{Q} \exp\left(-\frac{\Delta E_{Jp}}{kT}\right) = \frac{n}{Q} \exp(-p \cdot X) \tag{23}$$

where $X = \beta H/kT$ according to equation (19) and

$$Q = \sum_{p=-J}^{+J} \exp(-p \cdot X)$$

Thus equation (21) can be written as

$$\overline{m_J} = \frac{\sum_{p=-J}^{+J} \frac{p \cdot n}{Q} \exp(-p \cdot X)}{\sum_{p=-J}^{+J} \frac{n}{Q} \exp(-p \cdot X)} \tag{24}$$

$$= \frac{\sum_{p=-J}^{+J} p \cdot \exp(-p \cdot X)}{Q}$$

The exponential term can be evaluated by approximation through a power series expansion.

$$\exp(-p \cdot X) = 1 - p \cdot X + \frac{(pX)^2}{2!} - \ldots \tag{25}$$

At room temperature and for magnetic fields of 10,000 gauss or less, the quantity $p \cdot X$ is considerably less than unity. Hence the series is cut after the first two members.

$$\exp(-p \cdot X) = 1 - p \cdot X \tag{26}$$

Substitution into equation (24) yields:

$$\overline{m_J} = \frac{\sum_{p=-J}^{+J} p + X \cdot \sum_{p=-J}^{+J} p^2}{\sum_{p=-J}^{+J} (1 - p \cdot X)} \tag{27}$$

Since $p = m_J$ can assume the values $J, J - 1, \ldots, -(J - 1), -J$, then

$$\sum_{p=-J}^{+J} p = 0 \tag{28}$$

Furthermore, since m_J has $2J + 1$ possible values, the entire denominator becomes

$$\sum_{p=-J}^{+J} (1 - p \cdot X) = 2J + 1 \tag{29}$$

Finally, it may be shown that

$$\sum_{p=-J}^{+J} p^2 = \frac{J(J+1)(2J+1)}{3} \tag{30}$$

The substitution of equations (28), (29), and (30) into equation (27) yields:

$$\overline{m_J} = -\frac{J(J+1)}{3} X = \frac{g \cdot \mu_B \cdot H \cdot J(J+1)}{3kT} \tag{31}$$

From equations (5), (22), and (31) the paramagnetic contribution to the susceptibility is obtained:

$$\chi^{(V)} = \frac{N \cdot \mu_B^2 \cdot g^2 \cdot J(J+1)}{3kT} \tag{32}$$

The paramagnetic mole susceptibility is then, according to equations (11) and (13):

$$\chi_P^{(M)} = \frac{N_L \cdot \mu_B^2 \cdot g^2 J(J+1)}{3kT} = \frac{N_L \cdot \mu_{\text{eff}}^2}{3kT} \tag{33}$$

where $\mu_{\text{eff}} \equiv \sqrt{g^2 J(J+1)} \cdot \mu_B$ is analogous to equations (11) and (13).

Equation (33) has been derived under the assumption that J multiplet splittings λ are larger than kT, which in turn is larger than the Zeeman splittings ΔE, $\lambda \gg kT \gg \Delta E$. For $\lambda \ll kT \gg E$ the following is stated without derivation:

$$\chi_P^{(M)} = \frac{N_L \cdot \mu_B^2}{3kT}[4S(S+1) + L(L+1)] \quad (34)$$

$$= \frac{N_L}{3kT} \cdot \mu_{\text{eff}}^2$$

Finally, for $L = 0$, it follows from equation (33), as well as (34), that is, from $\lambda \gg kT$, as well as $\lambda \ll kT$, that

$$\chi_P^{(M)} = \frac{N_L \cdot \mu_B^2}{3kT}[4S(S+1)] = \frac{N_L}{3kT} \cdot \mu_{\text{eff}}^2 \quad (35)$$

From equations (33) to (35) it can be seen that the effective number of Bohr magnetons can be calculated fron the measured susceptibility:

$$n_{\text{eff}} = \sqrt{\frac{\chi^{(M)} \cdot 3kT}{N_L \cdot \mu_B^2}} = 2.84\sqrt{\chi^{(M)} \cdot T} \quad (36)$$

and then compared with the theoretical values:

$n_{\text{eff}} = \sqrt{g^2 \cdot J(J+1)}$ [in equation (33)]

$n_{\text{eff}} = \sqrt{4S(S+1) + L(L+1)}$ [in equation (34)]

$n_{\text{eff}} = \sqrt{4S(S+1)}$ [in equation (35)]

The paramagnetic susceptibility is always positive and temperature dependent (a fact that is expressed in the empirical Curie law):

$$\chi_P^{(M)} = \frac{C}{T} \quad (38)$$

Frequently, experimental data better fit a modified form of the Curie law, called the Curie-Weiss law, in which T is replaced by $(T + \theta)$ and θ is called the Weiss constant. Among the most frequently investigated inorganic compounds containing paramagnetic ions are the salts of the rare earths and the transition metals, the first nearly always following the described behavior and the latter scarcely ever fitting into the pattern. Reasons for these effects are given in the following section.

Deviations from Expected Magnetic Behavior

In the previous discussion, several assumptions were made that may cause measured magnetic behavior to deviate from the expected magnetic behavior.

The first assumption was that only the atomic or ionic ground states determine the value of the magnetic susceptibility. Actually, in some ions (e.g., Eu^{3+} and Sm^{3+}), low excited states do contribute. Thus they do not exhibit susceptibilities in accord with their ground state J value.

The other basic question is whether ions in solid substances (as used in this experiment) can be characterized by their S, L, and J values, which are defined for the undisturbed ionic state. Deviations due to this cause occur predominantly in transition metal compounds.

In rare earth salts the unpaired $4f$ electrons are nicely shielded from the influence of the surrounding crystal field by the closed shells of outer electrons, whereas the $3d$ electrons in transition metal ions are fully exposed. The well-known result is a complete decoupling of spin and orbital angular momenta, the resulting degeneracy of d orbitals being "broken" by the crystal field. It is evident that the orbital contribution to angular momentum can no longer be that of d electrons. The spin contribution, however, is retained, except for cases in which the high crystal field strength leads to spin pairing (low-spin compounds).

From the chemical point of view, the magnetic properties of transition metal compounds are especially interesting because they yield information regarding the spin state, as well as the geometric structure of complexes. Unfortunately, the magnetic behavior of these compounds exhibits a variety of patterns, which would necessitate a lengthy discussion. Therefore, a brief classification according to spectroscopic ground states will be presented in this section and particular cases will be treated in greater detail in Section B of this exercise.

The discussion will be based on the crystal field description given in Exercise 7C. In octahedral and tetrahedral crystal fields five different kinds of terms are encountered: A_1, A_2, E, T_1, and T_2. They can be related (at least for ground states) to the distribution of the appropriate number of electrons over two kinds of levels, e_g and t_{2g}, or, as they are sometimes called, d_γ and d_ϵ. Upon close examination, the A and E terms are found to be related either to configurations with electrons in e_g levels only or to those with filled or half-filled t_{2g} subshells and any possible number of electrons in the e_g subshell. On the other hand, T terms correspond to configurations of more or less than half-filled t_{2g} levels—again with any possible num-

TABLE 9-1
Crystal-field terms and possible configurations.

Terms	Configurations
A_1, A_2, E	e_g^n, $t_{2g}^3 e_g^n$, $t_{2g}^6 e_g^n$
T_1, T_2	$t_{2g}^m e_g^n$

Note: $m = 1, 2, 4, 5$; $n = 0, 1, 2, 3, 4$.

ber of electrons in the e_g levels. Table 9-1 lists terms and the corresponding configurations.

The three t_{2g} and two e_g orbitals originate from a set of five d orbitals, which are degenerate in the absence of a crystal field. The d_{xy}, d_{xz}, d_{yz}, $d_{x^2-y^2}$, and d_{z^2} orbitals, which were derived from and are equivalent to the five hydrogenlike orbitals characterized by $m_l = 2, 1, 0, -1, -2$, were discussed in Exercise 7C. As long as the five orbitals stay degenerate, each set can be transformed into the other. As soon as the crystal field interaction removes the degeneracy this is no longer the case. By examination, it can be seen that the three orbitals d_{xy}, d_{xz}, and d_{yz}, which form the t_{2g} set, are of the same form and can be transformed into each other by rotation. Since they are also degenerate, they behave like a set of three p orbitals with $m_l = 1, 0, -1$. Configurations t_{2g}^3 and t_{2g}^6, which correspond to half-filled and filled p shells, respectively, have a total angular momentum that vanishes ($L = 0$). All other configurations $t_{2g}^m e_g^n$ (see Table 9-1) possess the total angular momentum quantum number $L = 1$. They behave like P states.

In contrast to the t_{2g} orbitals, the $d_{x^2-y^2}$ and d_{z^2} orbitals, which form the e_g set, cannot be transformed into each other. They are degenerate in energy but otherwise are two orthogonal orbitals. They do not contribute to angular momentum ($L = 0$). It is evident that the effective orbital angular momentum quantum number $L = 0$ must be assigned to A and E terms, whereas T terms possess $L = 1$.

The first conclusion to be drawn from this is that all compounds with A and E ground states should display spin-only behavior according to equation (35). All T states should be adequately described by equation (34) with $L = 1$, since crystal field splittings are not too large and Zeeman splitting due to the magnetic field is even smaller. It can be expected that $\lambda \ll kT \gg \Delta E$. Unfortunately, equation (34) can only be considered a high temperature limit that can be approached from above or below depending on the number of d electrons. Experimental values can therefore be greater or smaller than those predicted by equation (34) but are generally closer to them than to values calculated from equations (33) or (35). (For further discussion see Section B of this exercise.)

Sometimes deviations from the "spin only" behavior are observed in compounds with an A or E ground state. These originate from the contributions of higher crystal field terms of the same multiplicity but of T character. Their magnitude is usually quite small. Section B of this exercise deals with these phenomena and with the magnetic properties of T terms in greater detail.

Experimental Method[1]

Description of the Gouy Balance

The measurement of diamagnetic and paramagnetic susceptibilities is conveniently carried out on a Gouy balance, which consists of a magnet (preferably an electromagnet) that generates a strong inhomogeneous field. The sample, contained in a long tube with a uniform cross section, is placed between the pole faces while suspended from one arm of an analytical balance (see Figure 9-3).

To correct for the diamagnetism of the glass, a piece of empty glass tubing extends from the bottom of the tube containing the sample and is identical to it (see Figure 9-4). The tubing is adjusted so that the plane between the filled portion and the empty portion is exactly level with the center of the pole faces. The sample is thus situated in a magnetic field that is inhomogeneous along the vertical axis and varies from H_0 at the plane dividing the tubing to H at the top.

The force dK, which acts on a small volume, dv, containing dn atoms in a field inhomogeneous in the y direction (see Figure 9-4), is

$$dK = M^{(m)} \cdot dn \cdot \frac{dH}{dy} \tag{39}$$

if the probe is surrounded by vacuum. According to equation (5)

$$M^{(m)} dn = \chi^{(V)} \cdot H \cdot dv \tag{40}$$

The introduction of equation (40) into equation (39) leads to

$$dK = \chi^{(V)} \cdot H \cdot \frac{dH}{dy} \cdot dv \tag{41}$$

[1]Refer to Brubacher and Stafford (1962).

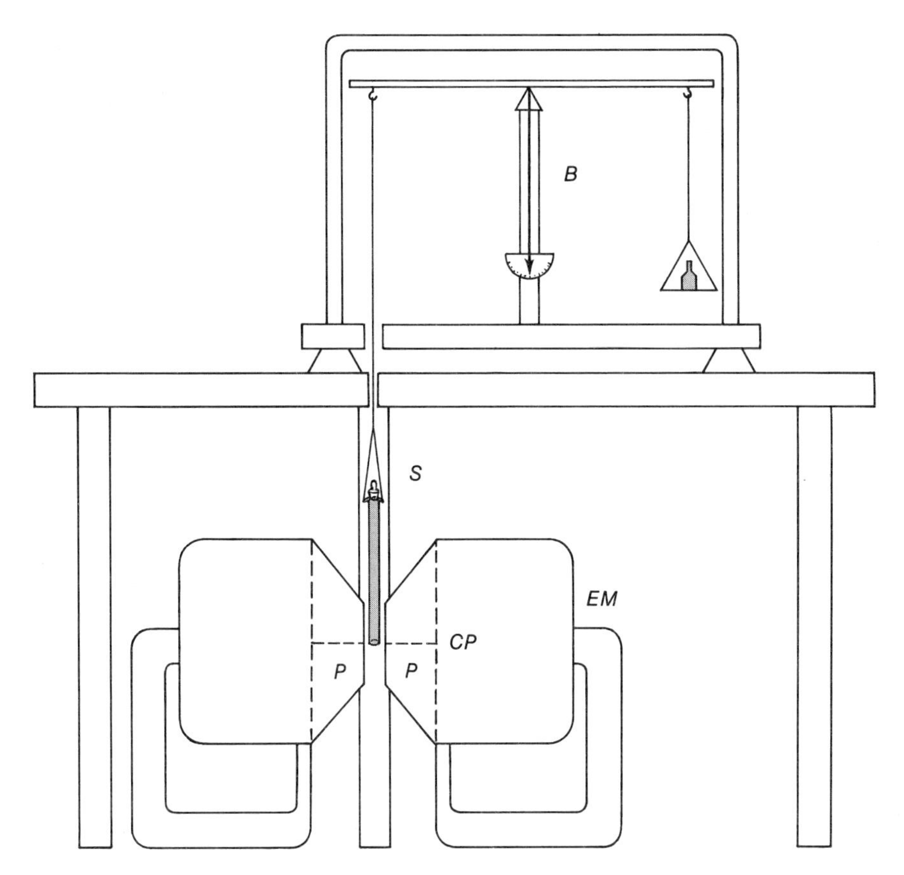

FIGURE 9-3
Schematic drawing of a Gouy balance: *B*, balance; *S*, shielding against air currents; *P*, pole faces; *CP* center of pole faces; *EM*, electromagnet. (See Figure 9-4 for details of sample tube and geometry of magnet.)

for a segment of dv in an extended probe. According to the symmetry of the tube, dv can be expressed by the inner cross-section A of the tube and the length dy necessary to enclose dv.

$$dv = A \cdot dy \tag{42}$$

For uniform cross section, equation (41) becomes, after the insertion of equation (42) and integration,

$$K = A \int_{H}^{H_0} \chi^{(V)} \cdot H dH \tag{43}$$

Actually, the tube is suspended in a gas (e.g., air) of susceptibility, $\chi_a^{(V)}$, and therefore the acting force is proportional only to the difference of the two susceptibilities. Assuming both susceptibilities to be uniform over the distance between the location of H and H_0, equation (43) can be expressed as

$$K = A(\chi^{(V)} - \chi_a^{(V)}) \int_{H}^{H_0} H dH \tag{44}$$

$$= \frac{A}{2}(\chi^{(V)} - \chi_a^{(V)})(H_0^2 - H^2)$$

If the sample extends sufficiently beyond the pole pieces (6–10 times the gap width), H^2 is negligible compared with H_0^2.

If the sample is suspended from an analytical balance, the apparent increase in weight, $g\Delta m$ (where Δm = weight in the presence of the field − weight without the field), equals this downward force:

$$g\Delta m = \frac{A(\chi^{(V)} - \chi_a^{(V)})}{2} H_0^2 \tag{45}$$

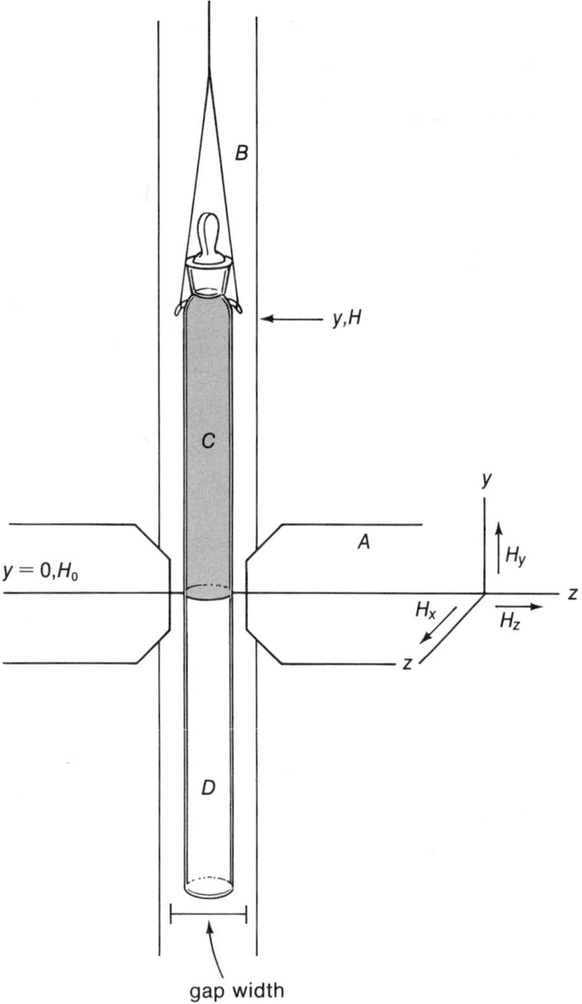

FIGURE 9-4
Arrangement of sample in the magnetic field: *A*, magnet (100-mm core); *B*, Lucite shielding (prevents draft). The front panel can be lifted for the insertion of the sample. *C*, sample tube (14–18 mm O.D., 150 mm long); *D*, glass tube for self-correction of glass diamagnetism (dimension equal to those of sample tube). Magnet has 75-mm pole pieces, 22-mm gap width.

where g is the gravitational constant and H^2 has been disregarded. Whether the susceptibility of the surrounding gas can also be disregarded depends on the precision of the measurements.

Measurements

Besides the increase in the weight of the sample upon the application of a field, to use equation (45) the cross-sectional area of the sample tube and the effective density of the substance must be known. For the calculation of both, obtain the volume of the tube by weighing it with and without water of a known temperature. The density of water as a function of temperature is tabulated in the Handbook of Chemistry and Physics. Calculate the cross-sectional area from the volume and length of the sample tube. (Sample tubes are marked to facilitate filling to the same height.) To determine the effective density of the sample, use its weight and the volume of the tube. The weighing with water can be the last step in the experiment to eliminate the drying time.

In general, the sample should be ground to a fine powder (with a mortar and pestle) to avoid anomalies due to anisotropy in the crystal and to facilitate the uniform packing that is essential. Uniformity in the size of the crystals is difficult to achieve and is usually unnecessary, provided the crystals are small. Magnetic anisotropy, that is, different paramagnetic susceptibilities along different axes, is quite frequently encountered in transition metal compounds. Such phenomena are correlated with the arrangement of different ionic and molecular species in the unit cell. The random orientation of small crystals in the sample still yields a homogeneous effective susceptibility.

To pack the sample, pour about 1 cm of powder into the tube, and tamp it down with a flat-ended glass rod. (Tapping causes the fine powders to settle and virtually guarantees denser packing at the bottom.) The uniformity of packing can be important. Therefore, to determine whether this is a limiting factor, data might be taken for two independently prepared samples of the same substance.

Be sure that the sample tube is suspended freely between the pole faces, that the support thread does not touch any shielding, and that the bottom of the filled part of the tube is at the center of the pole faces. Eliminate air currents, which affect weighing, by placing a shield around the sample.

Because magnetic susceptibilities depend on temperature, record the sample (room) temperature to ±0.5°C.

The field strength of an electromagnet is a function of the temperature of the iron core as well as the electric current. If an electromagnet is used over an extended period of time without external cooling, the resistence in the windings will cause the iron core to warm up. Therefore, the current should be turned on only as needed. If the iron core becomes noticeably warm, stop the experiment until the apparatus has cooled.

The field strength between magnet poles is determined by calibration. A suitable standard is mercury tetrathiocyanatocobalt, $Hg[Co(SCN)_4]$, which possesses exceptionally good packing properties and yields very reproducible results.

Measure Δm for the standard compound and two others obtained from the instructor at currents of $i = 2, 3, 4$, and 5 amp. Take several measurements at each current setting to verify their reproducibility and, as suggested previously, of a second independently prepared sample at one of the higher current settings.

Data

The difference in weight for a given sample with and without the magnetic field is in accord with equation (45).

$$\Delta m = \frac{A \cdot \text{const} \, (\chi^{(V)} - \chi_a^{(V)}) \cdot i^2}{2g} \tag{46}$$

where the constant is introduced through the proportionality of i and H in the linear current-field range of an electromagnet. A plot of Δm versus i^2 should produce a straight line connecting the origin and all points, if the system behaves correctly. Plots of this kind should be prepared as the measurements are taken, so that faulty behavior can be detected in time. Obviously, a straight line can be obtained only if $\chi^{(V)}$ is independent of the field strength. This is not the case if ferro- or ferrimagnetic impurities are present in the sample.

After the magnetism of air ($\chi_a^{(V)} = 0.03 \cdot 10^{-6}$) has been corrected for, the calibration with $Hg[Co(SCN)_4]$ ($\chi^{(V)} = 16.44 \cdot 10^{-6}$ at 20°C) yields the apparatus constant.

$$C_a = \frac{A \cdot \text{const}}{2g} \tag{47}$$

which in turn is used to calculate the susceptibilities of the test substances. If diamagnetic or very slightly paramagnetic test substances are encountered, water may prove to be a more suitable standard. The volume susceptibility of air-free water is $\chi^{(V)} = -0.721 \cdot 10^{-6}$.

Equation (46) shows a direct proportionality between Δm and $(\chi^{(V)} - \chi_a^{(V)})$, which could be exploited by the measurement of the standard and the test substance at identical current settings. Such a method does not yield reliable results because it is difficult to match current settings exactly, especially since the current must be changed only in the direction of increasing field strength to avoid obtaining inaccurate measurements because of hysteresis effects in the magnet.

In case the experimental results for the standard substance indicate that measurements have been performed outside the linear range of the current-field curve of the electromagnet, calculate H_0 from equation (45) for all current settings. Obtain a plot of H_0 versus i, from which H_0 for all settings of i can be read. Calculate the susceptibility of the test substance from equation (45), using the H_0 value taken from the curve.

From $\chi^{(V)}$, molar susceptibilities can be calculated using equations (12) and (13), and from these the effective number of Bohr magnetons can be obtained. However, n_{eff} can be calculated only after correcting for the additional diamagnetism of the total molecule.

As mentioned previously, all substances are diamagnetic in addition to any paramagnetism they display. Diamagnetic mole susceptibilities can be calculated approximately as the sum of the diamagnetic susceptibilities per gram ion, or gram molecule, of the constituent parts of the total complex.

$$\chi^{(M)}_{(\text{measured})} = \chi^{(M)}_{\text{param}} + \sum_{\text{ligands}} \chi'^{(M)}_{\text{diam}} + \sum_{\text{ions}} \chi'^{(M)}_{\text{diam}} \tag{48}$$

Then μ_{eff} and n_{eff} can be calculated from $\chi^{(M)}_{\text{param}}$.

A list of diamagnetic corrections is given in Table 9-2. Additional material may be found in

TABLE 9-2
Diamagnetic corrections.

Ion	$-\chi^{(M)}_{\text{diam}} \cdot 10^6$	Ion	$-\chi^{(M)}_{\text{diam}} \cdot 10^6$	Ion or molecule	$-\chi^{(M)}_{\text{diam}} \cdot 10^6$
Ce^{3+}	20	Hg^{2+}	38	OH^-	12
Co^{2+}	12	K^+	13	NO_3^-	20
Co^{3+}	10	NH_4^+	12	CNS^-	35
Cr^{2+}	15	Br^-	36	Acetate	32
Cu^{2+}	11	CN^-	18	Acetylacetonate	35
Cr^{3+}	11	Cl^-	26	Oxalate	34
Fe^{2+}	13	I^-	52	Ethylenediamine	46
Fe^{3+}	10	F^-	11	NH_3	18
Mn^{7+}	3	NO_2^-	10	H_2O	13

Lewis and Wilkins (1960), Adams and Raynor (1965), and Klemm (1940; 1941).

Errors

The precision of readings should be taken into consideration first. In this experiment masses are determined by the use of a balance, and the electric current is read from an ammeter. To what extent are these readings reproducible? Other factors include the accuracy with which the volume and the cross-sectional area of the tube can be determined. How reproducible are measurements taken on repeated packings of the same substance? Calibrations at the beginning and at the end of the experiment give an indication of the constancy of the magnetic field (heating of the magnet may have gone undetected).

Some sources of error may cancel each other as a result of the method of standardization. Are there any of this type?

Finally, the results must be compared with the literature values. An excellent compilation of magnetic susceptibilities has been made by Foëx (1957). Some are recorded in the *Handbook of Chemistry and Physics*. With care, an experimental result within 2% or less of the true value may be obtained for paramagnetic substances. If the discrepancy between the values you obtain and the literature values is much greater than the estimated uncertainty, there may be a real, systematic error to look for.

Impure solids may yield results that are greatly in error, particularly if the impurity is ferromagnetic. Trace ferromagnetic impurities may be detected by an apparent field strength dependence of the susceptibility.

Chemical Applications of Susceptibility Measurements

From the several experiments that follow, one or more shall be assigned to each student. Measurements under (1) should be performed by all students who have prepared chromous acetate.

1. Determine the magnetic susceptibility of $Cr(OAc)_2 \cdot 2H_2O$ as soon as possible after preparation (see Exercise 3B). Discuss your results in terms of the structure of chromous acetate. Compare these results with those obtained for $Cu(OAc)_2 \cdot 2H_2O$.

2. Determine the effective number of Bohr magnetons for $CuSO_4 \cdot 5H_2O$ and $Cu(OAc)_2 \cdot 2H_2O$. Compare the observed n_{eff} with that expected for Cu^{2+} with spin-only behavior. (Determinations in steps 1 and 2 can be combined.)

3. Determine the effective number of Bohr magnetons for $CeNH_4(SO_4)_2 \cdot 8H_2O$ and $CuSO_4 \cdot 5H_2O$. Each of these has only one unpaired electron; why do they have different effective moments? Calculate the expected number of Bohr magnetons under the assumption of L-S coupling for Ce^{3+} and in the form of the spin-only value for Cu^{2+}.

4. Determine the effective number of Bohr magnetons for (a) $Fe(NH_4)_2(SO_4)_2 \cdot 6H_2O$ and for $K_4Fe(CN)_6 \cdot 3H_2O$ and/or for (b) $Fe(NH_4)(SO_4)_2 \cdot 6H_2O$, $K_3Fe(CN)_6 \cdot 3H_2O$, and $[(C_2H_5)_4N]_2MnCl_4$. Explain your findings in terms of ligand field theory.

5. Determine the effective number of Bohr magnetons for $NiSO_4(NH_4)_2SO_4 \cdot 6H_2O$, for $[(C_2H_5)_4N]_2NiCl_4$,[2] and nickel dimethylglyoximate. Explain your findings in terms of ligand field theory.

6. Measure the susceptibility of $Co(en)_3Cl_3$ (see Exercise 1A for preparation and Exercise 7C for spectroscopy). The information obtained is helpful in the discussion of the spectroscopic properties of $Co(en)_3Cl_3$ and *cis*- and *trans*-$Co(en)_2Cl_3$.

B. PARAMAGNETISM OF TRANSITION METAL COMPLEXES

In the first section of this exercise it was shown that the paramagnetism of a transition metal complex depends on the nature of its crystal field ground state. All complexes with A and E ground states, upon first examination, were not supposed to possess any orbital angular momentum. Their paramagnetic susceptibility should thus exhibit spin-only behavior, which is adequately described by equation (33) in Section A of this exercise. Their effective number of Bohr magnetons would then be given by

$$n_{eff} = \sqrt{4S(S+1)} \quad (1)$$

For V, Ti, and Cr complexes of an A or E ground state, susceptibilities are indeed found quite close to the spin-only value, but there are striking deviations reported for other central ions, for example,

[2]For the preparation of bis(tetraethylammonium)tetrachloromanganate and bis(tetraethylammonium)tetrachloronickel see S. Y. Tyree, Ed., *Inorganic Syntheses*, Vol. 9 (New York: McGraw-Hill, 1967), p. 138.

Ni^{2+} in octahedral and Co^{2+} in tetrahedral complexes. Low-spin Co^{3+} complexes show a slight paramagnetism instead of being completely diamagnetic, and among the heavier central ions only Mn^{2+} and Fe^{3+} have the correct spin-only values.

These deviations from spin-only behavior can be explained by "mixing in" higher crystal field terms of the same multiplicity. The mixed-in term usually stems from the same free-ion parent term and possesses orbital angular momentum. The degree of mixing depends on the magnitude of the crystal field splitting $10D_q$, which separates the states, and the spin-orbit splitting λ, which decreases the separation between the states and introduces the splitting of the originally unsplit ground state. The number of Bohr magnetons can be written as

$$n_{\text{eff}} = n_{\text{s.o.}}\left(1 - a\,\frac{\lambda}{|10D_q|}\right) \qquad (2)$$

where a is a constant.

Since λ is positive for less-than-half-filled subshells and negative for all other configurations, a lowering of n_{eff} is found for the first half of the transition metal ions in the first row, and an increase of n_{eff} is found for the other half. The temperature dependence is the same as for ordinary paramagnetism.

The deviation from spin-only behavior depends on the magnitude of crystal field splitting. Thus a range of magnetic moments is expected for the same center ion in ligand fields of equal symmetry but different strength. For crystal field states of a d^5 configuration, which arise from ^6S terms, no states of sextet multiplicity are found near the ground state, since the ^6S term does not split in the crystal field. Previous reasoning indicates that no deviation from spin-only behavior is to be expected.

Unlike the A and E ground states, an orbital contribution to angular momentum is to be expected for all T ground states (see Section A of this exercise). The quantum number for the orbital angular momentum must be given the value $L = 1$. In contrast to the situation in free paramagnetic ions, where usually only the term of lowest J belonging to the state characterized by L and S is populated, several states of different $J (J = L + S, L + S - 1, \ldots, L - S)$ belonging to the crystal field state characterized by $L = 1$ and the appropriate S may contribute to the magnetic moment. Thus the condition $\lambda \gg kT \gg \Delta E$, where λ is the J multiplet splitting and ΔE is the m_J multiplet splitting, which usually holds for free paramagnetic ions, is fulfilled only at very low temperatures for crystal field states.

$$n_{\text{eff}(\lim T \to 0)} = \sqrt{g^2 \cdot J(J+1)} \qquad (3)$$

Equation (3) may be regarded as a limiting value for the effective number of Bohr magnetons of a first-row transition metal ion that possesses a T ground term in a crystal field of octahedral or tetrahedral symmetry. With rising temperature a point will be reached where $\lambda \ll kT \gg \Delta E$, and from then on n_{eff} can be represented by

$$n_{\text{eff}(\lim T \to \infty)} = \sqrt{4S(S+1) + L(L+1)} \qquad (4)$$

in accordance with equation (34) in Section A of this exercise. The actual value for n_{eff} at room temperature lies between the two limiting values, usually closer to the value predicted by equation (4). To know where these values may be expected, a more detailed discussion of J and m_J splittings of T terms originating from different d^n configurations is necessary.

The first step will be to classify T ground states according to their origin and configuration. Table 9-3 contains the four different T ground states, together with the respective configurations.

TABLE 9-3
Crystal-field terms and corresponding electron configurations in fields of octahedral and tetrahedral symmetry.

Term	Configuration					
	Octahedral				Tetrahedral	
	High-spin		Low-spin			
$^2T_2$	d^1	t_{2g}^1	d^5	t_{2g}^5	d^9	$e_g^4t_{2g}^5$
$^3T_1$	d^2	t_{2g}^2	d^4	t_{2g}^4	d^8	$e_g^4t_{2g}^4$
$^4T_1$	d^7	$e_g^2t_{2g}^5$			d^3	$e_g^2t_{2g}^1$
$^5T_2$	d^6	$e_g^2t_{2g}^4$			d^4	$e_g^2t_{2g}^2$

The magnetic properties of $^2T_2$ terms will be examined before briefly discussing the properties of the other terms. A ^2T state possesses $S = 1/2$ and $L = 1$. Such a state is split by $L - S$ coupling into the two states of $J = L + S = 3/2$ and $J = L - S = 1/2$. In contrast to a real ^2P state, where according to Hund's third rule the state of lowest J should have the lowest energy, this apparent ^2P state exhibits the $J = 3/2$ state as the lower one of the J multiplet. The $J = 3/2$ state is lowered by $\lambda/2$ and is fourfold degenerate according to the four values of m_J (3/2, 1/2, −1/2, −3/2). The state

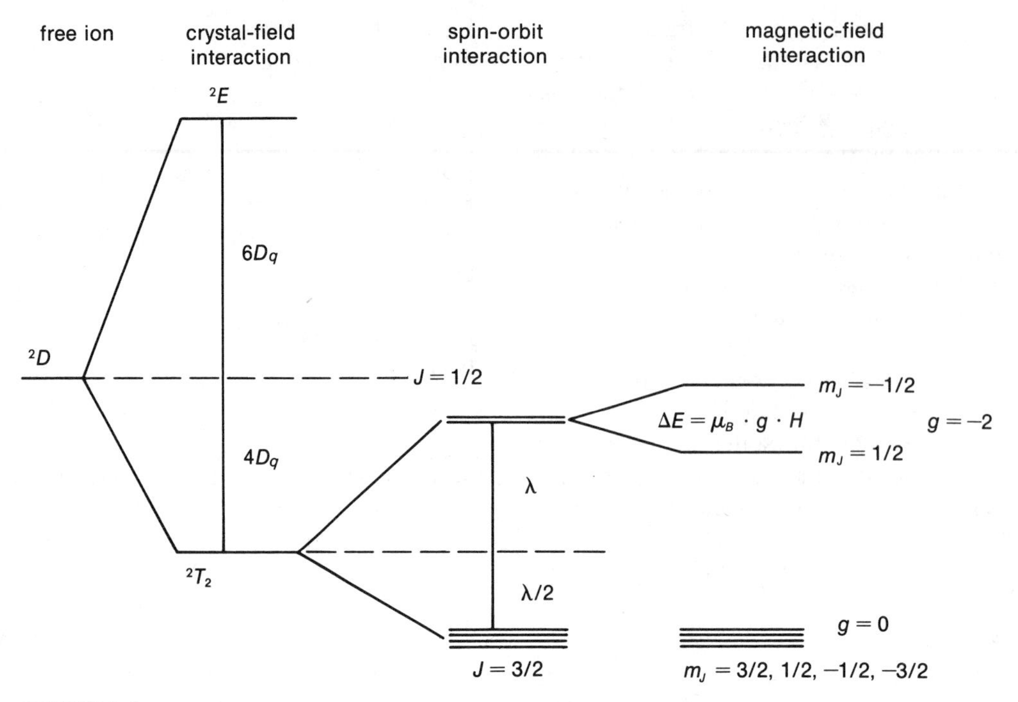

FIGURE 9-5
Schematic term splitting diagram for a d^1 configuration in an octahedral field. The diagram is also valid for a d^9 configuration in a tetrahedral field and a low-spin d^5 configuration in an octahedral field with the provision that λ becomes negative. The splittings due to L-S coupling and to magnetic-field interaction are exaggerated. The total degeneracy for $^2T_2$ is not shown. The orbital degeneracy is $M_L = 3$; the spin degeneracy is $M_S = 2$. Thus the level is sixfold degenerate.

$J = 1/2$ is the only twofold degenerate ($m_J = 1/2, -1/2$) and has been raised by λ to preserve the center of gravity. Figure 9-5 shows the splitting of the parent ^2D free-ion term due to the crystal field, $L - S$ coupling, and the magnetic field. Figure 9-5 also shows that the state with $J = 3/2$ is not split by the magnetic field because, according to equation (19) in Section A of this exercise, the energy difference ΔE between the state m_J in the magnetic field and the original degenerate state J is:

$$\Delta E = -\mu_B g H m_J \tag{5}$$

where g for d electrons in octahedral or tetrahedral crystal fields is:

$$g = 1 - (1/2)A + (2 + A)\frac{S(S+1) - L(L+1)}{2J(J+1)} \tag{6}$$

and A is equal to 1, except for low-field (high-spin) complexes with T_{1g} ground states. For these, A can vary between 1 and 1.5. This complication is due to the mixing of $T_1(F)$ and $T_1(P)$ terms. For each given central ion A remains constant.

For $^2T_2$, the splitting factor g takes the form:

$$g = +1/2 + 3\,\frac{S(S+1) - 2}{2J(J+1)} \tag{6a}$$

with $A = 1$ and $L = 1$. Thus $g = 0$ for $J = 3/2$ and $g = -2$ for $J = 1/2$. The $J = 3/2$ state is not split by the magnetic field, whereas the two components of $J = 1/2$ are removed by $\pm\frac{1}{2}\beta H$ from the degenerate state. A further splitting takes place because of the second-order Zeemann effect, which is not shown in Figure 9-5.

All of the states that were generated in the magnetic field are of very similar energies. An analysis of the change in population due to temperature change does not yield a simple expression from which to derive the temperature dependence of the magnetic susceptibility, as it did for paramagnetic susceptibilities. Therefore only very low temperature $T \rightarrow 0$ and very high temperature $T \rightarrow \infty$ will be considered. ($T \rightarrow \infty$ stands for

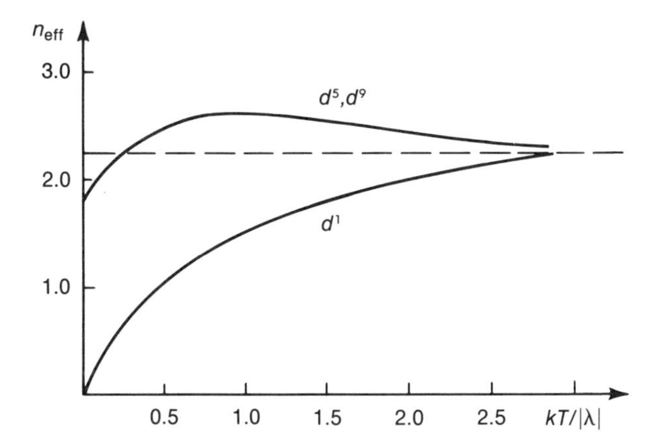

FIGURE 9-6
The temperature dependence of the magnetic moment of d^1 (octahedral), d^9 (tetrahedral), and d^5 (low-spin octahedral) configurations. The effective number of Bohr magnetons is plotted against $kT/|\lambda|$ because this ratio determines the population of states. [From Figgis and Lewis (1960).]

lim $kT \longrightarrow 10D_q$.) For $T \longrightarrow 0$, $n_{eff} = 0$ is obtained from equation (3) because only the lowest J level is occupied, for which $g = 0$. For $T \longrightarrow \infty$, $n_{eff} = \sqrt{4S(S+1) + L(L+1)} = \sqrt{5} = 2.24$ is obtained from equation (4). Figure 9-6 illustrates schematically how this limiting value is approached from below.

Tetrahedral d^9 and octahedral low-spin d^5 configurations involve more-than-half-filled shells. Therefore, λ changes sign; the lowest state is that of $J = 1/2$ and the higher that of $J = 3/2$. In Figure 9-5 this result can be obtained by a rotation of the J states around the $^2T_2$ term. This inversion results in a different temperature dependence of n_{eff}. Since the lowest state is now $J = 1/2$, with $g = -2$ according to equation (5), for $T \longrightarrow 0$, $n_{eff(\lim T \to 0)} = \sqrt{3} = 1.73$. For $T \longrightarrow \infty$, the limiting value $n_{eff(\lim T \to \infty)} = \sqrt{5} = 2.24$ is the same as for d^1. Between these extremes the magnetic moment rises at first, but soon begins to sink, as the nonmagnetic levels become available. Thus the limiting value is approached from above, as shown in Figure 9-6. Since d^1 complexes, on one hand, and d^5 and d^9 complexes, on the other, approach their room-temperature magnetic moments from different sides, d^1 moments are expected to be lower and d^5 and d^9 to be higher than the limiting value of $n_{eff} = 2.24$ Bohr magnetons. From this simple point of view the temperature dependence of n_{eff} is less for d^5 and d^9 than for d^1 and deviations should be smaller for the first than for the latter. There are exceptions to be found, of course, since any distortions of symmetry or electron delocalizations were not taken into account.

Of the configurations listed in Table 9-3, only two, besides those already discussed, are of interest in the context of this experiment. They are the configuration d^2 in an octahedral field, leading to a $^3T_1$ state, and the configuration d^6 in an octahedral crystal field, which is characterized by a $^5T_2$ ground state. Figure 9-7 shows the term scheme for d^2 (also valid for configuration d^8 in a tetrahedral field, and for a low-spin d^4 arrangement in an octahedral field). The sequence of J terms has to be inverted for d^8 and low-spin d^4 because for these λ is negative.

For a d^8 configuration the multiplet level with $J = 0$ becomes the ground state. Since this is so, it follows that $n_{eff(\lim T \to 0)} = 0$, according to equation (3). The high-temperature limit given by equation (4) is $n_{eff(\lim T \to \infty)} = \sqrt{10} = 3.16$. This limiting value is approached from above. For Ni^{2+} complexes the high temperature limit is reached at temperatures much higher than room temperature so that n_{eff} calculated from room-temperature susceptibility measurements is expected to be higher than $n_{eff(\lim T \to \infty)}$. The temperature dependence of n_{eff} is shown schematically in Figure 9-8.

Figure 9-9 shows the term diagram for tetrahedral d^4. This scheme must be inverted for d^6 because λ becomes negative for the half-filled shell. The lowest multiplet level has the property $J = 1$. To calculate the low-temperature limit for n_{eff}, the appropriate g for this state must be obtained from equation (5a), $g = 3.5$. Thus $n_{eff(\text{limit}\, T \to 0)} = 4.95$ is obtained from equation (3). The high-temperature limit is $n_{eff(\lim T \to \infty)} = \sqrt{26} = 5.1$, according to equation (4), and is approached from above. Figure 9-8 also contains a schematic drawing of the temperature dependence of $n_{eff}(d^6)$.

Although it is possible to understand the term arrangement in Figures 9-5, 9-7, and 9-9 by referring to Exercise 7A and C, and the limiting values for n_{eff} on the basis of the discussion in Section A of this exercise, an explanation for the approach of the high-temperature limit from above or below requires a more detailed statistical study.

In conclusion, the magnetic moments of substances with an A or E ground state are usually fairly close to spin-only values. Additional contributions stem from mixing in higher terms (borrowing orbital angular momentum). According to equation (1), this yields a constant additive term to the magnetic moment, the sign of which is positive for less than half-filled subshells and

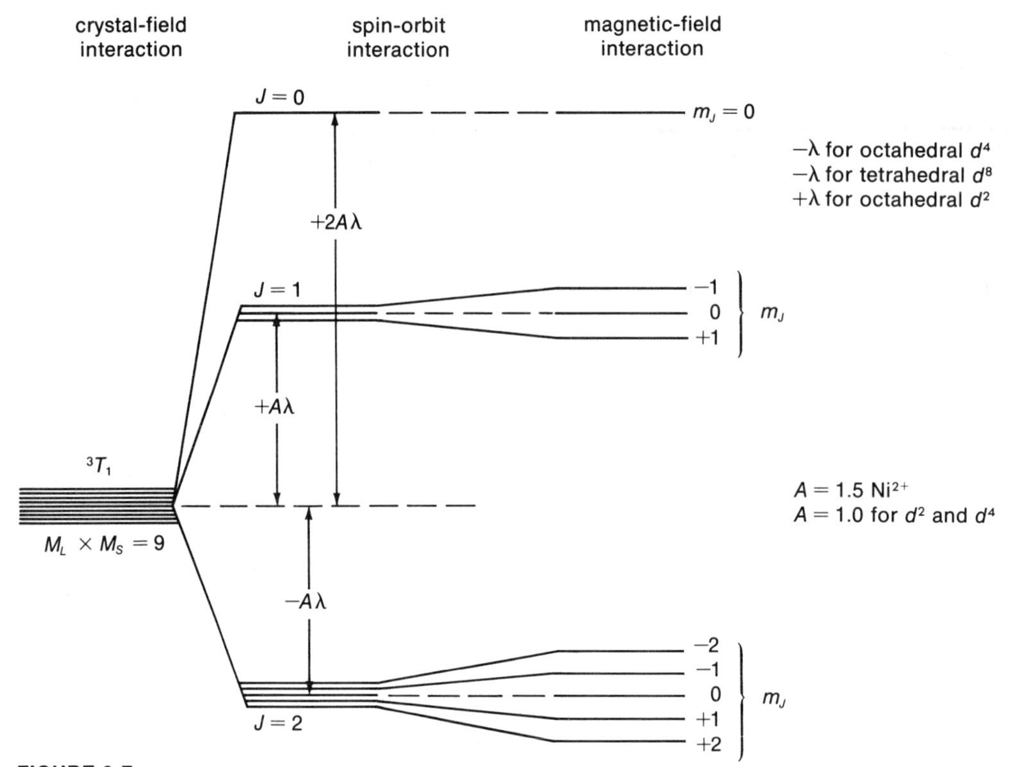

FIGURE 9-7
Splitting of the $^3T_1$ term originating from an octahedral d^2, a tetrahedral d^8, or an octahedral low-spin d^4 configuration.

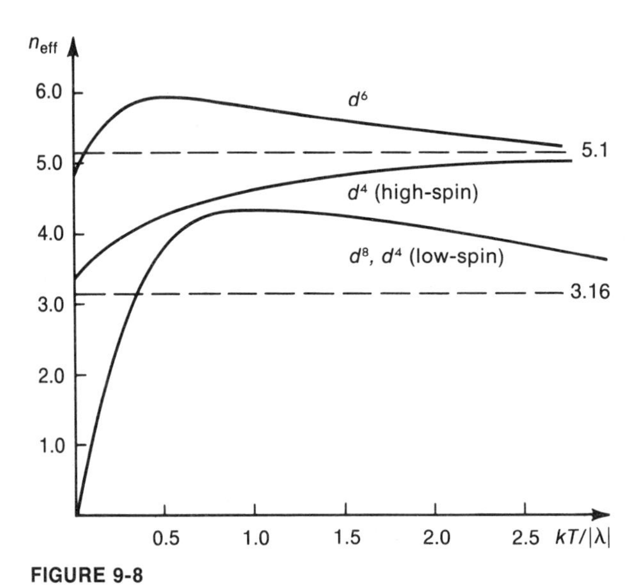

FIGURE 9-8
Temperature dependence of the magnetic moment for ions with a $^3T_1$ ground state (d^8 configuration in a tetrahedral field, and d^4 low-spin configuration in an octahedral field) or with a $^5T_2$ ground state (d^4 high-spin configuration in a tetrahedral field, and d^6 in an octahedral field). [From Figgis and Lewis (1960).]

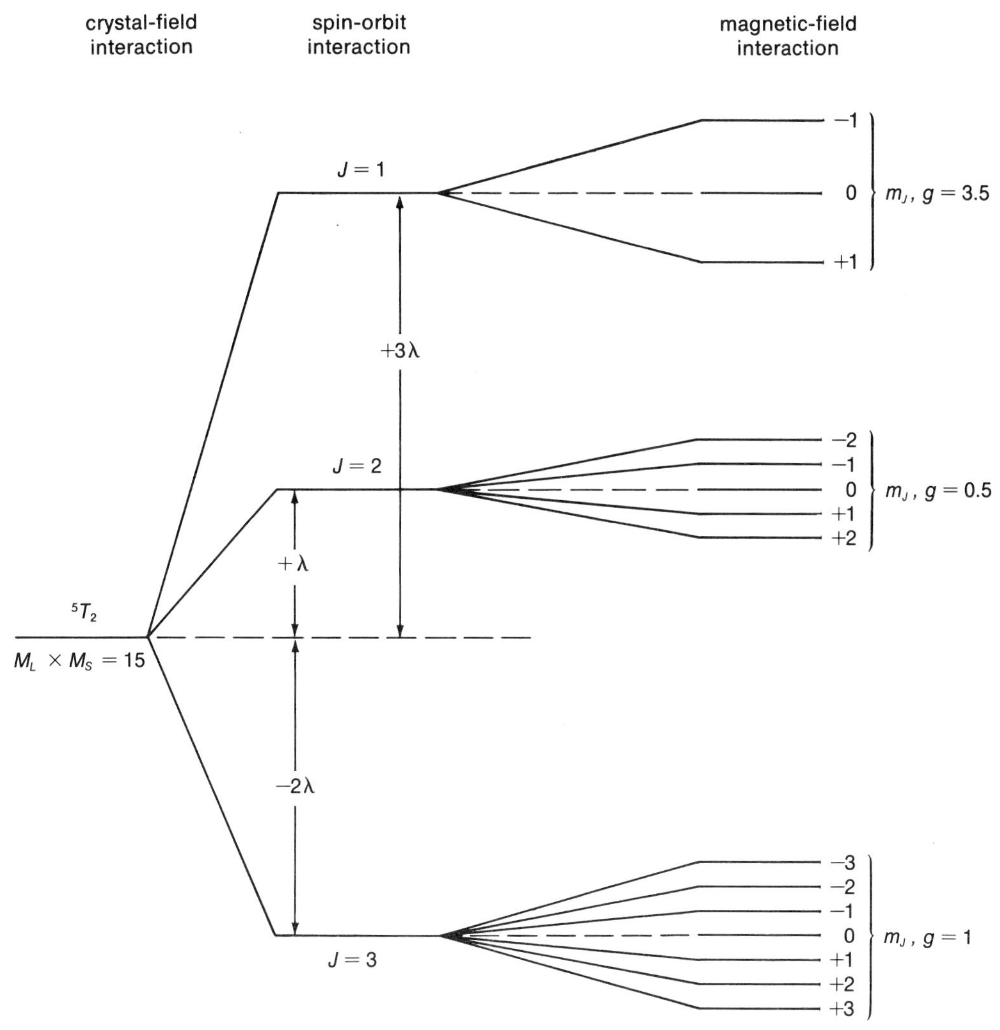

FIGURE 9-9
Schematic diagram of the splitting due to crystal-field, spin-orbit, and magnetic-field interactions of a $^5T_2$ level originating from a d^4 configuration in a tetrahedral crystal field. The splitting scheme must be inverted for a d^6 configuration in an octahedral crystal field, which leads to the same crystal field term because λ becomes negative.

negative for the others, and the magnitude of which depends on the strength of the crystal field. A Curie or Curie-Weiss law successfully describes the temperature dependence.

The magnetic behavior of substances with a T ground state, on the other hand, is characterized by a strong temperature dependence that cannot be described by a Curie-Weiss law, since the magnetic moment itself is temperature dependent. Lower and upper limits for the effective number of Bohr magnetons can be calculated by the characterization of these states by the total orbital angular momentum quantum number $L = 1$. Values for the magnetic moment at room temperature are usually comparatively close to the high-temperature limit. The sign of the deviation from the limiting value depends on the direction (above or below) from which it is approached.

Since the magnetic moment depends on the symmetry of the crystal field surrounding the paramagnetic central ion, susceptibility measurements constitute a sensitive method of structure determination for transition metal complexes. However, since it is difficult to predict the accurate magnetic moment at a particular temperature, measurement at just one temperature is insufficient. Instead, the characteristic differences in temperature dependence for a certain central ion in different crystal field symmetries should be exploited.

REFERENCES

Selwood (1954*a*; 1954*b*) and Bates (1951) provide general information on magnetic properties and techniques. Mulay (1963) discusses the method of measurement, and Cotton and Wilkinson (1966), Phillips and Williams (1966), Dorain (1965), and Sienko and Plane (1963) deal with the application of magnetic measurements to problems in inorganic chemistry.

Adams, D. M., and Raynor, J. B. 1965. *Advanced Practical Inorganic Chemistry*. New York: Wiley, p. 147.

Bates, L. F. 1951. *Modern Magnetism*. 3d ed. Cambridge: Cambridge University Press.

Brubacher, L. J., and Stafford, F. E. 1962. *J. Chem. Ed.* **39**, 574.

Cotton, F. A., and Wilkinson, G. 1966. *Advanced Inorganic Chemistry*. 2d ed. New York: Wiley, pp. 633–644 (general); pp. 662–673 (relation to crystal field theory); pp. 1056–1059 (lanthanides). Many specific examples are discussed in the text.

Dorain, P. B. 1965. *Symmetry in Inorganic Chemistry*. Reading, Massachusetts: Addison-Wesley, ch. 5.

Dunn, T. M., McClure, D. S., and Pearson, R. G. 1965. *Some Aspects of Crystal Field Theory*. New York: Harper & Row.

Figgis, B. N. 1966. *Introduction to Ligand Fields*. New York: Wiley.

Figgis, B. N., and Lewis, J. 1960. The Magnetochemistry of Complex Compounds. In *Modern Coordination Chemistry*. Ed. J. Lewis and R. G. Wilkins. New York: Wiley.

Foëx, G. 1957. *Tables des Constantes et Données Numerique. 7. Diamagnétisme et Paramagnétisme*. Paris: Masson.

Klemm, W. 1940. *Z. Anorg. Allgem. Chemie*. **244**, 377; 1941. **246**, 347.

Lewis, J., and Wilkins, R. G. 1960. *Modern Coordination Chemistry*. London: Wiley, p. 403.

Mulay, L. N. 1963. Magnetic Susceptibility. In *Treatise on Analytical Chemistry*. Vol. 4, Part 1. Ed. I. M. Kolthoff and P. J. Elving. New York: Wiley. A reprint of this chapter is available through the publisher.

Orgel, L. E. 1960. *An Introduction to Transition Metal Chemistry*. New York: Wiley.

Phillips, C. S. G., and Williams, R. J. P. 1966. *Inorganic Chemistry. II. Metals*. Oxford: Oxford University Press, pp. 191–192 (general); pp. 406–422 (*L-S* coupling); pp. 122 ff. (lanthanides).

Selwood, P. W. 1954*a*. *Magnetochemistry*. 2d ed. New York: Wiley.

———. 1954*b*. *Techniques of Organic Chemistry*. 3d ed. Vol. 1, Part 4. Ed. A. Weissberger. New York: Wiley, ch. 43.

Sienko, M. J., and Plane, R. A. 1963. *Physical Inorganic Chemistry*. New York: Benjamin.

EXERCISE 10

X-Ray Powder Diffraction of Crystalline Solids

A. DEBYE-SCHERRER-HULL METHOD

Introduction

For a long period in the history of science the only possible way to classify crystals was by their external symmetry. The idea that the observed symmetry was due to the atomic arrangement and order inside the crystal led to the concept of the point lattice, in which the crystal systems are characterized by their unit cells. The discovery of x-ray diffraction by crystals, which had been predicted on the basis of lattice symmetry, confirmed existing ideas and offered the scientist a means of studying the inner structure of matter. The following experiments serve as an introduction to the simplest of all x-ray investigations of crystals, the Debye-Scherrer-Hull powder diffraction method, dealing with its application to the identification and crystallographic characterization of crystalline solids.

Crystallographic Background

A crystal is an indefinite repetition in three dimensions of a parallelepiped called the unit cell. If the parameters of the unit cell are known, the total crystal structure is known. Unit cells can be constructed in many ways within one crystal, but they are always chosen so that they conform to the outer symmetry of the crystal and contain an integer number of molecules. The cell edges of lengths a, b, and c are placed parallel to the crystal axes x, y, and z. From the relationship between lengths a, b, and c and from the angles between cell edges α, β, and γ, the system to which the crystal belongs can be inferred. Crystal structures can be described by seven crystal systems and fourteen different unit cells, which define the fourteen Bravais lattices. Table 10-1 lists crystal systems and unit cells of corresponding Bravais lattices. As applied to the unit cell lengths a, b, and c, the unequal sign, $\neq$, means that the two are not equivalent. Usually, this is due to unequal lengths; however,

TABLE 10-1
Crystal systems and corresponding Bravais lattices.

Crystal system		Bravais lattice
Cubic	$a = b = c$, $\alpha = \beta = \gamma = 90°$	Primitive, body-centered, face-centered
Tetragonal	$a = b \neq c$, $\alpha = \beta = \gamma = 90°$	Primitive, body-centered
Orthorhombic	$a \neq b \neq c$, $\alpha = \beta = \gamma = 90°$	Primitive, body-centered, base-centered, face-centered
Rhombohedral	$a = b = c$, $\alpha = \beta = \gamma \neq 90°$	Primitive
Hexagonal	$a = b \neq c$, $\alpha = \beta = 90°$ $\gamma = 120°$	Primitive
Monoclinic	$a \neq b \neq c$, $\alpha = \gamma = 90° \neq \beta$	Primitive, base-centered
Triclinic	$a \neq b \neq c$, $\alpha \neq \beta \neq \gamma \neq 90°$	Primitive

two axes may be equal in length but not equivalent because of the symmetry of the atoms within the crystal. If the cell is primitive (or simple), the eight corners are occupied by atoms or molecules; if body-centered, the eight corners plus the center of the cell are occupied; if base-centered, the eight corners plus the centers of one pair of parallel faces are occupied; and if face-centered, the eight corners plus the center of all faces are occupied.

Lattice points in unit cells (sometimes also the spaces in between) are occupied by atoms, molecules, and simple or complex ions. Conditions imposed by ionic size and the demand of electroneutrality force many ionic substances to crystallize in typical model "structures." These structures are characterized by the coordination number for cations and anions. A unit cell of a certain structure is usually constructed so that these coordination numbers are clearly visible. As an example, the NaCl structure and the CaF_2 structure are shown in Figure 10-1. Both possess cubic face-centered unit cells. Two completely analogous unit cells have been drawn for CaF_2, the first clearly showing the coordination for anions, and the second showing the coordination of cations. The NaCl structure may be described as a lattice composed from a face-centered cation lattice interwoven with a face-centered anion lattice of the same magnitude.

A large number of univalent metal halides crystallize in the NaCl structure, but some are found to possess the CsCl structure also shown in Figure 10-1. Eight unit cells of the CsCl structure would have to be drawn to show the coordination of one chloride ion to eight caesium ions.

The interaction of an x-ray beam with a crystal can be explained by describing the corresponding point lattice as composed of stacked parallel planes, which contain the scattering centers. Many such planes can be found, and they are distinguished by the "Miller indices," which are the reciprocals of the separation of planes intersecting the crystallographic axes. The separations are measured in units of length of the cell edge parallel to the crystallographic axis. The Miller indices are integers obtained by clearing fractions in the reciprocals. Figure 10-2 shows how the Miller indices for an arbitrary plane are obtained. Some examples of lattice planes, together with their Miller indices, are given in Figure 10-3.

The distance d between planes can be expressed by means of the Miller indices. For orthogonal systems the derivation is simple. A given set of planes divides the x axis into lengths a/h, the y axis into b/k, and the z axis into c/l. Figure 10-4 shows the plane (A,B,C) and the origin (O), the segment (A,B) that the plane cuts from the x,y plane, and the planes perpendicular to the x,y plane, which contain the c axis.

The triangles AOB and AOS are similar and thus

$$\frac{OA}{OS} = \frac{AB}{OB}$$

or by substituting the corresponding lengths,

$$\frac{a/h}{s} = \frac{\sqrt{(b^2/k^2) + (a^2/h^2)}}{b/k}$$

or

$$\frac{1}{s} = \frac{hk}{ab}\sqrt{\frac{b^2}{k^2} + \frac{a^2}{h^2}} = \sqrt{\frac{h^2}{a^2} + \frac{k^2}{b^2}} \qquad (1)$$

Again, triangles SOC and SDO are similar and thus

$$\frac{OS}{OD} = \frac{SC}{OC}$$

number of atoms per unit cell		coordination number		structure
cation	anion	cation	anion	
1	1	8	8	CsCl
4	4	6	6	NaCl
4	8	8	4	CaF_2

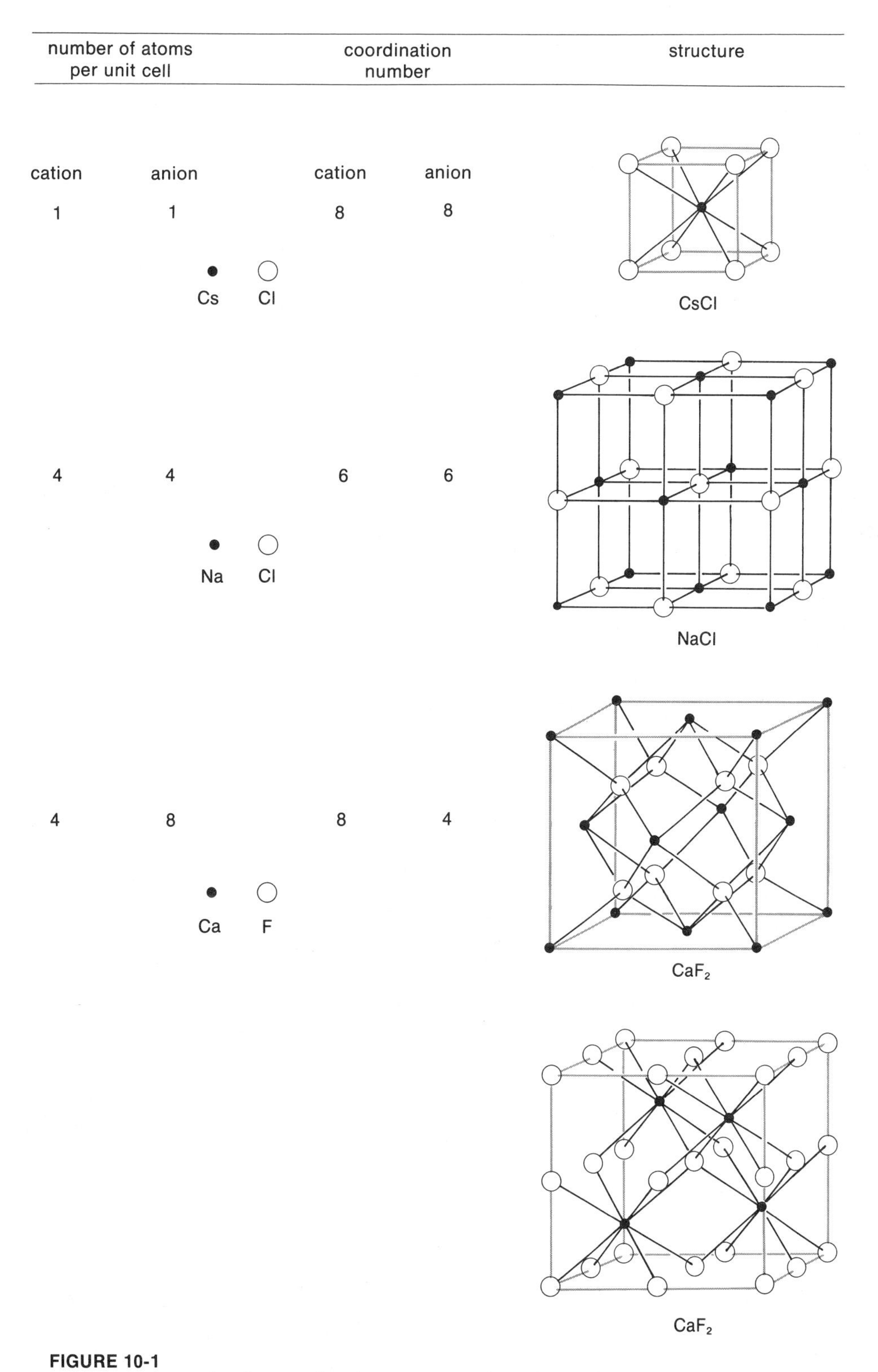

FIGURE 10-1
Unit cell and coordination number. In some cases (e.g., the NaCl structure) chemical bonds coincide with cell edges. Section D of this exercise, which deals with the spinel structure, gives further examples of unit cell representations.

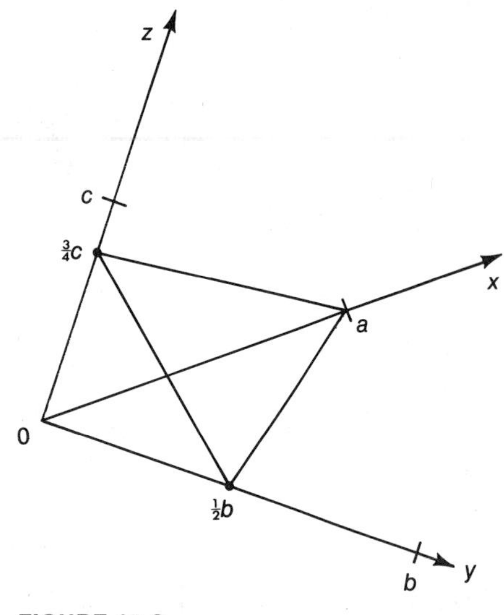

fractional intercepts of axis	x	y	z
	$1a$	$\frac{1}{2}b$	$\frac{3}{4}c$
reciprocals $(1/h, 1/k, 1/l)$	1	2	4/3
Miller indices (h, k, l)	3	6	4

FIGURE 10-2
Miller indices of a series of planes that cut the x axis in intervals a, the y axis in intervals $\frac{1}{2}b$, and the z axis in $\frac{3}{4}c$. The first plane is thought to contain the origin of the coordinates. (The axes need not be orthogonal.)

or by substituting the corresponding lengths,

$$\frac{s}{d} = \frac{\sqrt{c^2/l^2 + s^2}}{c/l}$$

or

$$\frac{1}{d} = \frac{1}{cs}\sqrt{\frac{c^2}{l^2} + s^2} = \sqrt{\frac{1}{s^2} + \frac{l^2}{c^2}}$$

Substitution of equation (1) leads to

$$\frac{1}{d} = \sqrt{\frac{h^2}{a^2} + \frac{k^2}{b^2} + \frac{l^2}{c^2}} \qquad (2)$$

Equation (2) is sometimes called the interplanar spacing equation. The derivation of similar equations for nonorthogonal systems is not as simple. Table 10-2 contains interplanar spacing equations for all orthogonal crystal classes.

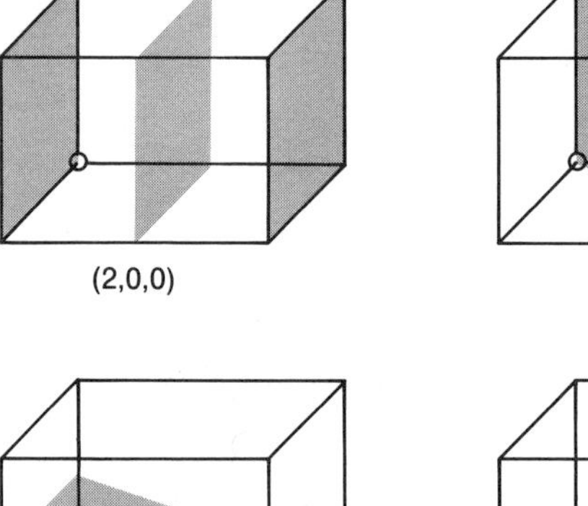
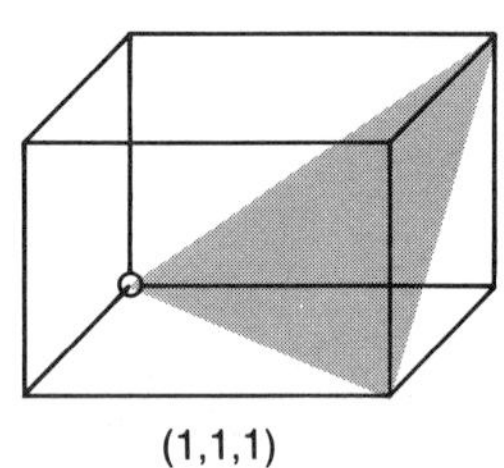

FIGURE 10-3
Lattice planes and their Miller indices.

The Debye-Sherrer-Hull Method

Physical Foundations

The wavelength of x-radiation is comparable to the interatomic distances or plane spacings in crystals. X-rays are diffracted by crystals in much the same way in which light is diffracted by gratings.

TABLE 10-2
Interplanar spacing equations for selected crystal systems.

Crystal systems	$1/d^2$
Cubic	$(1/a^2)(h^2 + k^2 + 1^2)$
Tetragonal	$(1/a^2)(h^2 + k^2) + (l^2/c^2)$
Orthorhombic	$(h^2/a^2) + (k^2/b^2) + (l^2/c^2)$
Hexagonal	$(4/3a^2)(h^2 + hk + k^2) + (l^3/c^2)$

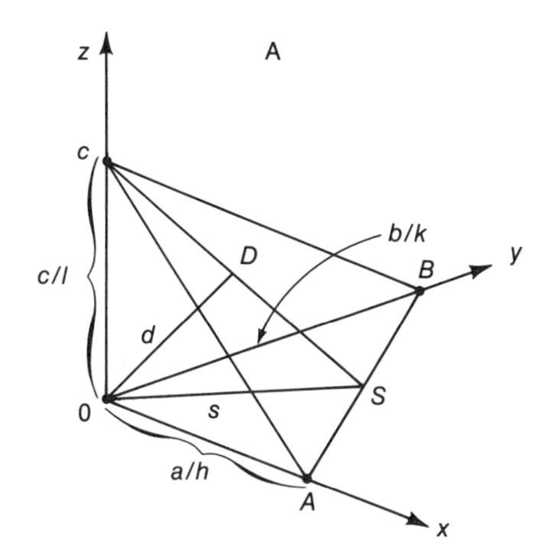

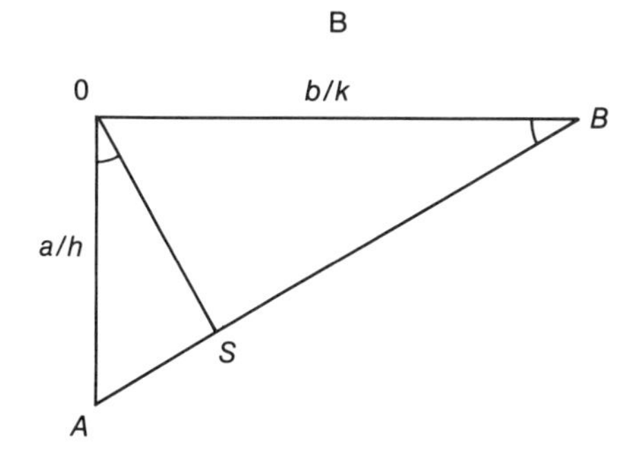

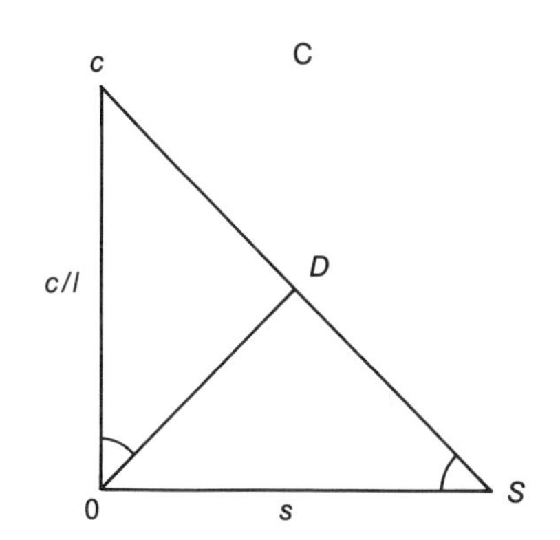

FIGURE 10-4
A. Crystal plane. B. Segment of (A) showing the *OAB* plane. C. Segment of (A) showing the *OCS* plane.

X-rays impinging on a crystal are scattered by the electron clouds around atomic nuclei. Each scattering center becomes the center of an emerging spherically symmetric wave. The only waves not extinguished by interference with neighboring waves in a single lattice plane are those for which the incident and refracted beams make the same angle with the plane. Since the x-radiation penetrates into deeper layers of the crystal, interference

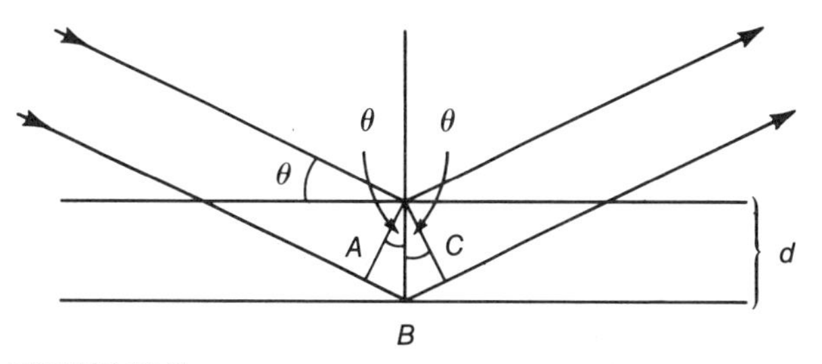

FIGURE 10-5
Diffraction of a well-collimated beam of monochromatic x-radiation on two lattice planes.

occurs between such reflected beams within an array of parallel lattice planes. Figure 10-5 illustrates the condition for constructive interference and the relations between observed diffractions and lattice planes. Constructive interference occurs whenever distances travelled by beams reflected from parallel lattice planes differ by an integer multiple of the wavelength.

From Figure 10-1 the difference in path length between reflections from two neighboring planes is read as:

$$AB + BC = 2AB$$

Trigonometry yields:

$$2AB = 2d \cdot \sin \theta$$

where d is the spacing between planes. With the condition $2AB = n\lambda$, where $n = 1, 2, 3, \ldots$, Bragg's equation is arrived at:

$$n\lambda = 2d \cdot \sin \theta \qquad (3)$$

which is the fundamental law of lattice diffraction and n is the order of diffraction. Whenever a parallel beam of monochromatic x-rays of wavelength λ impinges with angle θ on a set of parallel planes of distance d in a crystal, diffraction will be observed. According to equation (3), in principle diffraction is expected at multiple values of sin θ. Additional consideration of the vast number of possible lattices planes might lead to the prediction of complicated diffraction patterns. Fortunately, close examination reveals more simplicity.

For cubic crystals, d can be expressed as:

$$d^2 = \frac{a^2}{h^2 + k^2 + l^2} \qquad (4)$$

(see Table 10-1). The largest value for d can thus be predicted as $d = a$ for planes (1,0,0). Planes with the Miller indices (2,0,0) are spaced at exactly $d = a/2$. A glance at equation (3) shows that the second-order diffraction from (1,0,0) coincides exactly with the first-order diffraction from (2,0,0). Equivalent considerations hold for any other pair of planes. The smallest possible value of θ is thus to be found for first-order reflexion from (1,0,0). Any diffraction observed under a larger angle can subsequently be identified as a first-order diffraction from a set of planes with larger Miller indices, for example, smaller d. This can be done without fear of misinterpretation since all higher-order diffractions coincide with some first-order diffraction. The largest possible value for θ is to be found for $\sin \theta = 1$. Its introduction into equation (3) yields:

$$d = \frac{\lambda}{2} \tag{5}$$

The number of glancing angles under which diffraction can be observed is limited and consideration of first-order diffraction alone is sufficient. Equation (3) may be reformulated with the help of equation (4):

$$\frac{\sin^2 \theta}{h^2 + k^2 + l^2} = \frac{\lambda^2}{4a^2} \tag{6}$$

The ratio of the square of the sine of the glancing angle to the sum of the squares of the Miller indices for the diffraction set of planes always yields a constant. The range and the spacing of glancing angles are defined by the lattice constant a and the wavelength λ. For a given lattice constant, the choice of a wavelength that is too small leads to the observation of many lines [lattice planes of small d are still allowed, according to equation (5)], but these are crowded together in a range of small θ. The choice of a wavelength that is too large leads to the observation of a few lines in a narrow region close to $\theta = 90°$. The correct choice of wavelength results in a manageable number of glancing angles that are distributed over an accessible range. So far, the characterization of a crystal has been by the observation of glancing angles alone, without consideration of the fact that the intensity of the diffracted beam may change considerably for different lattice planes.

The intensity with which x-rays are scattered depends mainly on the electron density at the scattering site, with heavier elements scattering more than lighter elements. Planes that contain more scattering centers give rise to stronger diffractions than those of smaller population. In addition, beams diffracted from different sets of planes may interfere destructively and extinguish each other to a degree indicated by the scattering power of their constituents. These destructive interferences depend on symmetry and lead to so-called "systematic absences." A simple example is a body-centered cubic lattice. At the conditions for diffraction on the (1,0,0) planes, there is also diffraction on the planes lying between the (1,0,0) at $a/2$, but the waves emerging from the latter planes are out of phase with the diffraction from (1,0,0) by $\lambda/2$. Since both planes possess the same density of scattering centers, extinction will be complete if the scattering power is the same for both kinds of constituents in the planes. Thus no diffraction form (1,0,0) is observed. The same is true for diffraction by (1,1,1) planes. Diffraction of the beam is observed from (2,0,0), (2,2,2), and (1,1,0), since these go through all sites in the lattice. (See Figure 10-6.)

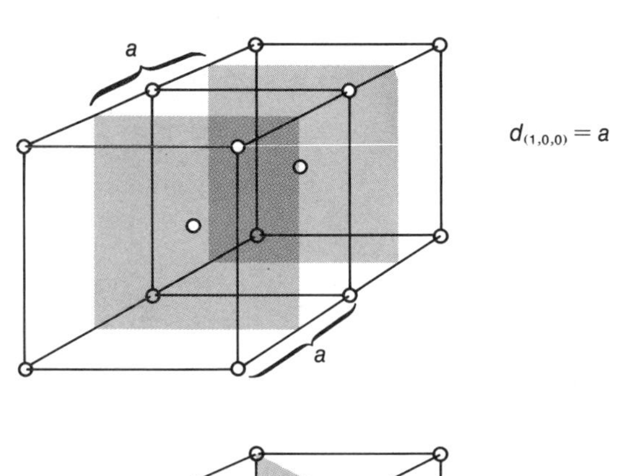

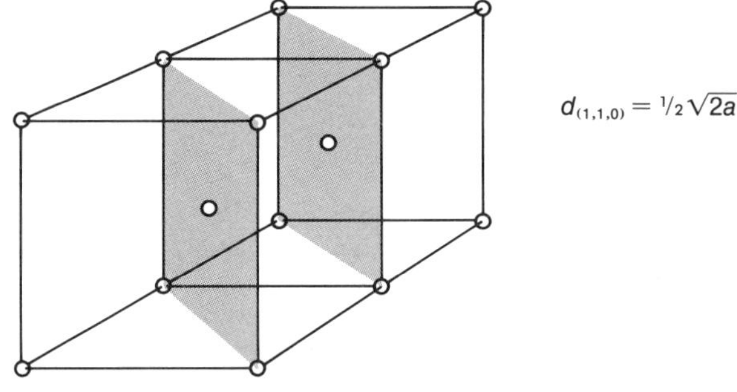

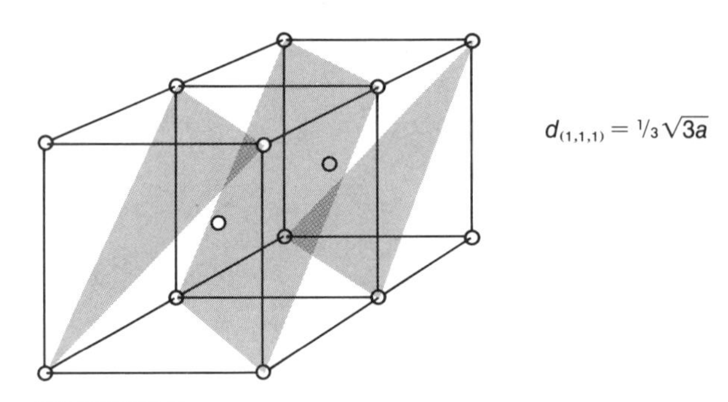

FIGURE 10-6
Planes in a body-centered cubic lattice.

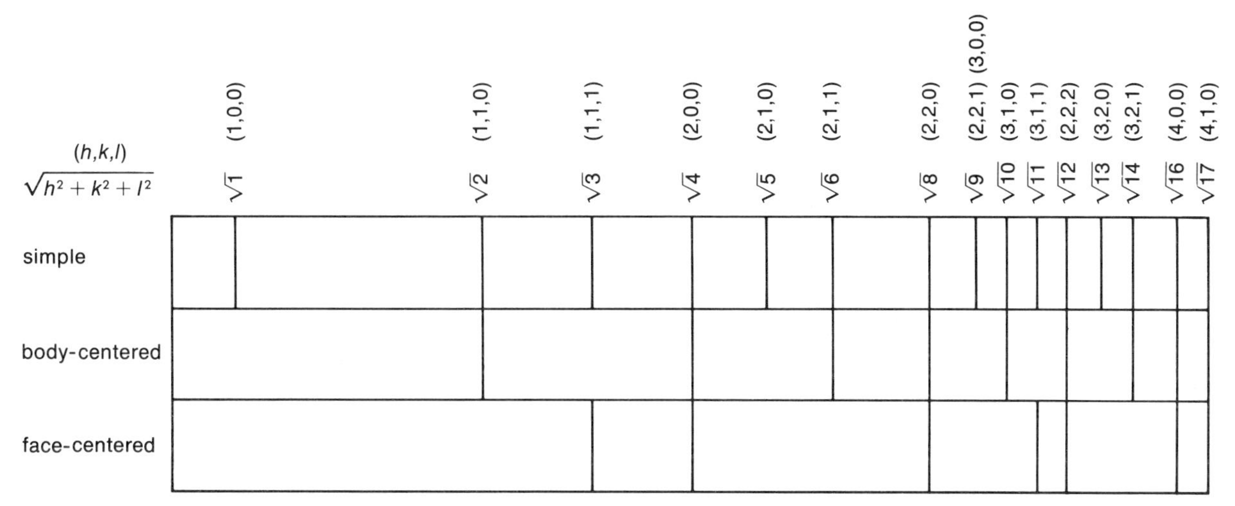

FIGURE 10-7
Allowed reflections from cubic structures.

In a simple cubic structure, planes always contain all lattice sites and thus all possible diffractions are observed. For face-centered structures, systematic absences occur.

According to equation (6), the sine of the glancing angle can be expressed as a constant times the square root of the sum of the squared Miller indices:

$$\sin \theta = \text{const } (h^2 + k^2 + l^2)^{1/2} \tag{7}$$
$$= \text{const} \cdot \sqrt{\text{integer}}$$

The integer values under the square root are restricted to those that can be obtained by the summation of three squares. Figure 10-7 shows the Miller indices of planes from which diffraction occurs for simple, body-centered, and face-centered cubic structures on a scale of log d. For body-centered structures, reflections are observed from planes for which $h + k + l$ is even. Diffraction from face-centered cubic crystals occurs only if all Miller indices are either odd or even. With experience, it is frequently possible to guess the lattice type from the regularities in the spacing of glancing angles.

Powder Pattern

The Debye-Scherrer-Hull technique is perhaps the most frequently employed powder diffraction method. With this technique lattice parameters and the symmetry of cubic, tetragonal, orthorhombic, and hexagonal crystals can be determined. The diffraction patterns of less symmetric crystal classes become too involved for evaluation.

A microcrystalline powder is brought into a collimated beam of monochromatic x-rays and is surrounded by a strip of photographic film. The film is exposed to reflections from all possible glancing angles, since the microcrystallites are randomly oriented in the sample. Thus for each set of planes, crystallites that form the correct angle with the x-ray beam can be found. If all orientations of crystallites under diffraction conditions are present in the sample, the diffracted beam forms a cone, which intercepts the film strip and looks like an arc on the developed film (see Figure 10-8). Samples are rotated in the apparatus to obtain full cones of diffracted radiation.

The relationship between the glancing angle θ and the arcs on the film is shown in Figure 10-9. The distance S between two corresponding arcs is related to 4θ, as illustrated in Figure 10-9. The arc length S depends on the radius R of the film circle.

$$S = 4 \cdot \theta \cdot R \tag{8}$$

and θ is given in radians:

$$\theta = \frac{S}{4 \cdot R} \tag{9}$$

or in degrees:

$$\theta = \frac{180}{4\pi R} S$$

The camera of commercial x-ray apparatus is frequently constructed so that the length of the filmstrip is either 180 mm (R = 57.3 mm) or 360 mm (R = 114.6 mm). In either case, θ can be

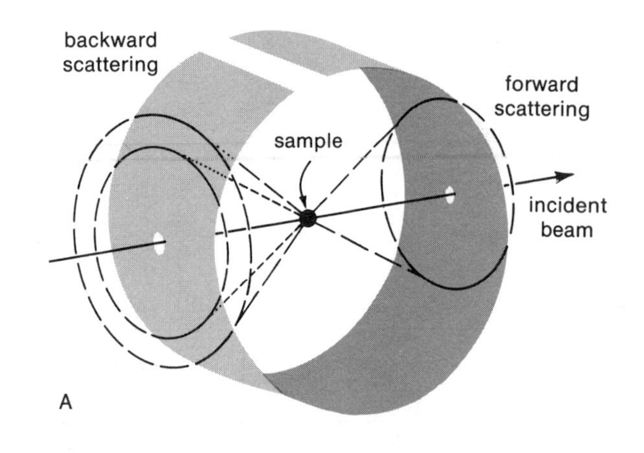

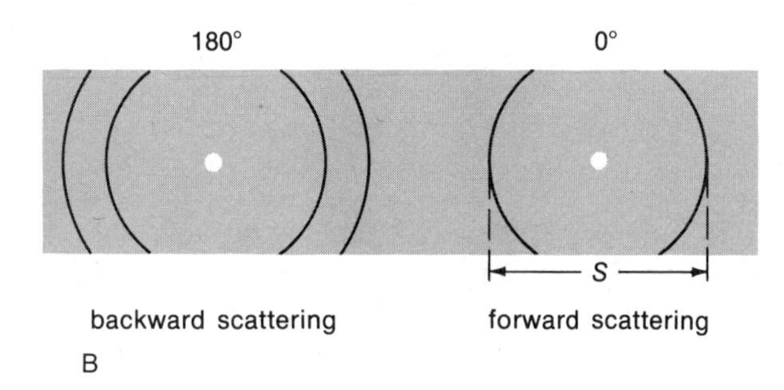

FIGURE 10-8
A. Cones of diffracted radiation. B. Developed film.

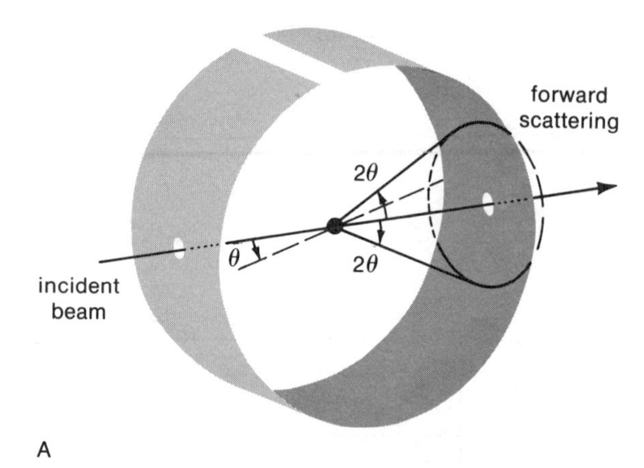

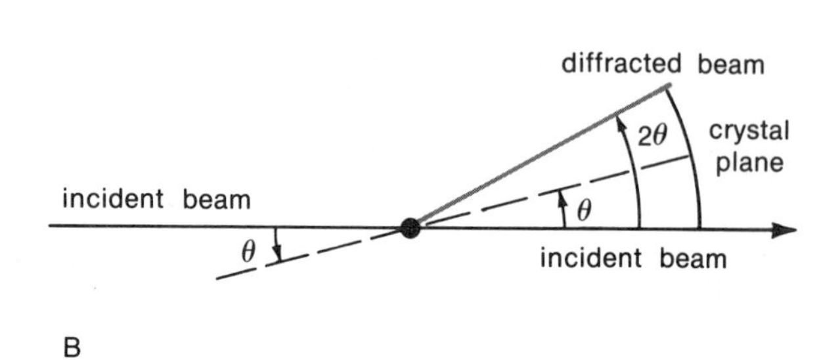

FIGURE 10-9
A. Production of arcs due to forward scattering. The broken line indicates the crystal plane. Rotation of the plane around the incident beam leads to a full cone of diffracted radiation. B. Production of a segment of the arc by one orientation of the diffracting crystal plane. The angle between the diffracted beam and the incident beam is seen to be $2 \cdot \theta$.

read from the film directly. For $R = 57.3$ mm. equation (9) reads: $\theta = S/2$ (with S in millimeters, θ in degrees). For $R = 114.6$ mm: $\theta = S/4$ (with S in millimeters, θ in degrees).

Apparatus

A diffractometer consists of an x-ray source, the necessary electronic equipment, and the camera.

X-Ray Source

X-rays are produced by the bombardment of a "target" with high-energy electrons. The electrons are boiled off a cathode and then accelerated in a field of 30,000 V. They possess enough energy to knock electrons out of the K shell of the target. After an electron has been removed in this way, another electron from the L or a higher shell takes its place, giving up its excess energy in the form of x-radiation during the transition. Thus each target yields a spectrum characteristic of its inner energy levels. Transitions occur mainly between L and K and between M and K shells. X-radiation from the first transition is called K_α and from the second, K_β. In addition to the spectrum, a continuous background of radiation, called "white radiation," is produced by the slowing down of impinging electrons in multiple interactions with the electrons of the target. The white radiation cuts off sharply at the wavelength corresponding to the maximum energy of the impinging electrons. Figure 10-10 shows a typical spectrum of a target.

Monochromatic radiation is needed for diffraction experiments. Since it is very difficult to construct a monochromator for x-rays, the output of the target is filtered. X-ray absorption spectra are characterized by increasing absorption as the wavelength at which K electrons of the absorbing element are excited is approached and by a sharp absorption edge as longer wavelengths are approached. Figure 10-10 shows the absorption spectrum of a suitable filter superimposed on the target emission spectrum. The filter has been

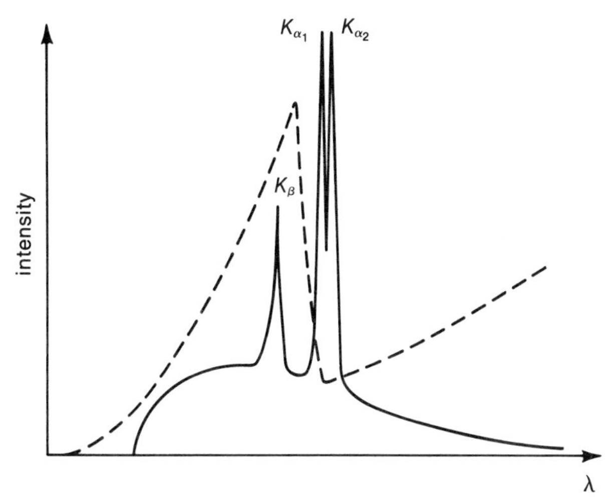

FIGURE 10-10
X-ray spectrum of a target bombarded by electrons. K_{α_1} and K_{α_2} are due to transitions between the two sublevels of the *L* shell and the *K* shell, and *K* stems from transitions between *M* and *K*. Lines due to transitions between higher levels are found at longer wavelengths. The dotted line represents the absorption spectrum of a filter.

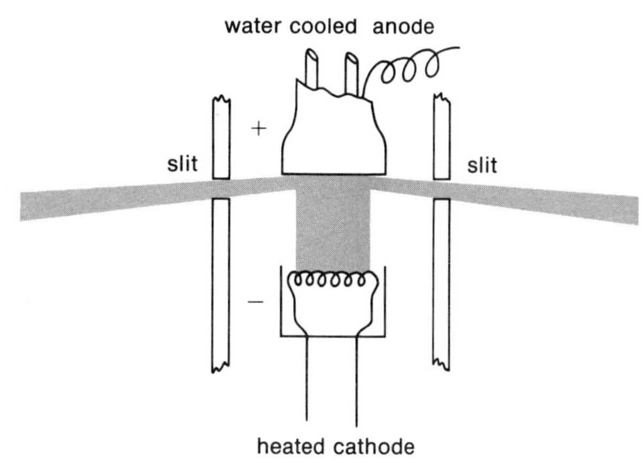

FIGURE 10-11
Schematic view of an x-ray tube.

chosen so that the absorption edge cuts off just before the wavelength of *K* radiation. Thus satisfactory monochromatic radiation is produced. Table 10-3 lists targets, filters, and wavelengths suitable for diffraction experiments. The beam emerging from the target can be collimated only by slits and metal tubes. Figure 10-11 shows an x-ray tube with the beam passing through the exit slit with a very small angle.

TABLE 10-3
Target elements, filter elements, and wavelengths for x-ray sources.

Target element	$\lambda_{K\alpha}$ (Å)*	Filter element
Mo	0.7107	Zr
Cu	1.5418	Ni
Ni	1.6591	Co
Co	1.7902	Fe
Fe	1.9373	Mn
Cr	2.2909	V

* $\lambda_{K\alpha}$ is a weighted average of $\lambda_{K\alpha 1}$ and $\lambda_{K\alpha 2}$.

Camera

Figure 10-12 shows the side and front views of an x-ray camera. The circular wall supports the film. One of the side walls can be removed to mount the sample and the film. The camera is attached very close to the source, so that no radiation escapes. The unreflected beam is absorbed in a filter of lead glass equipped with a fluorescing layer, which permits the observation of beam adjustment.

The centering knob aligns the specimen in the radiation path. The finger holds the film against the wall. The sample itself must be in the shape of a rod, about 0.5 mm in diameter and 10 mm long. This can be accomplished by (1) mixing the sample with an amorphous binder and extruding it; (2) coating the outside of a thin, glass fiber with sample powder (gelatin or plastic cement is used to attach the powder to the fiber); or (3) confining the powder in a thin-walled glass capillary. The last method is the most convenient, and it is suitable for moisture- or air-sensitive samples, since capillaries can be filled and sealed in inert atmospheres.

Procedure

Thoroughly grind the sample in an agate mortar. Fill the capillary with powder; check whether the powder has settled uniformly by looking at it through a magnifying glass in bright light. Then break the capillary at approximately one centimeter from the sealed end, and place the open end into the specimen holder. (Practice filling and breaking the capillary with a piece of used capillary.) The specimen holder is situated on a metal plate that is magnetically attached to the pulley-operated rotor (see Figure 10-12). Rotation increases the number of powder particles that can assume reflecting positions in the path of the incident beam, thereby producing smoother arcs

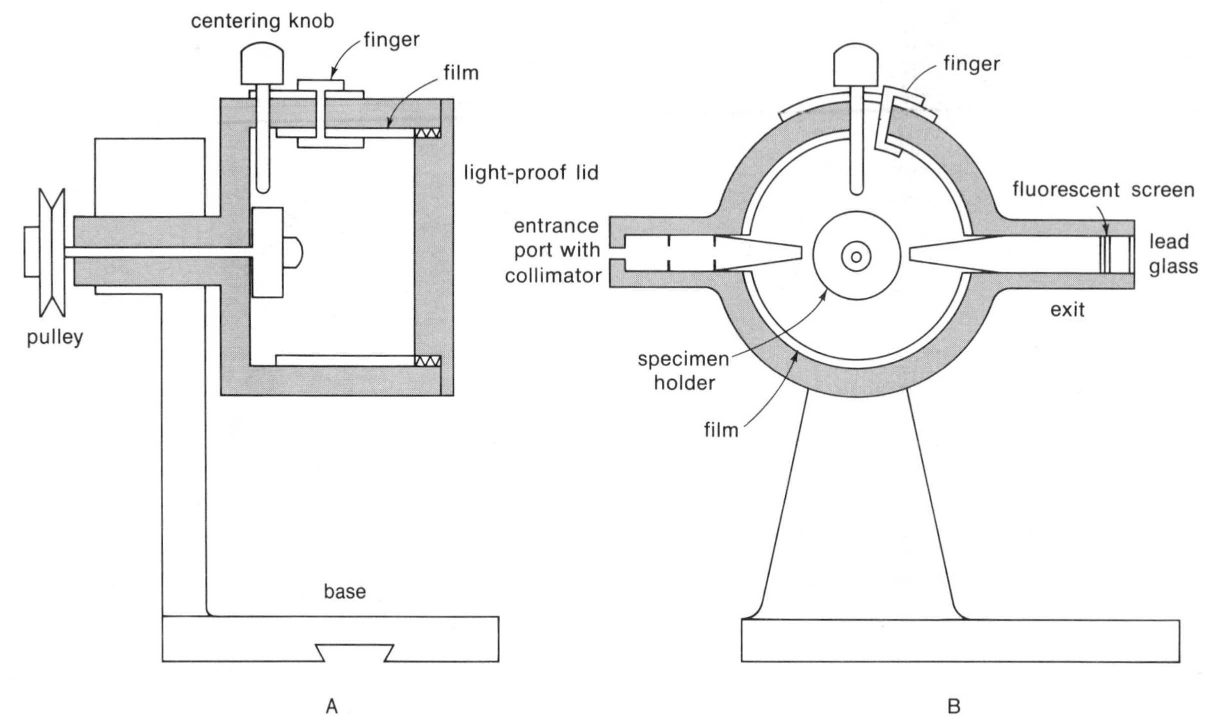

FIGURE 10-12
An x-ray camera: (A) side view and (B) front view.

than if the specimen were not rotated. Position the specimen so that it remains in the beam throughout rotation by moving the metal plate on the face of the rotor, using the plunger operated by the centering knob. During this procedure, observe the specimen through the x-ray collimator.

The specimen is illuminated by a light shining through the hole at which the exit port assembly will be placed later. Place a magnifying glass in the collimator to make viewing a bit easier. The specimen is centered satisfactorily when it remains visible through the collimator and is as stationary as possible throughout a complete rotation. Then take the camera to the darkroom, and cut and punch the film to fit the camera. X-ray film is less sensitive to ambient light than ordinary photographic film; nevertheless, care should be taken to minimize exposure to the red light normally used in darkrooms. Figure 10-13 illustrates a typical film cutter.

To prepare a filmstrip 360 mm long, insert the film until the end is against the stop and cut (the holes are punched at the same time). If 180 mm are desired, insert the film until the end of the strip is against the bolt in hole #1 and cut. Then place the bolt in hole #2 and punch the hole for the beam with the left puncher. Insert the cut strip into the film cutter from the opposite side and punch the second hole with the same puncher.

At this point it is advisable to put marks a known distance apart on the film, so that possible shrinkage can be detected. Marks are introduced by using a single-edged razor blade along the parts containing the puncher, the film being held in position by the puncher. This method produces a high degree of uncertainty, because the razor's edge must be exactly vertical to the plane of the film. It is much more desirable to calculate shrinkage from forward and backward scattering after development of the film. The place where the x-ray beam passed through entrance and exit hole can be determined precisely from the position of arcs around each hole. If no shrinkage has occurred, the distance between both centers of the beam should be exactly 90 or 180 mm.

Slide the film into the camera (being careful not to break the specimen at that time!) and tamp it to the wall by moving the "finger" (see Figure 10-12) to the right and then to the left. Reassemble the camera in the darkroom and take it to the diffractometer for mounting. Close the entrance port with a finger during transport, so that the film is not exposed to stray radiation. The diffractometer is turned on by the instructor. Students are not

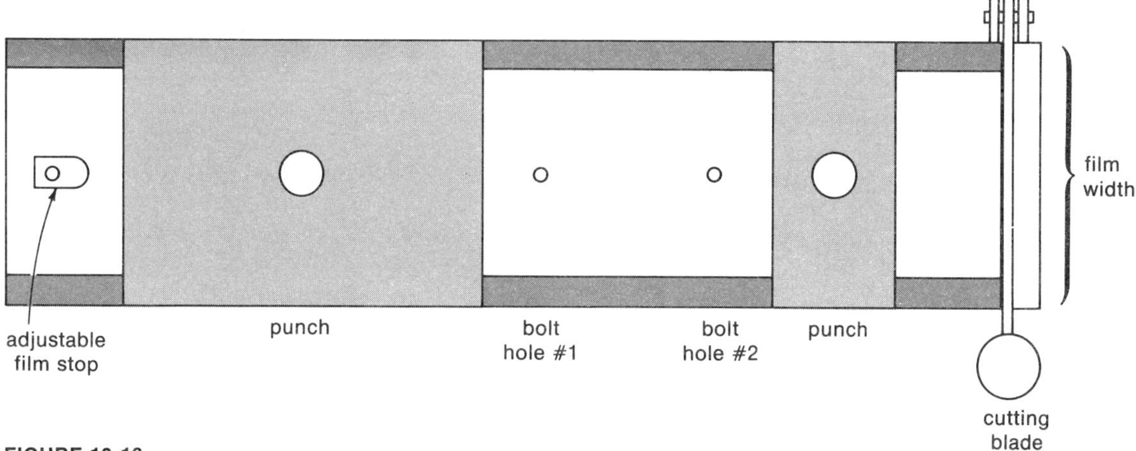

FIGURE 10-13
Cutter used to obtain 180-mm or 360-mm filmstrips for x-ray cameras.

allowed near a diffractometer while x-rays are generated. With some diffractometers, two cameras may be operated simultaneously, allowing two samples to be run at the same time. Cameras of large diameter are preferred for the determination of lattice parameters, because the relative error in the determination of arc distances is reduced.

Exposure time depends on the nature of the crystal. For the determination of lattice parameters of simple substances in large diameter cameras, three hours will usually do. Wash the film with distilled water before drying after development, so that no spots on the film can interfere with the measurement of arcs. Measure the distance S between arcs belonging to the same circular reflection with a scale or transparent millimeter paper on an illuminated surface. Commercial film-reading apparatus affords optimal reading accuracy.

Experimental Technique and Accuracy of the Result

A few experimental conditions that directly influence the attainable accuracy shall be mentioned here. Refer to chapters 14–16 of D'Eye and Wait (1960) for thorough coverage of the problem.

Sources of experimental uncertainty can be divided into two groups: those connected with the choice and condition of the source, and those connected with the quality and mounting of the sample. The factors bearing on the source choice, on the attainable range of reflections, and on the precision of angle determination were discussed earlier in this exercise and will be taken up again in Exercise 15 in Error Analysis and the Selection of Experimental Conditions. In addition to these considerations, which take into account only the conditions for refractions, the probability of the absorption of x-radiation by the sample must be given some thought. The discussion of filter materials for sources included the absorption behavior of different substances. For these materials, the absorption coefficient increases very steeply just before the K edge is reached. If the sample absorbs heavily, refraction can occur only from a small region of planes and the refracted beam will have a very small intensity. Another complication is that fluorescence of the sample may occur if the exciting wavelength is sufficiently near an absorption maximum. The fluorescent radiation is emitted at random and thus blackens the entire film and obscures the refraction lines. Atoms with atomic numbers two or three units less than the atoms in the target tend to absorb the radiation strongly.

The condition of the sample also influences the appearance of refraction halos. If crystallites are too small (the powder is amorphous), conditions for refraction cannot be met because the crystalline regions are too small to give rise to interference. If crystallites are too big, not all possible positions in the beam can be assumed, and halos do not appear in full. Correct centering is very important as well. Eccentricity of the specimen with respect to the axis of the film cylinder causes the displacement of reflections on the film, which makes reading of arc distances difficult.

The resolution of the K doublet offers a convenient means of obtaining an estimate of the reading accuracy. Wavelengths of the two components are: $K_{\alpha_1} = 1.54050\ \mathring{A}$, and $K_{\alpha_2} = 1.54434\ \mathring{A}$, for a copper target.

B. ANALYSIS OF A BINARY MIXTURE OF CRYSTALLINE SOLIDS BY X-RAY DIFFRACTION

Procedure

Obtain a sample, either a single compound or a binary mixture, from your instructor. Take the x-ray powder pattern of this unknown, measure the diffraction halos, and obtain the d values according to the procedure described in Section A of this exercise. Try to find out which two of the following salts are contained in the sample: NaF, NaCl, NaBr, KCl, KBr, $NaBrO_3$, $(NH_4)_2SO_4$, $CaCO_3$, $BaSO_4$, $HgCl_2$, and $CaSO_4$. All samples require an exposure time of 8 hours.

Data

Crystalline solids may be identified by their powder patterns because no two solids produce exactly the same diffraction pattern, if both the spacing and the intensity of lines are considered. A further aid to identification is that the pattern of a mixture is simply the superposition of the patterns of the separate compounds. The large number of possible compounds makes it necessary to have an idea of what the substance might be. The component parts of a mixture can be determined only if the mixture contains just a few compounds in almost equivalent proportions, so that lines distinctly attributable to each compound can be found. Patterns of unknown crystals can be identified either by direct comparison with powder patterns of known substances, or by matching d values and intensities with tabulated values. In the first case, the standard patterns must be obtained with a camera of the same size, using the same source and comparable exposure times. The second method can be used if compilations of x-ray data are available. The most comprehensive tabulation is found in the Powder Diffraction File published by the Joint Committee on Powder Diffraction Standards. The file consists of an index and the original, plus supplemental, sets of file cards. The index is arranged in two parts: one is alphabetic and the other, numerical. The alphabetical index should be used if a choice of substances is given. A typical card from the file is shown in Figure 10-14.

On the upper left side of the card, d values and the relative intensities for the three strongest lines, as well as for that of smallest θ, are given. Relative intensities are defined by setting the intensity of the strongest line equal to 100. The small five-digit numbers at the extreme upper left identify the card in the file on which complete powder data are given. The first digit is the file number; the last four, the card number. The observed powder data are tabulated in the columns on the right. The lower left side contains information about the type of scattered radiation, the crystal symmetry, and the references. The name and chemical composition of the substance are on the upper right.

In the numerical index each substance is listed three times. The spacing of the three strongest lines is called d_1, d_2, and d_3. In the first group, cards are listed with the spacings tabulated in this order; in the second group, the order of the spacings is d_2, d_1, and d_3; and in the third, d_3, d_1, and d_2. In all of these groups, cards are arranged in order of the decreasing value of the second spacing listed. The reason for this is that relative intensities can be altered by experimental circumstances. For example, a mixture may contain two compounds that have one line in common. The relative intensity of this line will then fit neither of the two patterns. The numerical index card also contains the serial number of the substance in the complete data file.

To identify an unknown crystalline substance, the serial numbers of compounds that have the same d values for the three strongest lines as the diffraction pattern of the unknown substance are obtained from the numerical index. By referring to the file cards and comparing the complete powder data, the unknown is identified when its pattern can be matched with one in the card file. For mixtures of two components, the six strongest lines must be matched. Since the process becomes quite complicated, it is less time-consuming and serves the purpose of instruction to hand out a list of possible compounds. Information about these is then taken from the alphabetical index. (Another useful compilation is in the U.S. National Bureau of Standards, Circular 539.)

For most qualitative purposes it is not necessary to use a numerical scale to represent relative intensities. The following relationships of visual estimates with numerical values have been suggested by the Joint Committee on Powder Diffraction Standards:

2829					
d	2.69	1.89	1.19	3.11	KF
1-1069					
I/I_1	100	80	25	15	Potassium Fluoride
1-1050					

Rad. 0.709 Filter
Dia. Cut off Coll.
I/I_1 d corr. abs?
Ref. Davey. Phys. Rev. 21,143 (1923)

Sys. Cubic** S.G. 0^5_H ${}^FM$ ${}^3M$
a_0 2.664 b_0 c_0 A C
a β γ Z 4
Rf. Wy

Ea nwβ 1.3629* E γ Sign
2 V D 2.534 mp 857 Colorless
Ref. C.C.

*AT 25°
B.P. 1502
**NaCl TYPE
Stock

d A°	I/I_1	hkl	d A°	I/I_1	hkl
3.11	15		0.94	5	
2.69	100		.90	5	
1.89	80		.89	8	
1.61	10		.84	8	
1.54	20		.80	8	
1.33	10		.71	8	
1.22	8				
1.19	25				
1.09	15				
1.02	5				

FIGURE 10-14
Sample card from Powder Diffraction File. [By permission of the Joint Committee on Powder Diffraction Standards.]

Strongest line = 100
Very strong = 90
Strong = 70 to 80
Medium = 50 to 60
Weak = 30 to 40
Faint = 20
Very faint = 10

C. DETERMINATION OF LATTICE PARAMETERS FOR CUBIC OR TETRAGONAL CRYSTALS

The basic task in this determination is to associate measured d spacings with the Miller indices for the reflecting planes. This procedure is called "indexing." Once a sufficient number of lines have been indexed, the symmetry of the lattice and the lattice constants can be calculated. This discussion will be confined to methods for indexing cubic and tetragonal crystals only. Crystals from these two classes can be indexed within a reasonable amount of time.

Cubic Crystals

From equation (6),

$$\frac{\sin^2 \theta}{h^2 + k^2 + l^2} = \frac{\lambda^2}{4a^2}$$

the proper sums, $h^2 + k^2 + l^2$, can be found for each $\sin^2 \theta$, which make the right side a constant. This can be done very easily with a slide rule. Mark all values for $\sin^2 \theta$ with pencil on the D scale of a slide rule. Then set the C scale so that an integer coincides with each value of $\sin^2 \theta$. Each integer is equal to the appropriate value of $h^2 + k^2 + l^2$. It is not necessary to try to fit all integers beginning with 1. Some cannot be matched because of systematic absences of reflections; others, for example, 7, 15, and 23, cannot be expressed as the sum of squares of three integers. It is also possible that some reflections are due to impurities. To achieve correct indexing, a few rules should be kept in mind. All strong lines must meet integer values on your slide rule; only faint reflexions can be disregarded. The smallest possible set of integers is the correct one.

The lattice symmetry can be inferred from the systematic absences. It is possible to observe, although very faintly, lines that should be absent. Preferably, the lattice constant a should be calculated from large angle scattering, and, if possible, from backscattering. (Refer to Exercise 15 for a discussion of error propagation.)

Arrange the data and the intermediate results in the form of a table. List, for example, S, the distance between arcs, θ, $\sin \theta$, $\sin^2 \theta$, d, $(h^2+k^2+l^2)$, h, k, l, and intensity. Number all arcs in order of increasing θ. From intensity and d values, the identity of the substance can be obtained.

Tetragonal Crystals

The interplanar spacing equation is:

$$d^2_{hkl} = \frac{1}{[(h^2 + k^2)/a^2] + (l^2/c^2)}$$

No easy, straightforward method such as that used for cubic crystals can be found. It is usually best to begin with a systematic search for those reflexions originating from planes parallel to the c axis. In this case, $l = 0$ and equation (6) reduces to:

$$\frac{\sin^2 \theta}{h^2 + k^2} = \frac{\lambda^2}{4a^2}$$

following the introduction of the Bragg equation (3) with $n = 1$. These reflexions behave like those from a cubic lattice, but they are widely spread out and are difficult to find. The only possible values for $h^2 + k^2$ are 1,2,4,5,8,13,18 . . . , and not all of these may be found because of systematic absences. Once a tentative value for a has been obtained in this way, an attempt is made to match the rest of the lines to:

$$\frac{c^2}{l^2} = \frac{1}{d^2} - \frac{a^2}{h^2 + k^2}$$

Successful matching leads to constant c values. Since at first the estimate of a is relatively inaccurate, values for c will scatter considerably. The best values for a and c are obtained from large angle scattering by plotting l^2/d^2 against $l^2/(h^2 + k^2)$. With correct indexing a straight line of the form

$$\frac{l^2}{d^2} = \frac{a^2 \cdot l^2}{h^2 + k^2} + c^2$$

will be obtained from which the best values for a and c can be calculated. The indexing of tetragonal substances demands much more time than that necessary for indexing cubic crystals.

Obtain a sample from your instructor and determine its x-ray powder pattern by exposing it for 3 hours. Measure the diffraction halos, and calculate d values according to Section A of this exercise. Index the lines as described herein. From the systematic absences decide upon the correct Bravais lattice. Calculate the unit cell length(s).

D. STRUCTURE OF COBALT FERRITE

Ferrites crystallize in the spinel structure, which is best described as a cubic close-packed (c.c.p.) array of oxygen ions with the metal ions occupying the holes between oxygen ions. Figure 10-15 illustrates two layers of oxygen ions in a c.c.p. arrangement.

One octahedral and two tetrahedral holes per oxygen atom are shown in Figure 10-15. Because of the stoichiometry of spinels (their general formula is AB_2O_4, where A is a doubly charged and B a triply charged ion), only one-half of the octahedral holes and one-eighth of the tetrahedral holes can be occupied by a metal ion. In normal spinels the energetically most favorable structure is reached if all triply charged ions occupy octahedral holes (B sites) and all doubly charged ions occupy tetrahedral holes (A sites). This normal structure is given the symbol $A[B_2]O_4$. Figure 10-16A–C shows the arrangement of A and B sites within the c.c.p. oxygen lattice. The segment

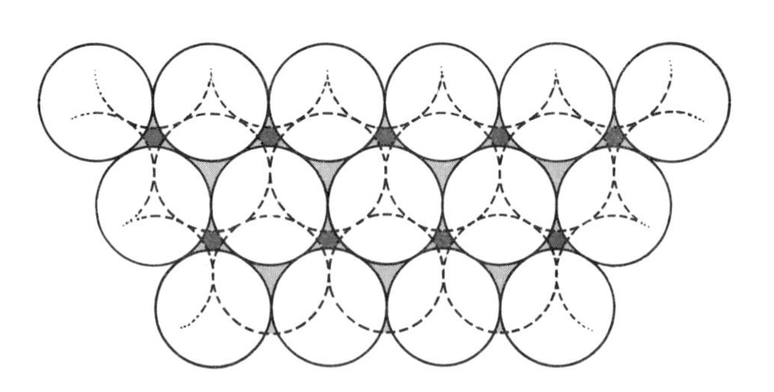

FIGURE 10-15
Cubic, close-packed array showing octahedral and tetrahedral holes that can be occupied by ions.

is large enough to enable you to find the unit cell, the symmetry of which ought to agree with the result of the x-ray powder pattern to be prepared in this experiment. In some cases, the total energy of the lattice is decreased if one-half of the triply charged ions are displaced from their B sites by the doubly charged ions. The resulting structure is called an inverted spinel and is indicated by the symbol $B[AB]O_4$. The x-ray powder pattern does not reveal whether $CoFe_2O_4$ crystallizes as an inverted spinel. Because of the large scattering power of cations, diffraction halos mainly indicate metal positions. However, since total electron densities of Co^{2+} and Fe^{3+} are very similar, the results for both structures would be indistinguishable.

Whether or not an inverted structure can be expected depends on the energy gained or lost upon the transfer of an A-site ion from the tetrahedral crystal field to the octahedral crystal field prevailing at the B site, together with the accompanying change of the displaced B-site ion from octahedral to tetrahedral crystal field symmetry. Thus the sum of changes in crystal field stabilization energy (CFSE) (Figgis 1966; McClure 1965) for the A and the B ion determines whether the displacement is possible. The crystal field stabilization energies are given by

$$\text{CFSE (oct)} = -[4n - 6(N - n)]D_q(\text{oct}) \quad (1)$$

for octahedral symmetry and

$$\text{CFSE (tetr)} = -[6(N - n) - 4n]D_q(\text{tetr}) \quad (2)$$

for tetrahedral crystal field symmetry. The quantity N is the total number of d electrons that the ion possesses, of which n electrons occupy the t_{2g} levels and $N - n$ electrons are distributed over e_g levels. Equations (1) and (2) are valid for high-spin arrangements only. Crystal field theory predicts that $|D_q(\text{tetr})| = 4/9|D_q(\text{oct})|$, which is found to be in sufficient agreement with experimental evidence. Experimental evidence indicating an inverted or normal structure is obtained through the measurement of saturation magnetization.

Procedure

Obtain the x-ray powder pattern of cobalt ferrite, which you prepared in Exercise 2B, or which you obtained from your instructor. To obtain a sharp pattern from a ferrite, it is necessary to use a target element other than the normally employed Cu. A Cr target and an exposure period of 5 hours, using a large-diameter camera (see Section A of this exercise), produce good results. Use the diffraction pattern for indexing (see Section B of this exercise), and calculate the volume of the unit cell.

Next, measure the density of the ferrite powder. Weigh an empty, stoppered pycnometer (preferably a small, wide-mouthed bottle with a fitted glass stopper from which a capillary protrudes); then partially fill it with ferrite and weigh it again. Fill the rest of the pycnometer with water, place it in a thermostat regulated at a temperature slightly higher than room temperature, and pump off any air that remains absorbed by the powder. After bubbles have ceased to appear, disconnect the pump, fill the pycnometer with water to the rim, and let it equilibrate in the thermostat. Wipe off the excess water and weigh. Clean the pycnometer, fill it with water, equilibrate the temperature in the thermostat, and weigh the pycnometer again after removing the excess water. From the density of the water at the thermostat temperature and the weight of the water contained in the pycnometer, calculate the volume of the pycnometer. Obtain the volume of water added to the ferrite from the weight of the pycnometer filled with water and ferrite and the weight with ferrite alone. With the correct water density, calculate the volume of water added to the ferrite and the volume occupied by the ferrite. Since the weight of ferrite is known, the density can be derived.

From the volume of the unit cell and the density of the ferrite, calculate the number of molecules contained in one unit cell. Check whether the unit cell that you selected from Figure 10-16 possesses the correct number of molecules.

From the elementary cell parameters, calculate the distances between the O and O, A site and O, B site and O, and B site and B site, using your model of the unit cell.

Based on the comparison of crystal field stabilization energies given in equations (1) and (2), predict whether $CoFe_2O_4$ is a normal or an inverted spinel.

In which kind of site do you expect to find contaminations of Fe^{2+}?

Estimating Residual Entropy: A Discussion and a Problem

Because one-half of the B sites in an inverted spinel are occupied by A ions, it may be expected to

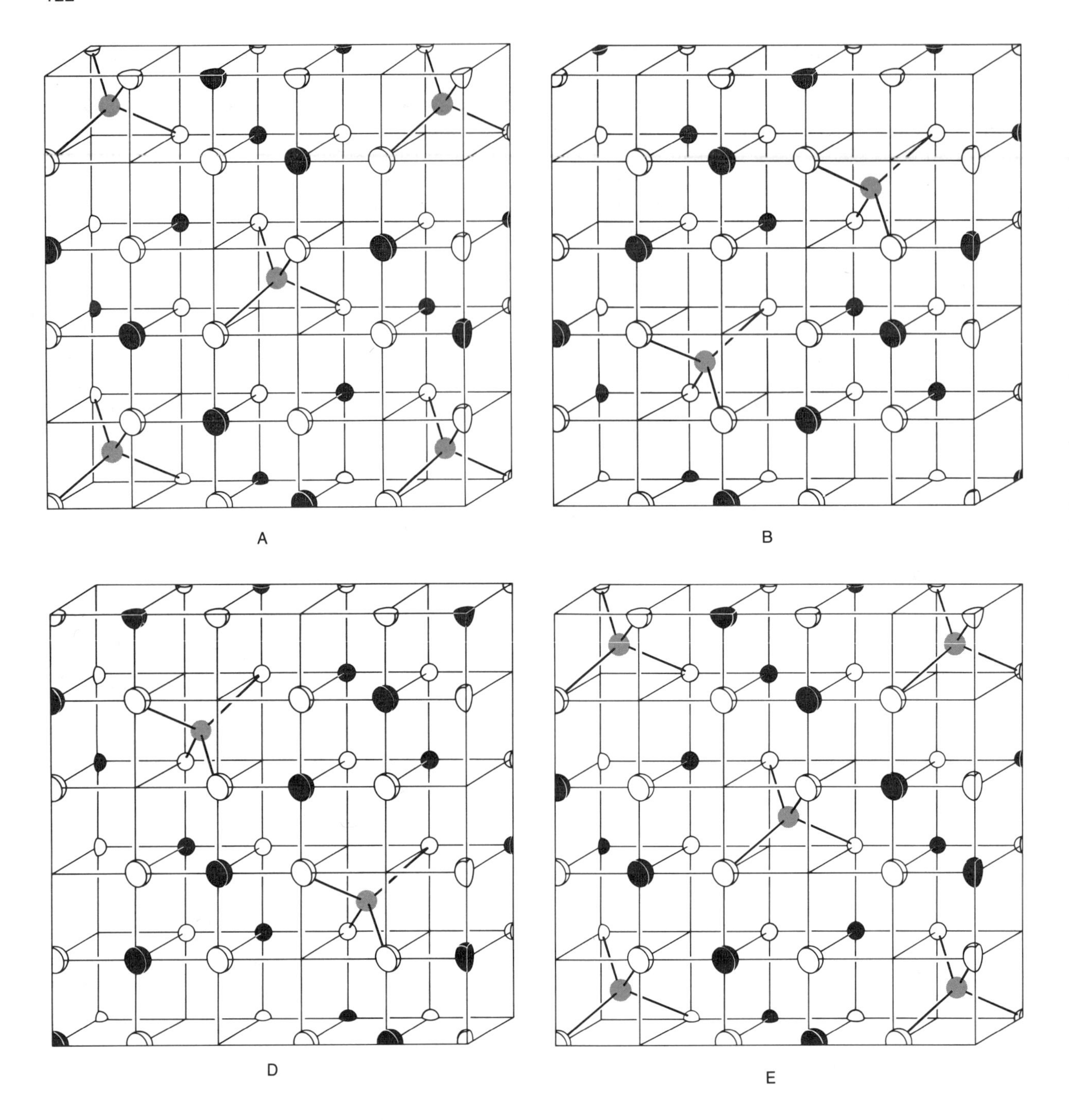

exhibit a residual entropy at absolute zero. At this temperature all contributions arising from the motions of ions around their lattice sites have faded. The residual entropy is thus solely due to the fact that there is a finite number of configurations (arrangements of ions on lattice sites) that comply with rules of near ordering, which are frozen in place as the temperature is lowered. Long-range order, tantamount to zero residual entropy, cannot be established because of the low mobility of the ions.

According to Boltzmann (Moore 1962), the residual entropy S is proportional to the logarithm of the number W of configurations compatible with the rules of near ordering.

$$S = k \ln W \tag{1}$$

where k is the Boltzmann constant. Obviously,

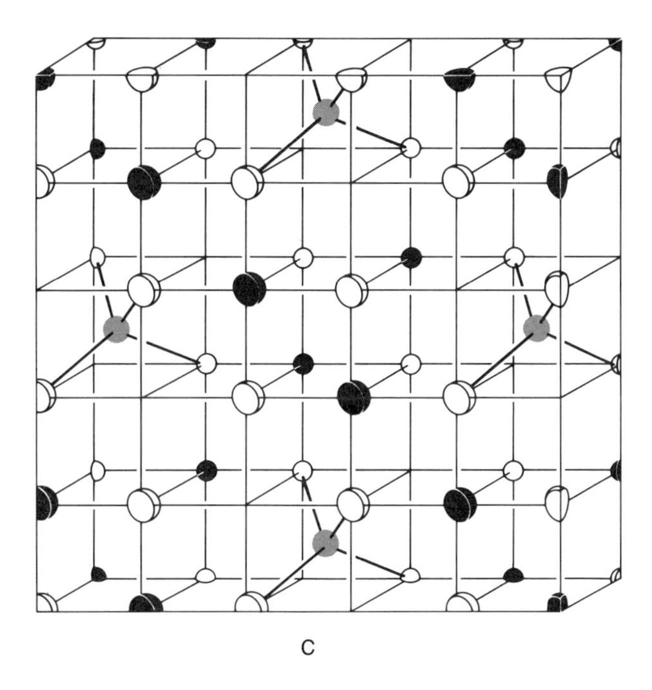

oxygen ion on front side

oxygen ion on back side

cation on B site in front

cation on B site in back

cation on A site

FIGURE 10-16
Segments cut out of a spinel lattice. The back side of any segment is equivalent to the front side of the following segment. Thus ions on network-crossing points are cut in half. The cubic, face-centered oxygen sublattice and the A-site sublattice are easily recognized. Detection of the B-site sublattice is more difficult (see the discussion on residual entropy in Section D of this exercise. Lines connecting A-site cations with the adjacent oxygen have been added to visualize the tetrahedral symmetry.

there is only one possible arrangement of O^{2-} ions on the O^{2-} sublattice and of one-half of the B ions on the A-site sublattice, since each of these sublattices contains identical ions only. However, this is not so for the B-site sublattice, one-half of which is occupied by A ions and the other half by B ions. The B-site lattice is shown in Figure 10-17. The B-site sublattice can be derived from the complete lattice (as shown in Figure 10-16A–E) by beginning at part D (omitting the top and right-hand rows), proceeding to part E (bypassing part A because parts E and A show the same lattice segment), and continuing to parts B and C. The number of configurations that are compatible with an even distribution of A and B ions over the B sites (near ordering) may now be estimated.

The B-site lattice, as depicted in Figure 10-17, is made up of two sets of tetrahedra. The tetrahedra in one set point to one side, and in the other set, to the other side. Each tetrahedron is connected at its four corners to members of the other set. The near-ordering rule may then be rephrased to state that each tetrahedron should host two A ions and two B ions. Calculate the number W of ways in which N_L A ions and N_L B ions can be distributed over N_L tetrahedra so that two corners of each tetrahedron are occupied by A ions and two by B ions, with the restriction that the tetrahedra are connected. The problem can be solved by analogy to the way in which Pauling (1960) estimated the residual entropy of ice. A slightly different version of his line of reasoning follows.

In hexagonal ice, each water molecule is engaged in two hydrogen bonds as a proton donor and in two as an acceptor. The regular ice lattice may be regarded, for the sake of simplicity, as a lattice of O^{2-} ions, each surrounded by four protons; two protons are in positions close to the O^{2-} ion, and two are far from it. This situation is expressed by Bernal and Fowler (1933):

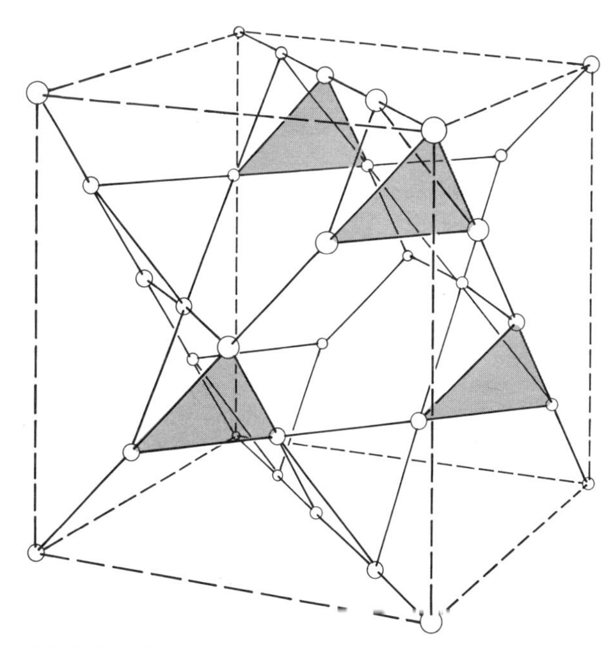

FIGURE 10-17
Spinel B-site sublattice (O^{2-} and A sites have been omitted). [From Anderson (1956).]

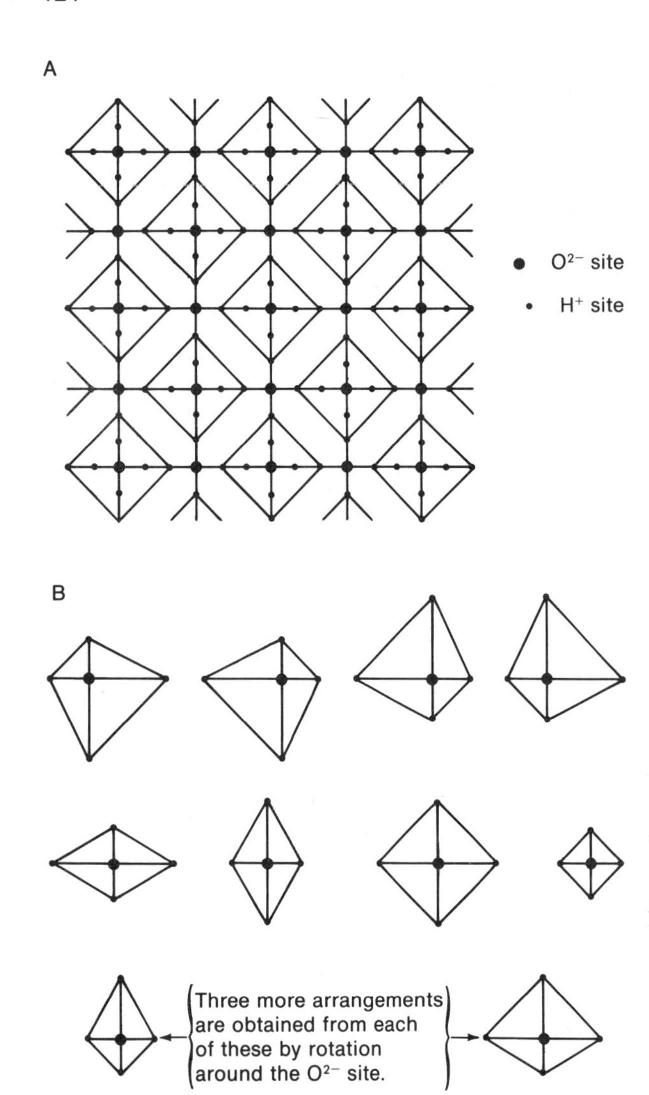

FIGURE 10-18
A. Two-dimensional ice lattice: only one-half of the proton sites are occupied; the placement of four protons around each alternate O^{2-} fills the lattice. B. Of the 16 possible arrangements of four protons around one O^{2-} ion, there are only six compatible with the condition that two protons should occupy a site near the O^{2-} ion and two far from it.

1. Each O^{2-} ion has four protons around it. There is just one proton between two neighboring O^{2-} ions.
2. Each O^{2-} ion has two protons that are close to it and two that are in positions far from it.

A two-dimensional representation of the ice lattice, sufficient for the purposes of this discussion, is shown in Figure 10-18 in which the two allowed proton positions between neighboring O^{2-} ions are indicated.

To calculate the number of possible proton arrangements that are compatible with the Bernal-Fowler rules, select every other O^{2-} ion (in the center of each square in Figure 10-18A) and position two protons close to each ion and two far from it. For any given O^{2-} ion in one of the squares this can be done in six different ways. For one mole of an ice crystal the number of these arrangements is $6^{N_L/2}$. The Bernal-Fowler rules are now fulfilled for every other O^{2-} ion, but not throughout the crystal. All arrangements that violate the Bernal-Fowler rules for O^{2-} ions outside the squares must be eliminated. If all of the above arrangements are truly independent of each other, all of the $2^4 = 16$ possible arrangements of four protons around a given O^{2-} ion that is not in a square will occur with equal frequency. Of these only six conform to the Bernal-Fowler rules, as mentioned earlier. Thus only a fraction of 6/16 of the proton arrangements around any one of the $N_L/2$ alternate O^{2-} ions is allowed.

This brings the number of allowed proton arrangements to:

$$W = 6^{N_L/2} \cdot \left(\frac{6}{16}\right)^{N_L/2} = \left(\frac{3}{2}\right)^{N_L}$$

and amounts to a residual entropy of:

$$S = R \ln \frac{3}{2}$$

per mole of ice. This entropy will persist even at absolute zero temperature when all other contributions to entropy will have vanished (third law of thermodynamics). The reason is that, despite a perfectly ordered oxygen sublattice and the observance of the Bernal-Fowler rules, one mole of an ice crystal can be made up by $(3/2)^{N_L}$ different proton arrangements.

Problem

Calculate the residual entropy of an inverted spinel in a manner similar to that used for ice in the preceding discussion. Figure 10-19 shows a suitable two-dimensional lattice, which is comparable to Figure 10-18. The insides of the tetrahedra (O^{2-} sites in Figure 10-18) are empty in Figure 10-19. There is only one possible place between these empty sites. Distribute the total number of A and B ions on B sites over one-half of the tetrahedra, in accord with the rules of near ordering. Draw the possible arrangements that comply with the rules. Draw all of the configurations obtainable on the alternate tetrahedra. Calculate the residual entropy of one mole of an inverted ferrite.

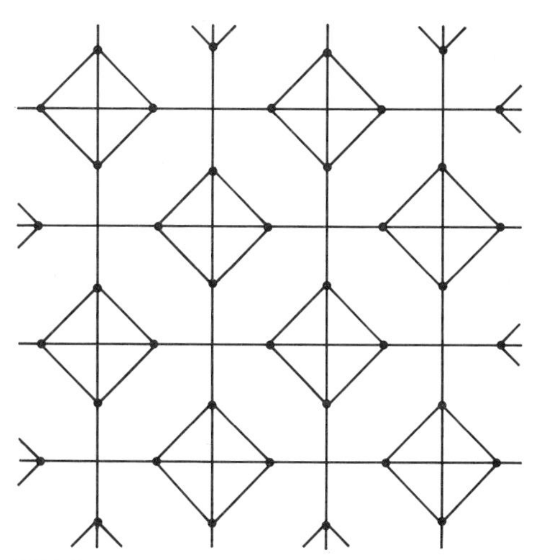

FIGURE 10-19
Two-dimensional B-site lattice. Occupation of four sites around each alternate network-crossing point fills the lattice.

REFERENCES

Anderson, P. W. 1956. *Phys. Rev.* **102**, 1099.

Azaroff, L. V., and Buerger, M. J. 1958. *The Powder Method in X-Ray Crystallography.* New York: McGraw-Hill.

Bernal, J. D., and Fowler, R. H. 1933. *J. Chem. Phys.* **1**, 8.

D'Eye, R. W. M., and Wait, E. 1960. *X-Ray Powder Photography in Inorganic Chemistry.* New York: Academic Press.

Figgis, B. N. 1966. *Introduction to Ligands Fields.* New York: Wiley.

McClure, D. S. 1965. The Effects of Inner Orbitals on Thermodynamic Properties. In *Some Aspects of Crystal Field Theory*, by T. M. Dunn, D. S. McClure, and R. G. Pearson. New York: Harper & Row.

Moore, W. J. 1962. *Physical Chemistry.* Englewood Cliffs, New Jersey: Prentice-Hall, p. 625.

Pauling, L. 1960. *The Nature of the Chemical Bond.* 3d ed. Ithaca, New York: Cornell University Press, p. 467.

PART 

ANALYTICAL METHODS

EXERCISE **11**

Chromatography

A. GAS-LIQUID CHROMATOGRAPHY

Gas-liquid chromatography has proven to be an extremely useful tool for analytical and preparative chemists. Its great advantage lies in the fact that, provided the right column has been chosen, fast separation of otherwise hard-to-separate mixtures can be achieved, even if only small samples of the product are available. Within certain limits the gas chromatograph can be used at the same time to analyze the sample both qualitatively and quantitatively.

Apparatus

Figure 11-1 is a schematic drawing of a gas chromatograph. Injection is made by means of a microliter syringe through a rubber membrane (*D* in Figure 11-1) at the entrance of the column. After the sample (also called the solute) has passed the column (*E*), it is detected in a thermal conductivity cell (*F*), and the result is recorded on a strip-chart recorder (*J*). The mode of operation of the detection system is discussed in another part of this exercise.

The flow rate of the carrier gas is measured at the point where the gas leaves the column by means of soap bubbles, which the gas forces along a graduated tube. The flowmeter can alternatively be attached to the outlet of the reference gas stream, if an adjustment of reference flow to carrier-gas flow becomes necessary.

The injector, column, and detector can be thermostated separately in some gas chromatographs. A constant temperature in the column is most important. Ask your instructor about the operation of the control units of the gas chromatograph.

A thermal conductivity cell consists in principle of heated wires arranged in a Wheatstone bridge. The temperature of these wires is determined by the rate of heating and the rate of loss by thermal conduction to the surrounding gas. In turn, the resistance of the wires depends on the temperature and thus, under fixed conditions, on the thermal conductivity of the surrounding gas. Figure 11-2 shows a representative bridge and cell arrangement for a gas chromatograph. In this case all of

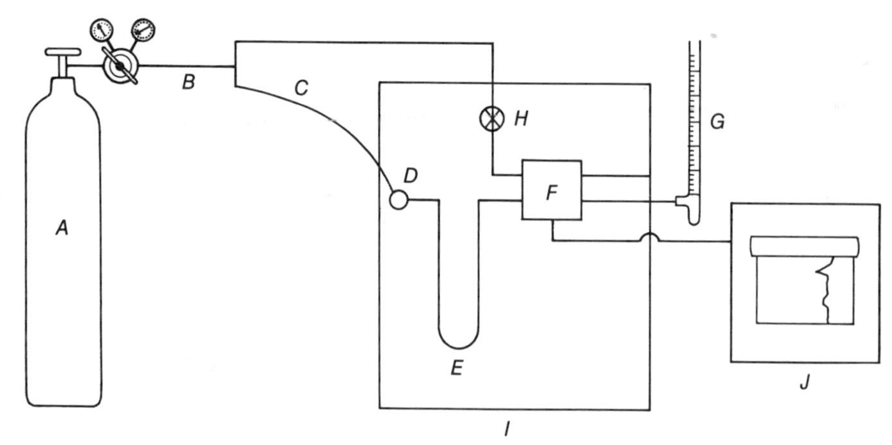

FIGURE 11-1
Elements of a simple gas chromatograph: *A*, source of carrier gas; *B*, stream division to obtain separate carrier-gas and reference streams; *C*, capillary section providing a smooth flow of carrier gas; *D*, injection system closed by rubber membrane; *E*, chromatographic column; *F*, detector; *G*, flow meter; *H*, throttle to adjust the reference flow rate to the carrier-gas flow rate; *I*, thermostat (or oven); *J*, recorder.

the resistances in the sample and reference flow are equally great. If the gas flow in both channels is exactly the same, ideally the bridge should be equilibrated. Small deviations from equilibrium can be corrected by means of the coarse and fine zero adjustment (see Figure 11-2). After the initial balancing of the bridge, each resistor possesses a constant additional part because of the coarse and fine resistor setting. These parts should be kept very small, and can thus be disregarded. The voltage difference, ΔV, between points a and b, which is delivered to the recorder by the way of the attenuator, is then given by:

$$\Delta V = \Delta V_0 \frac{S^2 - R^2}{(S + R)^2} = \Delta V_0 \frac{S - R}{S + R} \qquad (1)$$

where ΔV_0 is the voltage initially applied. If the carrier gas fills both channels of the cell, $S = R$ holds. Thus $S - R$ is directly proportional to the difference in thermal conductivity in the two channels.

An additional advantage of the arrangement shown in Figure 11-2 is that it allows equal current for both sides of the bridge. This guarantees an equal rate of heating for all resistors.

The ratio $\Delta V : \Delta V_0$ is of the order of 10^{-4} under operating conditions. Equation (1) is rewritten as follows:

$$\Delta V = \Delta V_0 \left\{ \frac{S - R}{2R + S - R} \right\} = \frac{\Delta V_0}{2R} (S - R) \qquad (2)$$

which holds since $S - R$ is negligible in comparison with $2R$.

Since S and R are proportional to $1/T$ and, on the other hand, $T \propto 1/\lambda$, where λ is the thermal conductivity, both S and R must be proportional to the thermal conductivity of their surrounding gas. The proportionality constant depends on cell geometry and flow rate. ΔV_0 and R remain constant throughout an experiment. Thus R refers to the carrier-gas thermal conductivity and S to that of the gas mixture in the sample channel of the cell. Equation (2) becomes:

$$\Delta V = a(\lambda_M - \lambda_C) \qquad (3)$$

where a is a proportionality constant depending on cell geometry, flow rate, applied voltage, and magnitude of resistances; λ_C is the carrier gas thermal conductivity; and λ_M can be expressed as the sum of the thermal conductivity λ_i of the separated compound i and of λ_C, both weighted by their momentary mole fractions in the cell.

The mole fraction of substance i at time t in the sample channel of the detector is:

$$x_{i,t} = \frac{n_{i,t}}{n_C + n_{i,t}}$$

Since $n_{i,t}$ is always small compared with n_C,

$$x_{i,t} = \frac{n_{i,t}}{n_C}$$

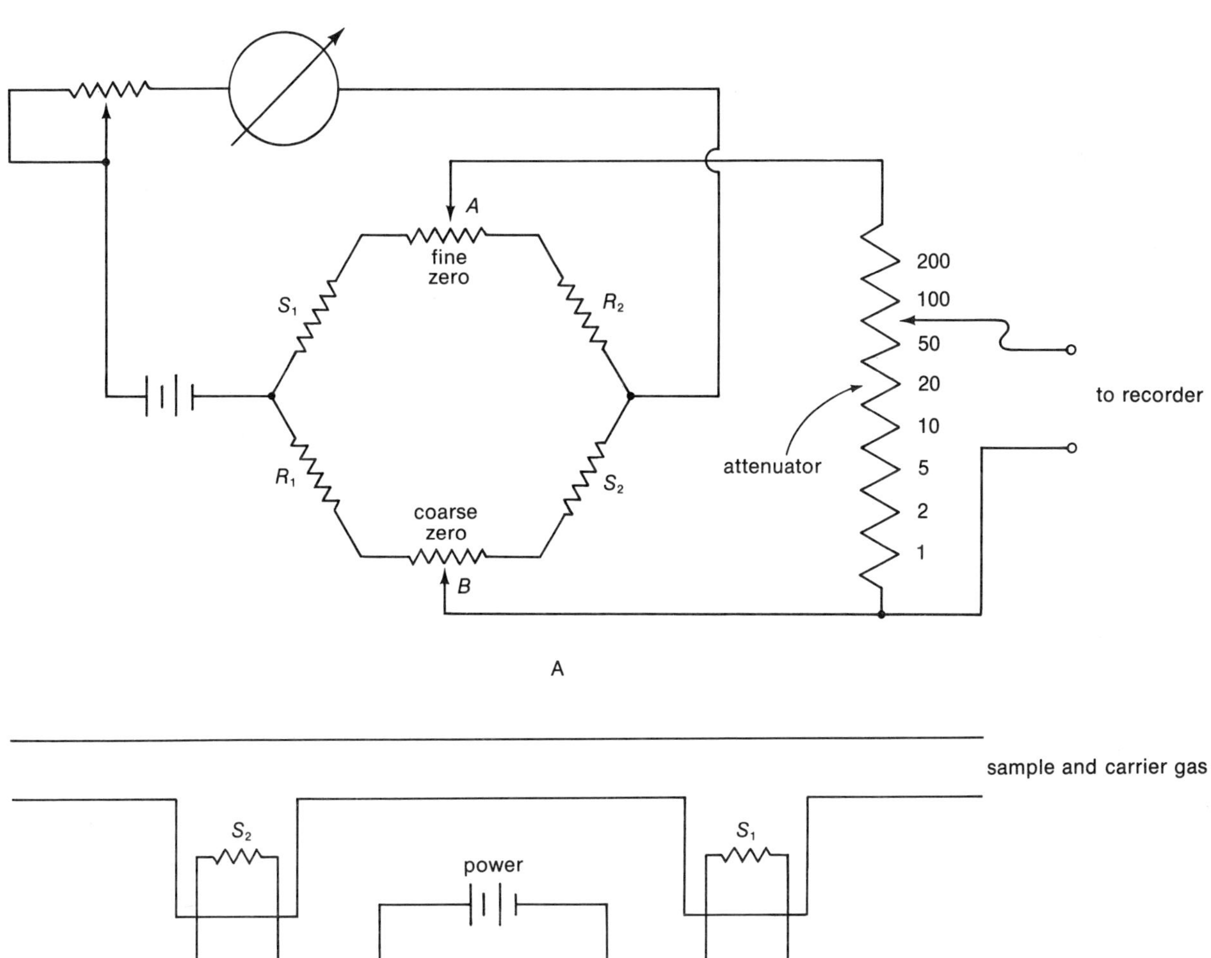

A

sample and carrier gas
S_2
S_1
power
coarse zero
fine zero
R_1
R_2
reference carrier gas

B

FIGURE 11-2
Thermal conductivity detector as used in a gas chromatograph: (A) the Wheatstone bridge and (B) the cell arrangement.

and the thermal conductivity of the mixture is:

$$\lambda_M = x_{i,t}\lambda_i + (1 - x_{i,t})\lambda_C$$

The voltage difference ΔV_t at time t is therefore a function of the mole fraction $x_{i,t}$ in the sample channel and of the difference of thermal conductivities of the carrier gas and the solute.

$$\Delta V_t = a x_{i,t}(\lambda_i - \lambda_C) \qquad (4)$$

The recorder plots ΔV as a function of time, and the total area A_i under a peak originating from a particular substance i in the sample channel is:

$$A_i = \int_0^t \Delta V_t dt = a \frac{n_i}{n_C} \Delta\lambda_i \qquad (5)$$

where $\Delta\lambda_i = \lambda_i - \lambda_C$, and n_i is the total number of moles of substance i in the original sample. ($a \cdot \Delta\lambda_i$ is sometimes called the detector response for substance i.) In addition to the variables already mentioned, a depends on the gas temperature. It is therefore reasonable to determine the constant by calibration with known substances under experimental conditions.

If all substances in the original sample have very similar thermal conductivities or if their thermal conductivities are all very small compared with that of the carrier gas, a knowledge of the peak areas alone suffices for the calculation of mole fractions in the original sample. The mole fraction of substance i in the original sample is

$$x_i = \frac{n_i}{\sum_{j=1}^{m} n_j},$$

if m substances are present. Therefore

$$x_i = \frac{A_i}{\Delta\lambda_i} \sum_{j=1}^{m} \frac{\Delta\lambda_j}{A_J} \qquad (6)$$

can be easily deduced from equation (5).

If the difference between the thermal conductivities of the carrier gas and any of the substances in the sample is of an order of magnitude that is higher than the differences between thermal conductivities of the substances, equation (6) reduces to equation (7).

$$x_i = \frac{A_i}{\sum_{j=1}^{m} A_j} \qquad (7)$$

Equation (5) showed that the detector response (the sensitivity of detection) depends on the choice of the carrier gas. It should possess a thermal conductivity as different as possible from those of the substances to be detected. Hydrogen and helium are the most desirable carrier gases in this respect, and helium is the safer of the two. In cases where one of these gases is used, equation (7) is almost always applicable. However, equations (6) and (7) can hold only if all of the substances emerge from the column in a reasonable length of time.

Separation Process Based on Repeated Equilibration

In this section a measure of column performance will be derived under the assumption that the solute repeatedly equilibrates between the liquid phase and the gas phase in the column. The column is tightly filled by the so-called stationary phase, which consists of a solid support evenly coated by a solvent suited to the sample. The solid support is not supposed to take part in the partitioning process. The carrier gas and the solute transported by it constitute the mobile phase. A gaseous or liquid sample introduced into the column is partly dissolved in the solvent to form the liquid phase and partly carried on by the carrier gas as the mobile or gas phase. Part of the sample in the liquid phase is evaporated and taken along by fresh gas to another portion of the solvent where it dissolves again. A sample therefore migrates through the column by repeated dissolution and evaporation. Depending on the fraction of the solute remaining in the gas phase, some substances travel faster than others and separation is finally achieved this way. However, the small volume originally occupied by the sample increases in this process, and separation can be partly negated by spreading. Even if a sample would never dissolve in the solvent, it would spread by diffusion. This can be easily demonstrated by injecting air into a column. In addition, irregularities in column packing, the formation of channels in which the substance can travel faster, or adsorption on the exposed parts of the solid support increase the width of the sample zone. Figure 11-3 illustrates the spreading of a sample during its transport through the column. The degree to which a substance spreads as it migrates through the column is a measure of the column performance. In the derivation that follows,

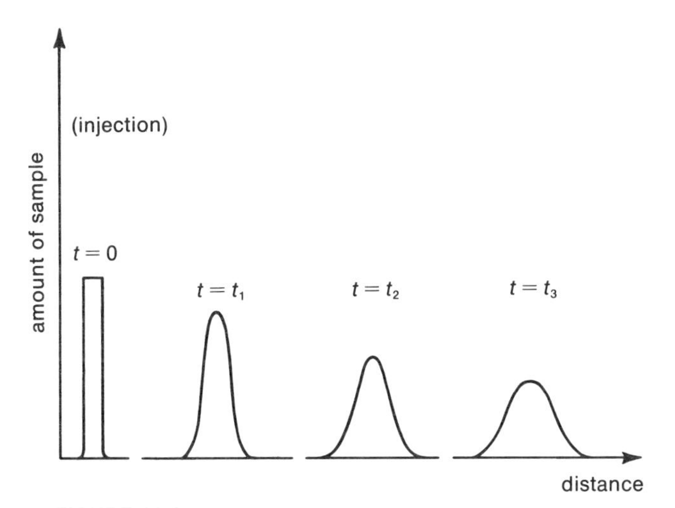

FIGURE 11-3
Spreading of solute during migration through column.

spreading is based solely on repeated equilibration. (Other sources of spreading will be taken into account later in the exercise.)

Theoretical Plates

The migration of the sample through the column and the spreading of a peak due to evaporation and dissolution of the solute can be described quantitatively based on the following assumptions:

1. The column can be divided into small, discrete volumes of equal size called theoretical plates. A theoretical plate is defined as the distance along the column that has the same effect as one discrete plate (equilibration) in, for example, a pseudocountercurrent operation. At the "outlet" of each plate, the equilibrium distribution between the gas phase and liquid phase is realized. Note carefully that this does not imply that equilibrium conditions are attained throughout the theoretical plate. (Why?) However, it is a useful model because the theory may be treated as if equilibration were realized throughout.
2. The actual continuous flow of carrier gas can be represented by pulsed flow. Each pulse contains the gas-phase volume ΔV_g of one theoretical plate.
3. The solute contained in each pulse of carrier gas is allowed to equilibrate with the liquid phase before the next pulse carries the gas phase into the next theoretical plate.
4. The partition coefficient k of the solute between the gas phase and the liquid phase is constant throughout the column (i.e., independent of concentration and pressure).
5. The whole sample is charged at time t_0 onto one theoretical plate.

The fraction of all solute molecules in one plate, which are in the gas phase, can be written as:

$$z = \frac{c_g \Delta V_g}{c_g \Delta V_g + c_l \Delta V_l} = \frac{\Delta V_g}{\Delta V_g + k_p \Delta V_l} = \frac{\Delta V_g}{\Delta V_p} \qquad (8)$$

where ΔV_l is the liquid volume in one plate, $c_g = (n_{i,t})_g / \Delta V_g$ is the concentration of solute molecules in the gas phase, and $c_l = (n_{i,t})_l / \Delta V_l$ is the concentration of solute molecules in the liquid phase of one theoretical plate. The partition coefficient is $k_p = c_l / c_g$, and $\Delta V_p = \Delta V_g + k_p \Delta V_l$ is the effective plate volume.

According to equation (8), the fraction of solute in the liquid phase of the same plate must be

$$y = \frac{c_l \Delta V_l}{c_g \Delta V_g + c_l \Delta V_l} = 1 - z \qquad (9)$$

According to point 5 of the assumptions, the complete sample is injected into one plate to which the plate number 0 is assigned for the sake of convenience (see Figure 11-4). If a new volume of carrier gas is added, fraction z is transported into the next plate and equilibrates there. The fraction y remains in plate number 0 and equilibrates with the new volume of carrier gas. The addition of a second volume ΔV_g leads to a fraction y^2 in the first plate; zy is transported into the second plate, but this plate lost z^2 to the third plate; and $2zy$ remains in the second plate. Figure 11-4 illustrates this process more clearly. It also demonstrates that the distribution of the solute after the passage of n volumes of carrier gas can be represented by the binomial expansion of $(y + z)^n$. The fraction of the solute in plate r is then given by the rth term of the expansion:

$$Q_{r,n} = \frac{n!\, y^{n-r} z^r}{r!\, (n-r)!} \qquad (10)$$

If n, r, and $n - r$ are large (a condition that is usually fulfilled in gas chromatographic processes by the time the solute leaves the column), $n!$, $r!$, and $(n - r)!$ can be approximated by Stirling's formula:

$$x! = (2x)^{1/2} x^x e^{-x} \qquad (11)$$

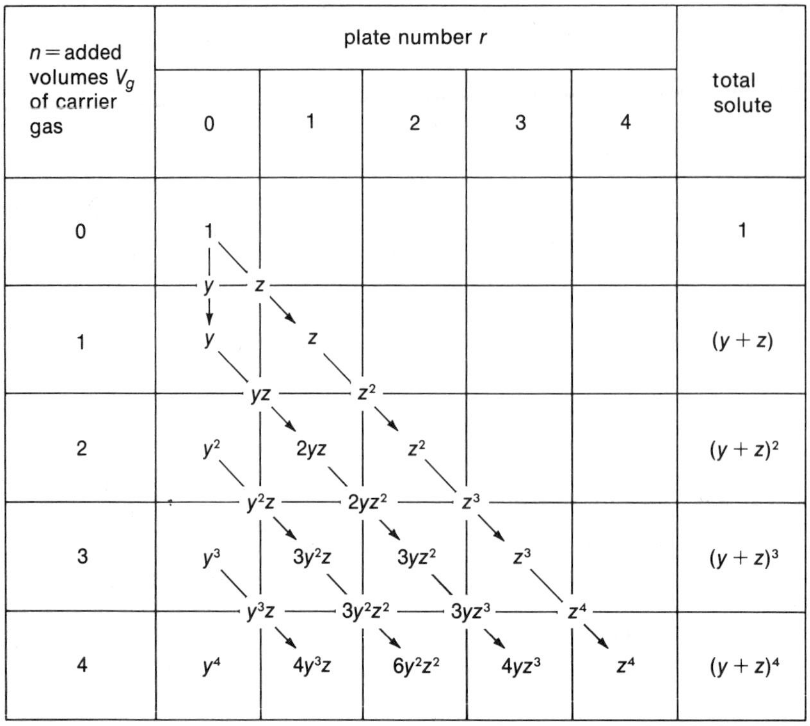

n = added volumes V_g of carrier gas	plate number r					total solute
	0	1	2	3	4	
0	1					1
	y	z				
1	y	z				$(y+z)$
	yz	z^2				
2	y^2	$2yz$	z^2			$(y+z)^2$
	y^2z	$2yz^2$	z^3			
3	y^3	$3y^2z$	$3yz^2$	z^3		$(y+z)^3$
	y^3z	$3y^2z^2$	$3yz^3$	z^4		
4	y^4	$4y^3z$	$6y^2z^2$	$4yz^3$	z^4	$(y+z)^4$

FIGURE 11-4
Solute distribution after the passage of zero to four volumes of carrier gas.

A new auxiliary variable u, which relates r to n, is introduced:

$$r = nz + u(nzy)^{1/2} \tag{12}$$

This measure effects a shift in the origin of the function and a constriction of the abscissa so that the mean value of u is zero and its variance is 1. The physical meaning of the term nz becomes obvious if equation (8) is rewritten as:

$$nz = \frac{n \cdot \Delta V_g}{V_p} = \frac{V}{V_p} \tag{13}$$

where $n \cdot \Delta V_g = V$ is the volume of carrier gas that has left the plate under consideration and nz is the number of effective plate volumes of gas that have left it. A series of considerations, including those already mentioned, leads to the reformulation of equation (10):

$$Q_n(r) = \frac{1}{(2\pi nzy)^{1/2}} \exp\left[-\frac{(r-nz)^2}{2nzy}\right] \tag{14}$$

This approximation is valid if

$$\frac{(r-nz)^3}{n^2} \xrightarrow[\text{limit } n \to \infty]{} 0$$

which later will be seen to hold in this case. The complete derivation of equation (14) is too long to be repeated here. Refer to standard texts of probability theory or von Mises (1964).

Equation (14) describes the distribution of the sample in a column after n gas volumes have been introduced. The largest concentration of the solute is then found in plate $r = nz$. From an experimental point of view, the distribution of the solute as a function of n at a fixed plate number N (the end of the column) is of interest, since the volume of carrier gas that has left the column at any given time is proportional to n by way of $V = n \cdot \Delta V_g$ (see equation 13). The quantity V is measurable and relates to the time that has elapsed by the constant volume flow rate $F = V/t$. Therefore a distribution over n should describe the observed elution curve. However, a simple exchange of n and r in equation (10) would not be sufficient to reformulate the distribution function, since n must necessarily be larger than r at the time that the solute appears at the end of the column. But the replacement of r by zn and n by r/z will facilitate the reformulation and leave the maximum of the distribution function at the correct place.

The new Gaussian function is:

$$Qr(n) = \frac{1}{[2\pi r(1-z)]^{1/2}} \exp\left[-\frac{(zn-r)^2}{2r(1-z)}\right] \quad (15)$$

Again, the maximum of a peak appears at the end of the column in the rth plate at $r = z \cdot n_{max}$. Since $z \ll 1$, the following may be written:

$$Qr(n) = Q_{max} \exp\left[-\frac{(n-n_{max})^2 z}{2n_{max}}\right] \quad (16)$$

The height of the observed elution peak is given by Q_{max}, the width is governed by the variance $\sigma^2 = n_{max}/z$. The distance Δn between the two points at which the tangents on the points of inflection intersect the n-coordinate is:

$$\Delta n = 4\sigma = 4\left(\frac{n_{max}}{z}\right)^{1/2} \quad (17)$$

For an elution curve measured in eluted volume or elapsed time this distance is, respectively:

$$\Delta V = \Delta V_g \cdot \Delta n = 4\Delta V_g\left(\frac{n_{max}}{z}\right)^{1/2} \quad (17a)$$

and

$$\Delta t = \frac{\Delta V}{F} \quad (17b)$$

(see Figure 11-5).

The number of theoretical plates $N = r$ may now be calculated from two easily measurable quantities: the distance between points of inflection [see equation (17a)], and the retention volume V_r. The latter is the volume of eluant that has left the column when the maximum of the elution peak appears.

$$V_r = n_{max} \cdot \Delta V_g \quad (18)$$

A combination of equations (17a) and (18) leads to the desired expression for N:

$$V_r^2 = n_{max}^2 \Delta V_g^2$$

$$\Delta V^2 = 16\Delta V_g^2\left(\frac{n_{max}}{z}\right)$$

and the following is obtained by division:

$$N = n_{max} \cdot z = 16\left(\frac{V_r}{\Delta V}\right)^2 \quad (19)$$

If time rather than volume was recorded, with $t_r = V_r/F$:[1]

$$N = 16\left(\frac{t_r}{\Delta t}\right)^2 \quad (20)$$

For a column of a given length L the height equivalent to a theoretical plate, HETP, is given by

$$\text{HETP} = \frac{L}{N} \quad (21)$$

For a given substance and column the performance is best where HETP is at a minimum. This topic is discussed later in this experiment.

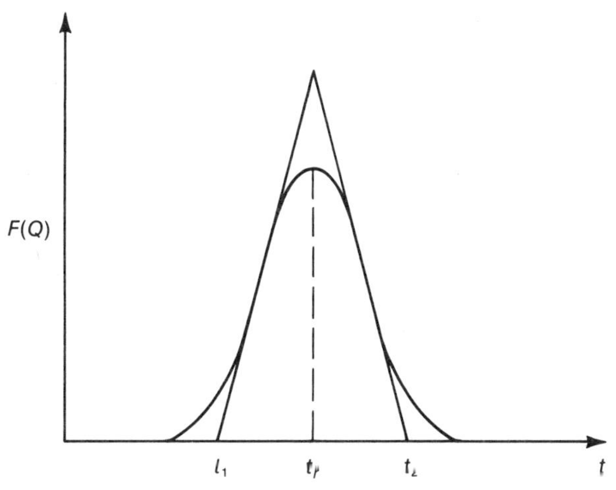

FIGURE 11-5
Distribution of the sample over time (equivalent to volume): t_1 and t_2 are the points at which the tangent at the inflection points cuts the time axis (thus $\Delta t = t_2 - t_1$); t_r is the retention time.

Operational Approach to the Theory of Gas-Liquid Chromatography

The treatment based on the model of the theoretical plate introduces some problems. The most important is the underlying emphasis on equilibrium considerations for a process in which nonequilibrium prevails. Although the theoretical-plate model was designed to allow the treatment of a theoretical problem as if equilibrium existed without implying its actual existence, the definition and concept of the theoretical plate has been distorted in some writings to the point where the original idea is often lost. As a result, chemists doing research on gas chromatography have tended to reject the theoretical-plate model in favor of a strictly opera-

[1]For different ways to calculate N see Purnell (1962). This is especially advisable if you are dealing with asymmetric peaks.

tional approach that avoids equilibrium concepts.

In the operational approach, column efficiency is measured in terms of a parameter h, which is a measure of the rate of spreading of the solute band with the distance traversed in the column. By definition:

$$h = \frac{d(\sigma^2)}{dl} \tag{22}$$

where σ^2 is the variance of the elution peak (σ = standard deviation), l is the distance, and h is referred to as the HETP, but only because of the historical significance of this terminology. Without the historical background h might be defined as the "index of column efficiency." The use of the variance of the peak in defining h is most convenient because of the additivity of variances. The length of the column L is related to h by the definition

$$L = hN \tag{23}$$

where N is the "number of theoretical plates." If h is independent of l, equation (22) leads to

$$h \cdot l = \sigma^2 \tag{24}$$

Since experimental observations are usually made as the material leaves the column where

$$l = L \tag{25}$$

the following may be written:

$$hL = \sigma^2 \tag{24a}$$

and combining equations (23) and (24a) yields:

$$N = \frac{L^2}{\sigma^2} \tag{26}$$

These quantities may be related to the experimental observables t_R and Δt (see Figure 11-5) as follows:

$$L = at_R \tag{27}$$

$$4\sigma = a\,\Delta t \tag{28}$$

where a is a proportionality factor. Thus

$$N = 16\left(\frac{t_R}{\Delta t}\right)^2 \tag{29}$$

Equation (29) is identical to the result obtained with the theoretical plate model equation (20). However, it is obtained by operational definitions without the need for a model of the physical processes taking place in the column.

Factors Controlling Column Efficiency

When a solute is injected into a column, the concentration versus the distance (axial) profile in the gas phase is initially a sharp impulse function (see the first peak in Figure 11-3). Regardless of whether the band is moving or not, it will spread because of diffusion (the flow of matter under the driving force of a concentration gradient). The band will spread symmetrically with time at a rate governed by the laws of diffusion (see Figure 11-3). The process is referred to as "axial diffusion" because it takes place along the axial direction in the column. The rate of spreading is proportional to the term.

$$2\gamma D_g$$

where D_g is the diffusion coefficient of the solute molecule in the gas phase and γ is a correction factor ($\gamma \sim 1$) that takes the intricacy of the diffusion path into account. The total band spreading due to axial diffusion when the solute leaves the column is proportional to the residence time in the column and is inversely proportional to the linear gas flow velocity, u. The axial diffusion contribution to the HETP is

$$h_B = \frac{2\gamma D_g}{u} \tag{30}$$

Note that the distance parameter is not included in equation (30) because of the way in which h is defined in (equation 24). A similar axial diffusion process occurs in the liquid phase. However, its contribution to h is negligible because diffusion is much slower in the liquid phase.

Band spreading is also influenced by the rate at which solute molecules are transported from the liquid phase to the gas phase and vice versa. The more time required for the solute molecules to leave the stationary liquid phase bulk and pass to the gas phase when a nonequilibrium distribution exists (excess in liquid), the more they will be displaced by the solute molecules travelling in the mobile gas phase. Similarly, the longer it takes for gas phase solute molecules to respond to a nonequilibrium distribution (excess in gas phase) and enter the liquid phase, the farther along the column they will move before entering the stationary liquid phase, placing them a greater distance ahead of the solute molecules that had been residing in the liquid phase. Clearly, the slower these transport processes are in relation to the rate of material transport in the mobile gas stream, the greater the band spreading. Once again, diffusion determines the rate of these transport processes. Although the rate of diffusion in the liquid phase is much slower than in the gas phase, under normal conditions of gas-liquid chromatography the time involved in transport from gas to liquid is comparable to that associated with the reverse process. This is because the thickness of the liquid-phase film is much less than the gas-phase dimensions. Therefore, both steps are important. The contribution to the plate height arising from a finite rate of diffusional transport from liquid to gas has the form

$$h_l = c\,\frac{k_p}{(l+k_p)^2}\cdot\frac{d_l^2}{D_l}\cdot u \qquad (31)$$

where c is a constant ($c \approx 8/\pi^2$), k_p is the partition coefficient (moles of the solute in gas phase over moles of the solute in liquid phase), D_l is the solute diffusion coefficient in the liquid phase, and d_l is the thickness (average) of the liquid. As might be expected from this discussion, the contribution to plate height increases as the gas flow rate and liquid film thickness increases, and decreases as the diffusion coefficient increases. A dependence on the partition coefficient is not surprising, since the effect in question should approach zero as $k_p \rightarrow 0$ (solute insoluble in liquid phase) and as $k_p \rightarrow \infty$ (no tendency of the solute to transport to gas phase), while a finite effect is expected for intermediate k_p values. The contribution to band spreading due to the finite transport rate from the gas phase to the liquid phase is

$$h_g = \frac{wd_p^2}{D_g}\cdot u \qquad (32)$$

where w is a very insensitive function of k (nearly a constant ~1), and d_p is the mean square packing diameter (Dg and u have been defined previously). The inverse dependence of this term on Dg is as expected. The dependence on d_p^2 manifests the reduction of the liquid phase surface area and greater inaccessibility of the liquid phase attending an increase in the solid support particle size. The liquid phase is found predominantly in the cracks and crevasses between solid support particles and in the pores of the support. Larger particle diameter means fewer such sites.

The final band spreading effect to be considered arises because the existence of solid support particles, column bends, and so forth, results in a distribution of path lengths that the solute molecules can follow in traversing the column (i.e., some molecules can take shorter paths than others). This term has the magnitude

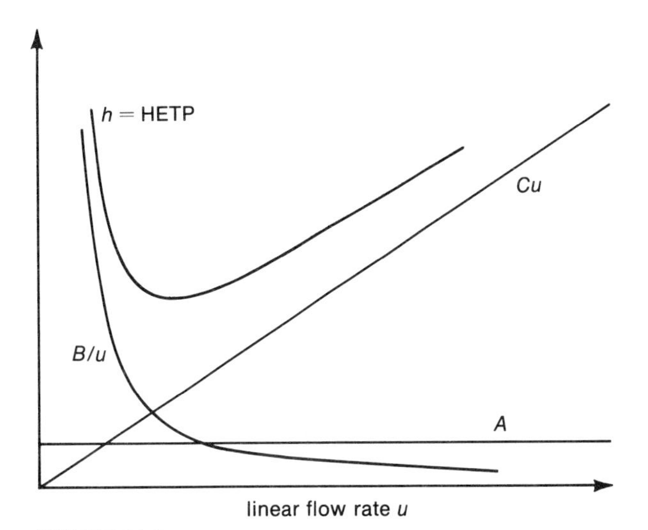

FIGURE 11-6
Schematic representation of van Deemter's equation. The different contributing terms are shown separately.

$$h_A = 2\lambda d_p \qquad (33)$$

where λ is a measure of packing irregularity (d_p has already been defined). Note that this term is independent of velocity. The h_A contribution is often referred to as the "eddy diffusion" term. By adding the various contributions to HETP (adding variances), the following is obtained:

$$h = h_A + h_B + h_g + h_l \qquad (34)$$

or

$$h = 2\lambda d_p + \frac{2\gamma D_g}{u} + \frac{c\cdot k_p\cdot d_l^2}{(1+k_p)^2 D_l}\cdot u + \frac{wd_p^2}{D_g}\cdot u \qquad (35)$$

This equation has the general form:

$$h = A + \frac{B}{u} + Cu \qquad (36)$$

a form originally proposed by van Deemter. Figure 11-6 plots the function h and contributing terms against the flow rate u. It is apparent that h has a minimum at a certain flow rate. This flow rate would then yield the optimum performance for a given column at a given temperature.

It should be recognized that the HETP will be a function of temperature because the parameters

D_g, k_p, D_l, and d_l are all temperature dependent. The effect of temperature on k_p is most important.

Experiments have revealed some shortcomings of these concepts, and some semiempirical modifications of equation (35) have been introduced. A problem arises because the linear velocity is not constant throughout the column because of the compressibility of the gas phase—that is, the pressure gradient varies with distance. Therefore, h is actually a function of distance. This problem can be circumvented somewhat by recognizing that

$$p_0 u_0 = pu \quad (37)$$

where p_0 and u_0 are the column outlet gas pressure and flow rate, respectively, while p and u are the corresponding quantities anywhere within the column (or at the inlet). Recognizing also that

$$D_g \propto \frac{1}{p} \quad (38)$$

the following may be written:

$$D_g = \frac{D_g^0}{p} \quad (39)$$

where D_g^0 is the solute gas phase diffusion coefficient when $p = 1$.

By combining equations (37) and (39), the axial diffusion term may be expressed as:

$$h_B = \frac{2D_g^0}{p_0 u_0} \quad (40)$$

Similarly, the following may be written:

$$h_g - \frac{wd_p^2 \cdot p_0 \cdot u_0}{D_g^0} \quad (41)$$

The h_l term is not modified so simply, but after appropriate theoretical arguments, the following is obtained:

$$h_l = \frac{c \cdot k_p \cdot d_l^2 \cdot j \cdot u_0}{(1 + k_p)^2 \cdot D_l} \quad (42)$$

where

$$j = \frac{3(\bar{p}^2 - 1)}{2(\bar{p}^3 - 1)} \quad (43)$$

and

$$\bar{p} = \frac{p_{\text{inlet}}}{p_{\text{outlet}}}$$

Finally, the "eddy diffusion term" has been the subject of much controversy and the existence of a flow-rate independent term in gas-liquid chromatography has been difficult to establish experimentally. It is argued that band spreading due to this effect is counteracted by cross-sectional diffusion (diffusion between flow streamlines). This not only reduces the eddy diffusion term, but makes it velocity dependent (faster flow velocity, less time for cross-sectional diffusion). To account for cross-sectional diffusion, the h_A term is written as

$$h_A = \frac{(2\lambda d_p)(w' d_p^2 p_0 u_0 / D_g^0)}{2\lambda d_p + (w' d_p^2 / D_g^0) p_0 u_0} \quad (44)$$

where w' is similar, but not identical, to w. This formulation of h_A takes into account the coupling of the eddy diffusion effect and cross-sectional diffusion, the latter being governed by the same physical processes as transport from the gas phase to the liquid phase, which accounts for the similarity in mathematical terms. Note that at high flow velocities (or small D_g^0)

$$h_A \cong 2\lambda d_p \quad (45)$$

while at low velocities (or large D_g^0)

$$h_A \cong \frac{wd^2 p_0 u_0}{D_g^0} \quad (46)$$

Equation (46) generally applies in gas-liquid chromatography and the h_A term is functionally indistinguishable from the h_g term [equation (40)] (i.e., experimentally, the existence of an h_A term is not apparent). The foregoing modifications lead to the result

$$h = \frac{(2\lambda d_p)(w' d_p^2 p_0 u_0 / D_g^0)}{(2\lambda d_p) + (w' d_p^2 p_0 u_0 / D_g^0)} + \frac{2\gamma D_g^0}{p_0 u_0} + \frac{wd_p^2}{D_g^0} \cdot p_0 u_0 + c \cdot \frac{k_p}{(l + k_p)^2} \cdot \frac{d_l^2}{D_l} \cdot j \cdot u_0 \quad (47)$$

Regardless of whether equation (35) or equation (47) is considered, it is apparent that the HETP will depend on the identity of the solute molecule since the parameters D_g^0, D_l and k_p are characteristic of the molecular species (the differences may be small with similar compounds; e.g., hexane and heptane). Thus, the solute molecule identity determines both the retention time (greater volatility = greater proportion of time in mobile phase = less time in column), and the HETP.

Separation and Identification of Substances

In a laboratory situation, several parameters affecting separation must be adjusted simultaneously. The first task is to find a suitable liquid phase. Monographs such as those by Berg (1963), Kaiser (1963), Pescole (1959), and Dal Nogare and Juvet (1962) contain lists of solvents that have proven to be useful in the separation of different classes of solutes. Next, column length, flow rate, and column temperature must be chosen. Liquid phase and column length have already been determined for this experiment. It is up to you to find optimal conditions for the separation of hexane isomers with respect to flow rate and column temperature.

To find optimal flow rates for the separation of several compounds, use the van Deemter curve (see Figure 11-6) for the hardest-to-separate compound. If the resolution permits, use a flow rate larger than the one for a minimum in HETP because of the resulting gain in time. If compounds appear too quickly, even at low flow rates, reduce the column temperature. If separation is too slow, raise the column temperature.

The identification of compounds becomes possible through the measurement of retention volumes or retention times, which are characteristic for any substance under constant column conditions. Strictly speaking, this means that a standard substance, suspected to be identical with one of the components in the sample must be run under exactly the same circumstances on the same column, if retention times are to be compared. Since all columns, although prepared in exactly the same way, are slightly different and even the same column shows a variation of behavior as it ages, the exact matching of retention times can be expected for two runs of the same substance on the same column only within a short time span (on the same day). The column tends to age quickly at elevated temperatures because of the "bleeding off" of the liquid phase.

Methods for the quantitative analysis of a sample have already been discusssed [see equations (6) and (7)].

Procedure

Find the optimal conditions for the separation of C_6 hydrocarbons on 5% hexadecane (weight percentage) on Chromosorb P. The qualitative and quantitative analysis of commercial *n*-hexane is attempted by comparing retention times and peak areas with those of available C_6 hydrocarbons.

The column consists of copper tubing, 2 meters long and 0.25 inch in diameter. Prepare the stationary phase by thoroughly mixing the dry solid support (Chromosorb P) with the required amount of hexadecane dissolved in reagent-grade ether. After the ether has evaporated, slowly fill the tubing with the powdery stationary phase and tap continuously to insure uniform settling. Coil the filled column so that it fits into the oven or thermostat. A newly filled column must be rinsed with carrier gas at a low flow rate for several hours, preferably the day before it is used in an experiment.

Operating Conditions

The following operating conditions have been found to be satisfactory when using a Barber-Coleman commercial gas chromatograph:

Injection port temperature	50°C (75°C)
Detector temperature	room temperature (50°C)
Detector current	150 ma
Attenuator setting (start)	10
Chart speed	1 inch/minute

The sample size is 0.5 microliter.

All of the following experiments are to be done with the column at room temperature. If sufficient time is available, select several to be repeated at higher or lower temperatures.

1. Inject a sample of 0.5 μl *n*-hexane (reagent grade) at flow rates of approximately 40, 60, 80, 100, 120, 140 cc/minute. Repeat the injection at two-minute intervals three or four times at each flow rate.
2. Calculate the HETP at each flow rate, and make the additional determinations required to find the optimum flow rate. Do all of the experiments at the optimum flow rate.
3. Determine the retention times for the five C_6H_{14} isomers and methylcyclopentane.
4. Prepare a calibration mixture by placing 0.5–1.0 ml of each isomer in a vial with a threaded cap. (Weight to 0.1 mg.) Determine the composition of this mixture (weight % or

mole %). Analyze the mixture by gas-liquid chromatography to determine the relative detector response for each component.

5. Identify the impurities in commercial *n*-hexane. Estimate the percent abundance of each isomer.
6. Identify the compounds in a mixture of C_6 hydrocarbons obtained from the instructor and calculate their relative amounts from the areas of peaks and data obtained in step 4.

B. ION-EXCHANGE CHROMATOGRAPHY

The use of ion-exchange procedures in analytical chemistry has increased tremendously in recent years. Ion-exchange chromatography has become an especially valuable aid in the separation of chemically similar ions and ionic compounds such as rare earth compounds and amino acids. In many cases separation is achieved comparatively quickly and at a very low cost.

Apparatus

50-ml buret for ion-exchange column
Automatic sample collector (optional)
Spectrophotometer with cells
*p*H meter
25-ml pipet
15-ml pipet
1-ml pipet
2-ml graduated pipet (0.1 ml)
60 test tubes
2 100-ml volumetric flasks
Mariotte flask
Glass wool

Chemicals

12 *M* HCl purified by ion exchange
6 *M* HCl prepared by diluting purified 12 *M* HCl
4 *M* HCl prepared by diluting purified 12 *M* HCl
Nitroso-R-salt solution, 1% in water (nitroso-R-salt is an abbreviation for 1-nitroso-2-naphtol-3,6-disulfonic acid disodium salt)
Concentrated sodium acetate solution, or 3 *M*.
Dowex 1 X 8 ion-exchange resin, 100–200 mesh
Fe^{2+} and Co^{2+} stock solutions

The goal of the experiment is to find optimal conditions for the separation of Fe^{2+} from Co^{2+} by elution ion-exchange chromatography. Krauss and Moore (1953) describe the separation of Ni^{2+}, Mn^{2+}, Co^{2+}, Cu^{2+}, Fe^{3+} and Zn^{2+} in the form of chloride complexes on an anion-exchange column. They also give data for the behavior of Fe^{2+} on such a column (Krauss and Moore 1952).

The metals form complexes of the general form

$$M^{n+} + mCl^- = MCl_m^{n-m}$$

with chloride ions. If $m > n$ the complex is an anion. Which complex exists in highest abundance depends on the concentration of chloride and cation, as well as on the equilibrium constants of the various reactions. At a particular chloride concentration, separation on the column will be largely due to the state of complexity attained by the different cations. Fortunately, not much has to be known about the various complex processes that occur during the transport of different metal ions in whatever state they may be. Plate theory enables us to predict the behavior of mixtures on a particular column rather well, after a few initial measurements have been made. However, since plate theory is based on equilibrium assumptions in a situation that is clearly nonequilibrium, the relations it provides will be used strictly in an operational sense.

The results of this experiment can be used in the analysis of cobalt ferrite, which was prepared previously. The magnetic properties of this compound depend heavily on the oxidation state of the iron. It is therefore very important to find the traces of Fe^{2+} that may be present in the bulk of Fe^{3+} and Co^{2+}. It is very difficult to find these traces without separation. The separation of Fe^{3+} from the mixture is of no concern in this experiment because Fe^{3+} is retained on the column and can be washed off with water after Fe^{2+} and Co^{2+} have been removed from the column.

Separation Process

Ion-exchange resins are usually composed of long strands of polystyrene cross-linked by divinylbenzene groups. They also contain functional groups, which are either basic or acidic. A cation exchanger usually contains sulfonate groups; an anion exchanger may possess quarternary amine groups, which interact with the ions to be separated. An anion exchange column can be pictured as a bed of pervious material containing immobile cations through which a solution carrying anions is percolated. At the beginning of the process the resin should be saturated with a solution containing just one kind of anion (concentrated HCl) to assure uniformity throughout the column and the resulting reproducible behavior. A bulk of a different anion introduced to the column will, after a sufficient period of time, achieve equilibrium distribution between the cation sites and the solution by the replacement of Cl^- ions. The addition of an eluant transports the ions in solution to regions solely occupied by Cl^- ions and is followed by the redistribution and further transport of the ions remaining in solution. Different kinds of anions travel at different speeds, depending on their success in the competition for cation sites. Separation will be achieved in this way if the column is long enough and the flow rate slow enough to allow sufficient time for the kinetic processes involved in redistribution. The formation of channels in the column upsets the process severely since the flow of the solution is much faster within these channels than in the remainder of the resin bed. Commercial resins therefore come in beads of nearly uniform size to allow better packing. The bead size is given in mesh numbers, a higher mesh number indicates smaller particle diameter.

The partitioning process does not take place on the surface of the beads only; ions actually enter the resin and their rate of exchange usually determines the time required for separation. Resins of high cross-linkage permit only slow "particle diffusion," which this process is called; but with small ions the loss in time is entirely offset by increased selectivity. Low cross-linkage is preferred in separations of large ions.

The presence of functional groups and exchangeable counterions in the beads leads to an uptake of water, accompanied by a swelling of the resin until the strain of the resin matrix balances the osmotic pressure. Highly cross-linked resins exhibit less swelling because of the increased rigidity of their structure. In Dowex resins the degree of cross-linking is expressed as the percentage of divinylbenzene added at the time of polymerization. Dowex 1 X 8 indicates the addition of 8%, not all of which may actually have formed cross-links.

Plate Theory

The performance of a column depends on the choice of the resin, the flow rate, the composition of the eluant, and the nature of the mixture to be separated. For a given column and eluant with a constant flow rate, the continuous nonequilibrium transport of a certain kind of ion through the column may be approximated by a step-by-step transport process, in which equilibrium is reached at each step. With this method column behavior towards different ions may be predicted to a certain extent and parameters may be calculated, which otherwise would have to be found by trial and error.

It is assumed that a number of equilibrations of ions between the eluant and the resin have been reached in the column before the bulk of the ions reaches the end of the column. The time required for each equilibration depends on the combined rates of the different transport processes in the column. It is not necessary to know these parameters. The length of column, Δx, in which equilibration can be achieved is given by the linear flow rate of the eluant, Δx = HETP (height of a theoretical plate) and a column of length L contains $p = L/\text{HETP}$ theoretical plates, which are numbered $r = 0,1, \ldots, p$. The eluant is added to this column in portions having the magnitude of the interstitial volume ΔV_i that is accessible to the eluant in one plate. The flow is practically continuous since equilibration is always reached after the volume has travelled the distance of a plate, according to the definition of theoretical plates. The volumes of eluant added to the column are counted by the number $n = 0,1,2, \ldots$

Upon the addition of each volume of eluant the fraction y remains in the plate:

$$y = \frac{c_r \Delta V_r}{c_r \Delta V_r + c_{\text{el}} \Delta V_i} = \frac{C}{C+1} \tag{1}$$

where $C = c_r \Delta V_r / c_{\text{el}} \Delta V_i$; whereas the fraction z

$$z = \frac{c_{\text{el}} \Delta V_i}{c_{\text{el}} \Delta V_i + c_r \Delta V_r} = \frac{1}{C+1} \tag{2}$$

is transported to the new plate. C is the distribution coefficient; c_r and c_{el} are the concentrations of solute in the resin phase and in the eluant, respectively; and ΔV_r is the volume occupied by the resin in one plate.

The fraction of solute in plate r after n volumes of eluant have been added, is then derived from the binomial expansion of $(y+z)^n$.

$$Q_{r,n} = \frac{n!\, y^{n-r} z!}{r!\,(n-r)!} \tag{3}$$

For large n, r, and $n - r$ and after the conversion to a distribution over volumes n for a given plate, $r = n_{\max} \cdot z$ (see Section A of this exercise for details), equation (3) can be approximated by a Gaussian distribution:

$$Q_r(n) = Q_{\max} \cdot \exp\left[-\frac{(nz - n_{\max}z)^2}{2n_{\max}z(1-z)}\right] \quad (4)$$

Equation (4) describes the distribution of the solute over the number n of volumes ΔV_i introduced onto a column of $p \equiv N$ plates, that is, the measured elution curve.

Equation (2) can be written as

$$z = \frac{\Delta V_i}{V_p} \quad (5)$$

with

$$V_p = \Delta V_i + \left(\frac{c_r}{c_{el}}\right)\Delta V_r \quad (5a)$$

where V_p is the effective plate volume.

The distribution of the solute over the volume leaving the column is a measurable quantity. To determine this quantity, z in equation (4) is expressed with the aid of equation (5), keeping in mind that the total volume, V, which has left the column after n plate volumes have been added, is $V = n \cdot \Delta Vi$. The volume, $V_{\max}$, at which the maximum of the solute appears at the end of the column is then $V_{\max} = n_{\max} \cdot \Delta Vi$. Equation (4) is now:

$$Q_p(V) = Q_{\max} \cdot \exp\left(-\frac{(V - V_{\max})^2(C+1)}{2V_{\max} \cdot C \cdot V_p}\right) \quad (6)$$

Since the plate number, N, is $N = n_{\max} \cdot z$, the following may be written:

$$V_{\max} = \frac{n_{\max} \cdot \Delta V_i \cdot V_p}{V_p} = N \cdot V_p$$

or

$$\frac{1}{V_p} = \frac{N}{V_{\max}} \quad (7)$$

which leads to:

$$Q_p(V) = Q_{\max} \cdot \exp\left(-\frac{(V - V_{\max})^2(C+1)N}{2V_{\max}^2 \cdot C}\right) \quad (8)$$

Equation (8) describes the elution curve of a solute with a distribution coefficient C leaving a column composed of N theoretical plates.

Calculation of the Number of Theoretical Plates

Equation (8) serves to evaluate the number of theoretical plates a column possesses. At the volume, Ve, where the concentration of the solute has fallen to $1/e$ of the maximum value, $Q_p(V) = Q_{\max}/e$, it holds that:

$$1 = \frac{(V_e - V_{\max})^2(C+1)N}{2V_{\max}^2 \cdot C} \quad (9)$$

and equation (9) can be solved for N, if C is known. The distribution coefficient C can be obtained from the volume at which the maximum of the solute appears, $V_{\max}$. The distribution coefficient C was defined as the amount of solute in the resin of one plate divided by the amount in solution in the interstitial volume of the same plate at equilibrium:

$$C = \frac{c_r \cdot \Delta V_r}{c_{el} \Delta V_i} \quad (10)$$

The introduction of equation (10) into equation (5a), which defines V_p, leads to:

$$V_p = (C + 1)\Delta V_i \quad (11)$$

and combining equations (11) and (7) yields:

$$V_{\max} = (C + 1) \cdot V_{in} \quad (12)$$

where $V_{in} = N\Delta V_i$ is the total interstitial volume of the column. The determination of V_{in} is described in the "Procedure" of this section of the exercise.

Calculation of Column Length

So far, one solute on its way through the column has been considered. The question of how to determine the length of column necessary to separate two substances, a and b, if their plate numbers N_a and N_b have been measured for the same column shall be taken up next. It is advisable to convert equation (8) into the form of the normal error function

$$y = \frac{1}{\sqrt{2\pi}} \exp\left(-\frac{x^2}{2}\right) \quad (13)$$

so that standard tables may be used (see, e.g., *Handbook of Chemistry and Physics*). Equation (8) then reads

$$\frac{Q_p(V)}{\sqrt{2\pi} \cdot Q_{\max}} = \frac{1}{\sqrt{2\pi}} \exp\left(-\frac{(V - V_{\max})^2(C + 1)N}{2 \cdot V_{\max}^2 \cdot C}\right) \tag{14}$$

A comparison of equations (13) and (14) shows

$$x = \frac{V - V_{\max}}{V_{\max}} \cdot \left\{\frac{N(C + 1)}{C}\right\}^{1/2} \tag{15}$$

The error function [equation (13)] is normalized so that

$$\int_{-\infty}^{+\infty} y \cdot dx = 1$$

or because of its symmetry around $x = 0$ the following may be written:

$$\int_{-\infty}^{0} y dx = \int_{0}^{+\infty} y dx = 0.5$$

If a cross contamination of 0.05% of substance a in b or b in a seems tolerable, then x_a must be found for which 0.9995 times the total amount of a has left the column. This means that the value of x_a must be determined for which the following relation holds:

$$0.5 + \int_{0}^{x_a} y_a dx_a = 0.9995 \tag{16}$$

In this case $x_a = 3.29$. On the other hand, only 0.0005 times the total amount of compound b is assumed to have left the column at this time.

$$0.5 - \int_{-x_b}^{0} y_b dx_b = 0.0005 \tag{17}$$

and $x_b = 3.29$.

From equation (15) and x_a and x_b the following two equations can be derived:

$$V_a = V_{a,\max} + 3.29\, V_{a,\max} \left\{\frac{C_a}{N_a(C_a + 1)}\right\}^{1/2} \tag{18}$$

and

$$V_b - V_{b,\max} \quad 3.29\, V_{b,\max} \left\{\frac{C_b}{N_b(C_b + 1)}\right\}^{1/2} \tag{19}$$

It follows that $V_a = V_b$ because the elution curves cross at this point.

The number of plates in a column of length L can be expressed as the number of plates per centimeter, P, times the length.

$$N_a = P_a \cdot L$$

and

$$N_b = P_b \cdot L$$

where $P = 1/\text{HETP}$. The introduction of these equations plus equation (12) into equations (18) and (19) and solving for L leads to

$$(L)^{1/2} = \frac{3.29}{C_b - C_a} \left[\left(\frac{C_b(C_b + 1)}{P_b}\right)^{1/2} + \left(\frac{C_a(C_a + 1)}{P_a}\right)^{1/2}\right] \tag{20}$$

from which L can be calculated. If other degrees of contamination seem tolerable, only the value for the limit of the integral over the error curve must be changed. In general, the limits of contamination have to be very low if the two substances are present in very different amounts; otherwise the substance present in the smaller quantity will be too heavily contaminated.

A safety margin of 10% is added to L, to account for experimental deviations. This should be done because a new column prepared to the calculated length frequently does not show the required number of theoretical plates because of differences in packing or in the resin.

In many cases, the separability of two substances on a given column can be altered drastically by changing the eluant. It is advantageous to elute the first component with an eluant that holds the second component back, and then to change to an eluant that washes the second component out relatively quickly. The change is made immediately after the first component has left the column.

Separation by Changing the Eluant

Let us assume that compound a is eluted from the column by eluant no. 1, which is replaced by eluant no. 2 right after the maximum of substance a leaves the column. Since the actual separation of substances a and b takes place under the influence of eluant no. 1 only, the required length, L, of column can be calculated, with the help of equation (20), either by choosing a very small degree of cross contamination or by increasing the safety margin. Either of these procedures is permissible because, as pointed out before, the objective is to arrive at a method to estimate optimal column conditions, rather than an exact calculation.

To decide whether a change in eluant will shorten the required time and decrease the volume to be collected decisively, the volume that leaves the column is calculated until the maximum of substance b appears. Compare the result to equation (12)—written for substance b and eluant no. 1. The indices now refer to compound and eluant.

$$V_{1,b,\max} = V_{\text{in}}(C_{1,b} + 1) \qquad (12a)$$

Equation (12a) states the amount of eluant to be collected until the maximum of compound b appears under the action of eluant no. 1. The calculation of $V_{1,2,b,\max}$, the volume necessary to wash out substance b if the eluant is changed from no. 1 to no. 2 when the maximum of compound a appears at the end of the column, follows.

First, obtain the length l_1 that the peak of b travelled under the influence of eluant no. 1. The amount of eluant necessary to move peak b through L cm to the end of the column is $V_{1,b,\max}$ ml. Of this amount only $V_{1,b,\max} - V_{\text{in}}$ ml have actually passed through this peak, since V_{in} ml remain in the column. Each milliliter of eluant no. 1 moves peak b $L/C_{1,b} \cdot V_{\text{in}}$ cm along the column; $V_{1,a,\max}$ ml transport the peak along the length l_1.

$$l_1 = \frac{L \cdot V_{1,a,\max}}{C_{1,b} \cdot V_{\text{in}}} \qquad (21)$$

By the same reasoning each milliliter of eluant no. 2 moves the peak $H/C_{2,b} \cdot V_{\text{in}}$ cm, and V_2 ml transport peak b along length l_2 to the end of the column.

$$l_2 = \frac{L \cdot V_2}{C_{2,b} \cdot V_{\text{in}}} \qquad (22)$$

If the total volume to be collected until the maximum of peak b appears is $V_{1,2,b,\max}$, from which $V_{1,a,\max}$ transported the peak along distance l_1, and if V_{in} did not move through the peak (because it was already contained in the column), then V_2 is given by

$$V_2 = V_{1,2,b,\max} - V_{1,a,\max} - V_{\text{in}} \qquad (23)$$

The sum of the two partial lengths l_1 and l_2 constitutes the total column length L.

$$l_1 + l_2 = L \qquad (24)$$

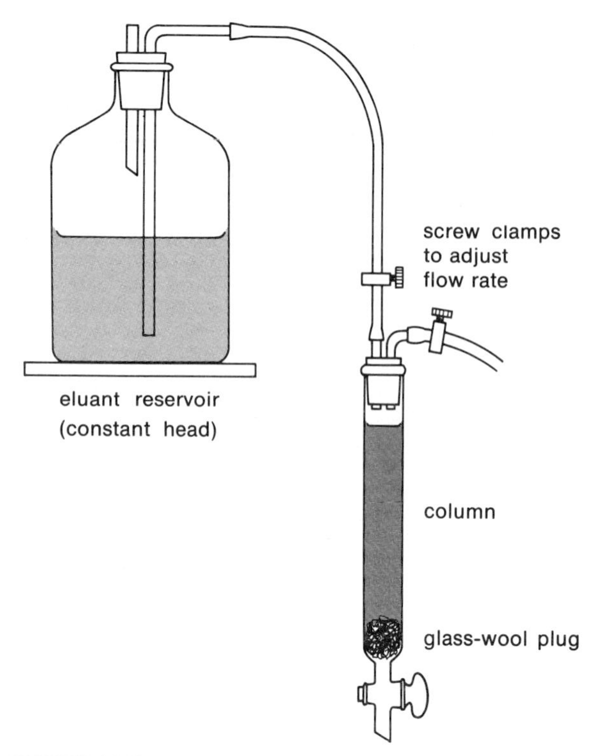

FIGURE 11-7
Column arrangement to obtain constant flow rate of eluant.

The introduction of equation (23) into equation (22) and the combination of equations (21) and (22) resulting in equation (24) yield the desired expression for the total volume to be collected before the maximum of peak b occurs.

$$V_{1,2,b,\max} = V_{1,a,\max} + V_{\text{in}} + C_{2,b}\left(V_{\text{in}} - \frac{V_{1,a,\max}}{C_{1,b}}\right) \qquad (25)$$

All calculations have been performed on the basis of a constant flow rate of the eluant!

Procedure

To prepare the column place a small piece of glass wool in the bottom of a buret. Place about 15 g of the resin into a beaker and slurry with distilled water. While it is in suspension, rapidly pour the slurry into the buret and wash the resin to the bottom with distilled water. Gently tap the buret to insure the uniform settling of the resin until the height of the resin is about 25 cm. Assemble the complete apparatus in accord with Figure 11-7.

To determine the interstitial volume of the column, run 20 ml of normalized 0.1 N HCl through the column. Drain to the surface of the resin; wait 10 minutes and wash again with 20 ml of 0.1 N HCl. Titrate the last 3 ml that leave the column against 0.01 N NaOH and repeat the washing with 0.1 N HCl until the solution leaving the column is 0.1 N HCl. Drain the solution to the resin level again. At no time during the operation of the column should the solution fall below the resin level. Dry the walls of the column with cotton swabs or paper tissues taking care not to touch the resin surface. Percolate 0.5 N sodium nitrate solution (*not normalized*) through the column until the eluant obtained from the column is neutral. Test frequently with pH paper; then obtain a final reading with a pH meter. Gather all of the solution that was removed from the column in a volumetric flask, dilute to the mark, and measure the pH of aliquot samples. The interstitial volume of the column is:

$$V_{\text{in}} = \frac{\text{milliequivalents of H}^+}{\text{normality of initial HCl}}$$

The volume of HCl that was eluted from the glass-wool plug and the lower part of the buret can be estimated from blank runs. The eluted milliequivalents of H^+ must be adjusted for this error.

Bring the column into the desired form by washing with three portions of 20 ml each of purified[2] 6 M HCl. The measurement of interstitial volume V_{in} cannot be performed at this stage since too much HCl would penetrate the resin phase (Donnan penetration). After the last portion of the 6 M HCl has been drained to the resin level, introduce 0.5 ml of Co^{2+} and 0.5 ml of Fe^{2+} stock solutions into the column. Insert the pipet deeply into the column to avoid contaminating the walls with the stock solutions. Drain to allow the sample to sink into the resin; then wash the walls above the resin with 1 ml of 6 M HCl. Drain to the resin level again and wait 10 minutes to let the sample equilibrate. A stock solution should contain approximately 2–3 mg of a metal per milliliter in concentrated, purified HCl. (The stock solutions are easily prepared from cobaltous chloride and iron powder. (Prepare two stock solutions and reduce the Fe^{2+} solution with hydroxylamine immediately before use.) Fill a Mariotte flask with 6 M HCl and attach it to the column as shown in Figure 11-7. Adjust the screw clamps to obtain a linear flow rate of 0.5 cm/minute. (Calculate the linear flow rate from the volume flow rate and the inner diameter of the tube.) Use an automatic sample collector to collect about 150 ml of solution in 3-ml portions, a process that takes about 5 hours if a normal 50-ml buret has been used as a column. Nevertheless, it is entirely possible to do the experiment without using a sample collector if 3-ml portions are collected only during intervals in which either Fe^{2+} or Co^{2+} are eluted from the column. However, it is necessary to know beforehand how much time is required to elute the various solutes. The first 15 ml will not contain any Fe^{2+}. This ion will emerge from the column after roughly 20–35 ml of eluant have passed through the column. The next 30 ml consist of pure eluant and then the Co^{2+} begins to appear in a broad band. This information is necessarily imprecise, since the actual experimental conditions vary appreciably and procedures have to be altered accordingly. Therefore the procedure is as follows: Discard the first 10-ml portion after checking briefly for any Fe^{2+}. Then collect consecutive portions of 5 ml each and check for iron by the procedures described herein. Three-milliliter portions are to be taken following the first sample that contained iron until all of it has left the column. Continue with 10-ml portions until 70 ml has been eluted. Then gather 5-ml portions, always checking for Co^{2+} until the first trace appears, switch to 3-ml portions, and continue until all of the cobalt has been eluted.

If a sample collector is used, check every fifth test tube for iron or cobalt. Once the two bands have been located, color has to be developed in all samples containing metal atoms. Discard the rest.

Prepare the column for reuse by washing with concentrated HCl. Percolate water through the column after it has been used twice to prevent a buildup of impurities (especially Cu^{2+}, Fe^{3+}, and Zn^{2+}). Check the influence of the concentration of the complexing agent (HCl) by rinsing the column three times with 20 ml each of 4 M HCl. Then introduce a 0.5-ml sample of Co^{2+} stock solution and wash it into the column as before. Elution will now be much faster, which makes it advisable to collect only 3-ml portions and to analyze each from the beginning until all metal has left the column.

Analysis

The method of analysis to be used in this experiment is spectrophotometry because many samples

[2]Purify by passing through a cation-exchange column to remove trace metals.

can be inspected in a comparatively short time. For the purpose of evaluating elution curves, precision can be sacrificed slightly if the method is fast and both components can be analyzed in one process, which is advantageous if peaks overlap. Nitroso-R-salt, which absorbs in the region below 5000 Å, develops a deep red-brown color with cobaltous ions and a bright green color with ferrous ions at a pH between 5 and 8. The color develops to its maximum intensity upon heating to about 80°C and is stable for more than 24 hours. The ferrous complex binds one mole of the salt and has an absorption maximum at 7200 Å, whereas the complex formed by one cobaltous ion and three dye molecules has an adsorption maximum at 4100 Å. A less pronounced shoulder on the absorption curve at 5100 Å is sufficiently strong to allow the detection of Co^{2+} at concentrations prevailing in this experiment. Thus Fe^{2+} and Co^{2+} can be detected at 7200 Å and 5100 Å, respectively, without the interference of excess nitroso-R-salt. At 5100 Å, however, both complexes absorb. For the determination of an elution curve any property that is directly proportional to the concentration suffices, since the concentration by itself is never used in any calculation. Thus absorbance at 7200 Å can be plotted without conversion against the eluted volume to yield the elution curve of Fe^{2+}. The same is true for absorbance at 5100 Å and the elution curve of Co^{2+}, as long as no absorption is observed at 7200 Å. In regions where peaks overlap, the absorbance of Co^{2+} at 5100 Å, A_{5100}^{Co}, is given by:

$$A_{5100}^{Co} = A_{5100} - A_{7200}\ \text{const} \qquad (26)$$

where const $= a_{5100}^{Fe}/a_{7200}^{Fe}$ and a_{λ}^{M} is the absorption coefficient. If the metals are completely separated by the column, the analysis of Fe^{2+} is performed at 7200 Å, and that of Co^{2+} at 4100 Å.

Pipet 15 ml of concentrated sodium acetate solution, followed by 0.5 ml of a 1% indicator solution, into each 3-ml portion collected during the run with 6 M HCl, in which metal ions are expected. If the red color of the cobalt complex is observed, add another milliliter of the indicator solution. Put all test tubes over a steam bath for at least 10 minutes, cool, and record the absorbance at 4100 Å and 7200 Å.

Treat the portions collected during the run with 4 M HCl in the same way, adding 10 ml of concentrated sodium acetate solution instead of 15 ml.

It is advisable to check some of the developed colored solutions for their pH.

If separation is not complete, record absorptions at 5100 Å and 7200 Å. If possible, use two spectrophotometers, one set at 7200 Å and the other at 5100 Å. Since the absorbance at 5100 Å is measured at a part of the spectrum that is slightly sloped, appreciable error can be introduced by small deviations in wavelength.

Prepare solutions 6×10^{-5} M in either Fe^{2+} or Co^{2+} and check whether the pH is between 5 and 8; if not, add sodium acetate. These solutions are for the determination of absorption coefficients. Take aliquot samples. Reduce the solution containing Fe^{2+} with hydroxylamine and add 1 ml of 1% nitroso-R-salt solution per milligram of metal. Add 3 ml of 1% nitroso-R-salt solution per milligram of metal to the sample containing Co^{2+} ions. Put both solutions on a steam bath. Wait for 10 minutes. Cool to room temperature and record the spectrum of each between 3500 Å and 7500 Å. Obtain the ratio of absorption coefficients at 5100 Å and 7200 Å from the spectrum of the Fe^{2+} complex. Check the ratio for different concentrations.

Handle all 3-ml portions[3] from a run that has not led to complete separation exactly as described herein, but obtain the absorption of each sample at both wavelengths (5100 Å and 7200 Å).

Data

Plot the elution curves for both runs as indicated in the preceding discussion of "Analysis." Calculate the length of the column necessary to separate Fe^{2+} and Co^{2+} to 0.01% cross contamination. What are the volumes at which maxima of the two peaks appear for the calculated column length? How much time does the experiment take? Is there an advantage in eluting Fe^{2+} with 6 M HCl and then eluting Co^{2+} with 4 M HCl? What would the total volume to be collected be and how much time would be required?

If time permits, run additional experiments to see whether their results agree with your calculations.

C. SEPARATION OF POLYPHOSPHATES BY PAPER CHROMATOGRAPHY

Apparatus

Jar for ascending chromatography (Wide mouthed, 2000-ml, solid-dispensing jars with cork stoppers

[3]Treat all larger portions with proportionally greater amounts of reagents.

and glass rods to suspend the paper will do. The corks should have bores that match the diameter of the rod. See Figure 11-8.)
Filter paper (W & R Ralston, Ltd., Genuine Whatman Filter Paper No. 1 and Schleicher and Schüll No. 589 Orange R filter paper have been used successfully.)
5 5-microliter disposable pipets (e.g., Drummond Microcaps, Drummond Scientific, Broomall, Pa.)
Spray bottle
Ultraviolet lamp
Drying oven at 80°C

Chemicals

0.01 *M* orthophosphate solution, as standard
0.05 *M* pyrrophosphate solution, as standard
Ebels solution:
750 ml isopropyl alcohol
50 g trichloroacetic acid
2.5 ml concentrated ammonia
Acid molybdate solution:
25 ml of 60% perchloric acid
5 ml concentrated hydrochloric acid
5 g ammonium heptamolybdate tetrahydrate
500 ml distilled water

Introduction

Paper chromatography is one of the most widely used separation techniques in chemistry. One reason for its success is its economy in terms of equipment, sample size, and actual working time. Nevertheless, relatively little is known about the processes underlying the separation mechanism, which is frequently very complex. A simple explanation is that the separation is due to the partitioning of the solute between solvent-saturated cellulose as the stationary phase and a migrating solvent as the mobile phase. Due to the partitioning, different solutes travel at different speeds and are finally separated.

Paper chromatography can be performed with an ascending or descending solvent. With an ascending solvent, the solute moves by capillary action upwards into the paper; with a descending solvent, it runs down from a suspended reservoir. In both cases, the *Rf* values for a particular solute are found to be the same for a given solvent. As in gas-liquid or liquid-liquid chromatography, the separation can be enhanced by consecutive partitioning with two suitable solvents. This is done in two-dimensional chromatography. The solute is developed in one direction with the first solvent and then in a perpendicular direction with the second solvent. Before elution, the paper is saturated by solvent vapor in a closed container, to obtain an equilibrated stationary phase. The container remains closed during elution to avoid loss of solvent to the atmosphere. Such losses may render the separation process less reproducible. Only small amounts of solute can be separated in one run, since the stationary phase is easily overloaded. Because of the small concentrations, elaborate techniques are frequently used to make the zones clearly visible. The solute is deposited on the paper in a dissolved form, so that it may easily penetrate the stationary phase. At the same time, as in all chromatographic processes, care should be taken to restrict the volume into which the sample is charged. Successful partitioning is assured by the slow movement of the solvent and by constant gas-phase composition. Both conditions are conveniently fulfilled for the ascending solvent in a closed jar. Polyphosphates of different chain lengths can be separated on certain filter papers with a solvent of Ebel's solution—a mixture of isopropylalcohol, trichloroacetic acid, and ammonia. The ease with which polyphosphates hydrolyze necessitates the use of a nonaqueous solvent. After chromatography is finished, the different zones are made visible by the conversion of some of the polyphosphates to a heteropolyacid—very probably $[P(Mo_3O_{10})_4]^{-3}$—and the reduction of the heteropolyacid, in the course of which the blue color of the developed spots begins to appear. The formation of heteropolyacid is preceded by the hydrolysis of the parent polyphosphates, since the orthophosphate anion is one of the reactants in spot development [see van Wazer (1958) pp. 559 ff.]. Chain and ring polyphosphates remain unresolved but can be separated by means of two-dimensional paper chromatography.

Procedure

The experiment can be run with tripolyphosphate and its hydrolyzate, prepared as described in Exercise 2C, or with commercial tripolyphosphate. In 3 hours of developing time, ortho-, pyro-, and tripolyphosphate are well resolved whereas higher polyphates remain partially unresolved. For the separation of higher polyphosphates (up to 5 phosphorus units) 6 hours and 14 inches of paper are required. The experiment described here takes 3 hours.

Cut a 7″ × 7″ square from the chromatographic paper. Draw a pencil line parallel to and 1 inch from one side. Make five pencil marks along the line at 1-inch intervals, beginning 1.5 inches from one end. Spot one of the marks with a 5-microliter sample of 0.01 *M* orthophosphate, one with 0.03 *M* pyrophosphate from standard solutions, one with 0.05 *M* tripolyphosphate, and two with hydrolyzate. It is worthwhile to obtain two samples of hydrolyzate, distinguished by different reaction times, to observe the progress of hydrolysis qualitatively. The size of the spots may be minimized by making the application in several portions at 15- to 30-second intervals.[4] Staple the paper to form a cylinder, the ends touching but not overlapping with the line of spots at the lower end. Attach a wire to the upper end with paper clips.

Suspend each chromatogram in a separate bottle above 1 cm of Ebel's acid solution (750 ml of isopropyl alcohol combined with a solution of 50 g of trichloroacetic acid in 250 ml of distilled water and 2.5 ml of concentrated ammonia). Run the middle portion of the wire through the hole in the cork stopper and twist it around the glass rod to keep the paper above the level of the solution. The arrangement is shown in Figure 11-8.

Allow several hours for equilibration, so that the paper is saturated with the solvent vapor. Then lower the paper into the solution, using the glass rod to close the hole, and let the chromatogram develop until the solvent front is about 1 inch from the top (3–4 hours). Mark the solvent front, gently remove the staples from the paper, and hang the chromatogram by the paper clips in a place that is free of contamination, such as an unused hood. Air-dry; then place in an oven at 85°C for 10 minutes for a final drying. Spray the chromatogram with an acid molybdate solution in the hood (5 ml of 60% perchloric acid, 1 ml of concentrated HCl, and 1 g of ammonium heptamolybdate tetrahydrate; dilute to 100 ml with distilled water). Figure 11-9 shows a spraying bottle specifically manufactured for chromatographic use.

Dry the chromatogram again at 85°C for 10 minutes. Finally, expose to ultraviolet light to develop the spots. CAUTION: Avoid looking directly into the lamp; ultraviolet light damages the eye irrevocably.

[4]Disposable micropipets can be used to advantage here.

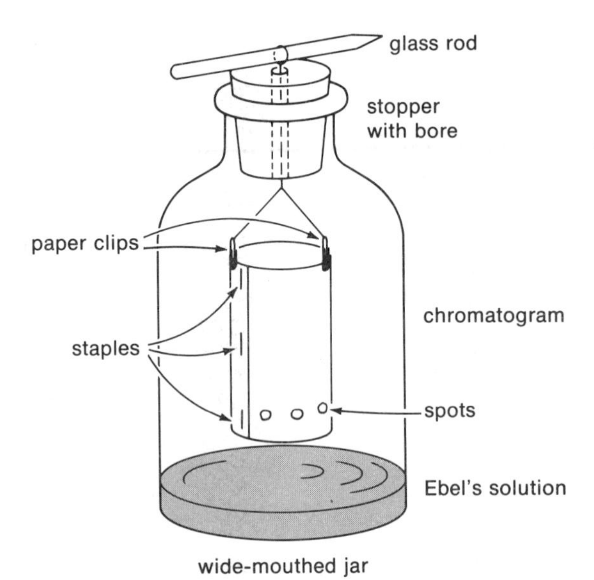

FIGURE 11-8
Chromatogram suspended above the solution in a jar. The glass rod can be used to close the hole in the stopper after the paper has been lowered into the solution for the development of the chromatogram.

Mark the positions of the spots using either the centers or the leading edges. Calculate the R_f values:

$$R_f = \frac{\text{distance solute moved}}{\text{distance solvent moved beyond initial solvent spot}}$$

Prepare a graph in which logarithm R_f is plotted against chain length n to show whether the following relationship holds:

$$\log R_f = -an + b$$

where a and b are constants.

Which components are present in the prepared or commercial triphosphate? How did hydrolysis change the composition? Did the preparation of tripolyphosphate proceed to completion? If end-group titration (see Exercise 12E) has been performed on the same substance, compare the two results.

Additional Work

The separation of ring and chain polyphosphates has been described by van Wazer and Karl-Kroupa

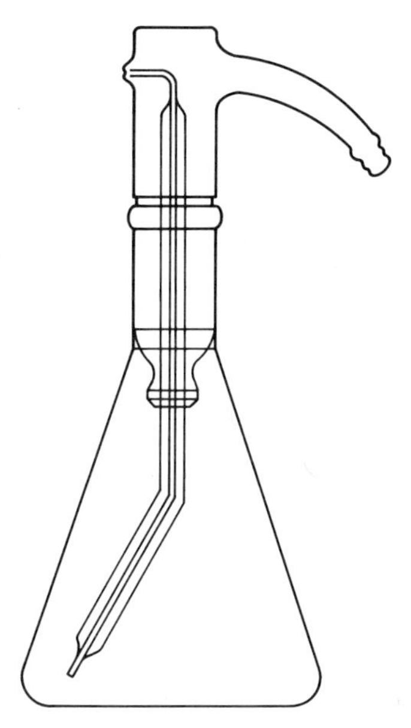

FIGURE 11-9
Spray bottle for use with indicator in paper chromatography.

(1956). The method is more time-consuming, but of special interest, if chromatographic results are to be compared with those of end-group titration (see Exercise 12E).

REFERENCES

Berg, E. W. 1963. *Physical and Chemical Methods of Separation.* New York: McGraw-Hill.

Dal Nogare, S., and Juvet, R. S. 1962. *Gas Chromatography.* New York: Wiley.

Giddings, Y. C. 1965. *Dynamics of Chromatography. I. Principles and Theory.* New York: Dekker.

Hanes, C. S., and Isherwood, F. A. 1949. *Nature* **164**, 1107. Paper chromatography.

Kaiser, R. 1963. *Gas Phase Chromatography.* London: Butterworth.

Krauss, K. A., and Moore, G. E. 1952. *J. Am. Chem. Soc.* **74**, 843; 1953. **75**, 1460. Ion-exchange chromatography.

Pescok, R. L. 1959. *Principles and Practice of Gas Chromatography.* New York: Wiley.

Purnell, H. 1962. *Gas Chromatography.* New York: Wiley, p. 105.

Rieman, W., III, and Sargent, R. 1961. In *Physical Methods of Chemical Analysis.* Vol. 4. Ed. W. G. Berl. New York: Academic Press.

van Wazer, J. R. 1958. *Phosphorus and Its Compounds.* Vol. 1. New York: Wiley.

van Wazer, J. R., and Karl-Kroupa, E. 1956. *J. Am. Chem. Soc.* **78**, 1772. Paper chromatography.

von Mises, R. 1964. *Mathematical Theory of Probability and Statistics.* New York: Academic Press, p. 273.

Westman, A. E. R., and Scott, A. E. 1951. *Nature* **168**, 740. Paper chromatography.

EXERCISE 12

Electrochemical Methods

A. INTRODUCTION TO ANALYTICAL APPLICATIONS

The analytical applications of electrochemistry are based on the concentration dependence of the various physical characteristics of an electrochemical cell. For the present purposes, an electrochemical cell is simply described as an electrolyte solution in which two electrodes (sometimes three) are immersed. To understand more fully the conditions under which the experiments in this exercise are performed, these physical characteristics will be reviewed before proceeding to a discussion of the experimental circumstances under which all but one of them contribute negligibly to observed behavior, a condition that must be met to render observations interpretable.

Physical Cell Characteristics

The best known cell characteristic is the "equilibrium cell potential," which is measured under static conditions (no current flowing).

Dynamic cell properties are described by the "cell impedance," which consists of two parts—the "bulk impedance" of the electrolyte, which is due to all of the forces hampering the migration of ions in the electric field, and the "interfacial impedance." The latter has two contributors—the "double-layer impedance" and the "faradaic impedance." The magnitude of the faradaic impedance depends on the rate at which ions, which can undergo charge transfer at the electrode, are transported toward the electrode and the rate at which charge transfer takes place. The electrical double layer around the electrode is formed by oppositely charged ions, drawn into the vicinity of the electrode without being able to react with it. In equilibrium they just neutralize the charge of the electrode and thus constitute (together with the electrode surface and its charge) a capacitance.

Electrochemical Methods

The following methods are exploited in the experiments in this exercise: potentiometry, conductometry, and voltammetry and polarography. They shall be described in terms of primarily observed cell characteristics.

Potentiometry

Conditions to be observed here are chemical equilibrium and the negation of any current flowing within the cell. The appropriate electronic circuitry is described in Sections B and C of this exercise.

The equilibrium cell potential is given by the Nernst equation

$$E_{\text{cell}} = E^0_{\text{cell}} - \frac{RT}{nF} \ln \frac{a^{\alpha}_{\text{ox}1} a^{\beta}_{\text{red}2}}{a^{\gamma}_{\text{red}1} a^{\delta}_{\text{ox}2}} \qquad (1)$$

where α, β, γ, and δ are the stoichiometric coefficients of the cell reaction

$$\delta \text{ ox}2 + \gamma \text{ red}1 = \beta \text{ red}2 + \alpha \text{ ox}1$$

and F is Faraday's constant, n is the number of electrons involved in the reaction, and a is an activity.

Frequently, the ratio of one redox couple (for instance, $a_{\text{ox}1}/a_{\text{red}1}$) is fixed because of the use of a standard reference electrode. Thus only the activity ratio of the remaining pair is measured.

An ion-sensitive electrode paired with a standard electrode is conveniently used to determine concentration. The best known ion-sensitive electrode is the glass electrode, in which an electrical double layer is formed at the solution-glass interface by a combination of ion-exchange and ion-penetration processes that primarily involve the hydronium ion. Under favorable conditions the resulting potential difference is proportional to the logarithm of the hydronium ion activity

$$E = E^{\circ} + \frac{RT}{nF} \ln a_{\text{H}^+} \qquad (2)$$

The proper linear response to pH is checked by standardization with buffer solutions (see Section E of this exercise).

Conductometry

The bulk impedance of the cell depends on the mobility and concentration of all of the ions present in the electrolyte solution (for theoretical background see Section G of this exercise). Thus information about the number and nature of ions can be obtained only if a few different species (one salt) are contained in the solution. The contributions of the double layer and of faradaic impedance are minimized by the application of high frequencies (1000 c/second) to large, "unpolarizable" electrodes. Their presence, nevertheless, shows up in the observed cell capacitance.

Polarography and Other Forms of Voltammetry

Here the objective is to observe the faradaic current. Its concentration dependence renders it suitable for quantitative determination, while the characteristic potential (half-wave potential) allows qualitative recognition of the electroactive material. Under these circumstances the interfacial impedance of one electrode is negligible; therefore a large unpolarized electrode is frequently combined with a small working electrode (for another method see Section C of this exercise). In addition, the voltage drop across the bulk ohmic resistance must be minimized and that is accomplished by large concentrations of the supporting electrolyte (10^{-1} to 10^{-2} M), compared with the concentration of the electroactive substance (10^{-3} to 10^{-4} M). At the same time the presence of a majority of electrically inactive ions impairs the migration of the electroactive species (the depolarizer) so that they arrive at the electrode by diffusion, which makes theoretical treatment much simpler. The double-layer charging current still remains a problem, which is solved in different ways depending on the actual experimental situation.

Several analytical applications of voltammetry are illustrated in Section C of this exercise, together with suitable electronic circuitry. Several polarographic methods are demonstrated in Section D. The electronics used in these experiments is based on operational amplifiers. Although some of the experiments can be performed alternatively on commercially available apparatus, a much more versatile polarograph is constructed by the student from operational amplifiers. The following experiment offers, therefore, an introduction to operational amplifier circuitry.

B. OPERATIONAL AMPLIFIER

Introduction

Electronic "gadgetry," both simple and sophisticated, is a ubiquitous commodity in the modern chemical research laboratory. A chemist should have some knowledge of electronics and instrument design for a variety of reasons. A thorough familiarity with the instrumentation employed in an experiment is needed to understand the experiment fully and be able to recognize subtle indications of instrument misbehavior or maladjustment. In the purchase of electronic equipment, a proper and intelligent evaluation of manufacturers' specifications often requires that the chemist be at least moderately conversant in electronic matters. Sufficient electronic knowledge to "troubleshoot" at least the simpler causes of instrument breakdown often saves the chemist considerable time relative to that required to effect repair through the manufacturer's representative. Commercial instrumentation, designed usually for routine applications, often proves to be inadequate for basic experimental research that is not routine. The modification of a commercial instrument to particular specifications or the construction of special-purpose instrumentation obviously requires a knowledge of electronics.

Frequently, construction and/or modification of special-purpose instrumentation is the most important contribution to the success of the chemist's research program. It is also the most demanding of the chemist because it constitutes an actual problem in instrument design and usually requires a sophisticated knowledge of electronics.

The purposes of this experiment are (1) to give the student some experience in reading electrical circuit diagrams, wiring electronic circuits, and testing them with the aid of both analog (oscilloscopes, recorders, oscillators) and digital (digital voltohmmeters) test equipment and (2) to acquaint him with the operational amplifier.

The operational amplifier is an extremely versatile device that frequently can be used to advantage in instrument construction and/or modification involving signals of frequency $\leq$ 100 KHz (approximately). Successful and intelligent use of the operational amplifier can be accomplished without detailed knowledge of basic electronics. A large variety of circuits useful in chemical instrumentation can be constructed around the operational amplifier. Indeed, almost any conceivable analog operation (amplification, integration, differentiation, multiplication, etc.) can be effected with operational amplifier circuits. A circuit designed to perform a particular analog function will most likely achieve "state-of-the-art" performance at a relatively low cost, if built with the aid of operational amplifiers. The versatility of this device is appealing because a half-dozen operational amplifiers make available many dozens of electronic operations if the necessary passive circuit elements (resistors, capacitors, etc.) are at hand. This obviates the relatively expensive practice of purchasing a separate "black box" for each electronic function that might be performed in day-to-day research.

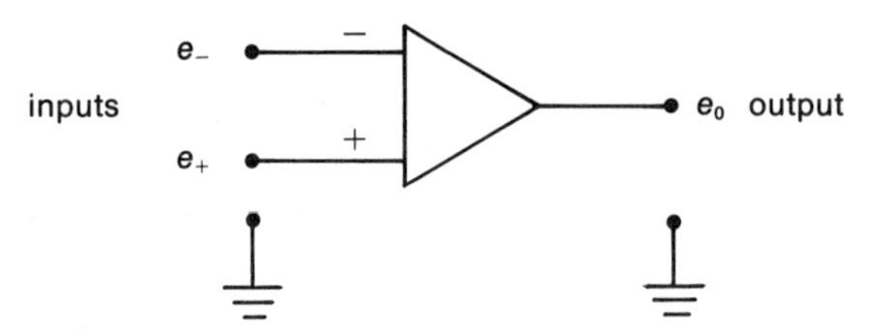

FIGURE 12-1
Operational amplifier.

This experiment consists of constructing and evaluating some basic operational amplifier circuits. The Philbrick/Nexus Model RP Operational Amplifier Manifold, which makes the various amplifier terminals available on a patchboard and allows the fast, convenient, solderless wiring of external components, can be used. The power supply necessary to activate the operational amplifiers ($\pm$15 volts, d.c.) is part of the manifold. Dual-beam oscilloscopes with a scope camera, *x-y* recorders, digital voltohmmeters, a.c. function generators, and d.c. voltage supplies are furnished to support this study.

In a later experiment, the Philbrick/Nexus Manifold will be used to construct a modern polarograph from operational amplifiers.

Properties of Operational Amplifier

For the purposes of this experiment the operational amplifier can be represented by the symbol shown in Figure 12-1, which shows two inputs, one designated by a minus (− input) and one designated by a plus (+ input). The input and output signals are referenced to the ground. The operational amplifier has the following important properties:

1. The response characteristic is given by the relationship

$$e_0 = \alpha(e_+ - e_-) \qquad (1)$$

where α ranges from 10^5 to 10^9 at d.c., depending on the specific amplifier model. In other words, the device is a very high gain, differential amplifier. The response is broad band with α rolling off to unity gain at about 10^6 to 10^8 Hz.

2. The input impedances are very high, typically 10^8 to 10^{12} ohms.
3. The output impedance is low (≤ 100 ohms); the amplifier current output is limited to 1–50 milliamps, although this limit can be extended to several amps by using current boosters.
4. The output voltage is limited to ± 12 volts with most solid-state versions and ± 100 volts with vacuum-tube models and special solid-state types.
5. The voltage offset is very low, typically 3μ volts to 1 millivolt, and the offset drift per hour or per degree is of comparable magnitude.
6. The current offset is very low, typically 10^{-8} to 10^{-14} amps and the current offset drift (per hour or per degree centigrade) is of comparable magnitude.
7. Power requirements are modest; for example, $+15$ volts and -15 volts are required for most solid-state versions.

The properties that are most important in making the operational amplifier unique in electronic signal manipulation are the very high gain and input impedance, which, in combination with negative feedback, yield desirable results. The circuit shown in Figure 12-2 demonstrates this.

The basic relations (based on Ohm's Law, etc.) are:

$$i_1 = \frac{e_i - e_-}{R_1} \tag{2}$$

$$i_f = \frac{e_- - e_0}{R_2} \tag{3}$$

$$i_1 = i_a + i_f \tag{4}$$

$$i_a = \frac{e_-}{Z_i} \tag{5}$$

where Z_i is the input impedance at the $-$input terminal. Equation (1) is also needed. For this case, since $e_+ = 0$,

$$e_0 = -\alpha e_- \tag{6}$$

Combining equations (2) and (6) gives

$$i_1 = \frac{e_i + (e_0/\alpha)}{R_1} \tag{7}$$

Equations (3) and (6) give

$$i_f = -\frac{e_0[1 + (1/\alpha)]}{R_2} \tag{8}$$

Equations (5) and (6) give

$$i_a = -\frac{e_0}{\alpha Z_i} \tag{9}$$

Substituting equations (7)–(9) in equation (4) gives

$$\frac{e_i + (e_0/\alpha)}{R_1} = -\frac{e_0}{\alpha Z_i} - \frac{e_0[1 + (1/\alpha)]}{R_2} \tag{10}$$

This can be rearranged to

$$e_0 = -e_i\left(\frac{R_2}{R_1}\right)\left(\frac{1}{1 + (1/\alpha) + (R_2/\alpha R_1) + (R_2/\alpha Z_i)}\right) \tag{11}$$

Under normal conditions it is apparent that

$$\frac{1}{\alpha} \ll 1 \tag{12}$$

$$\frac{R_2}{\alpha R_1} \ll 1 \tag{13}$$

$$\frac{R_2}{\alpha Z_i} \ll 1 \tag{14}$$

so that

$$e_0 = -e_i\left(\frac{R_2}{R_1}\right) \tag{15}$$

Equation (15) is precisely what would have been obtained by applying the assumptions that the input current to the amplifier (i_a) is zero and that the potential difference between the + and $-$ inputs is zero ($e_- = 0$) at the outset of the analysis of the circuit in Figure 12-2. The high gain and

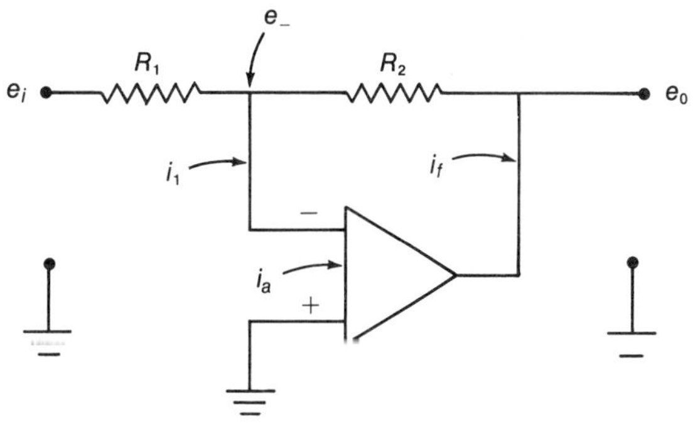

FIGURE 12-2
Inverting amplifier configuration.

input impedance of the operational amplifier validate these assumptions at d.c. and moderate frequencies (up to 5–50 kc) so that they usually can be employed in the analysis of operational amplifier circuits. The most interesting and important aspect of equation (15) is that it shows the circuit response to be independent of amplifier gain and input impedance. The drift in amplifier gain and input impedance with time, temperature, and so forth, is not manifested in the response of the circuit. Accuracy is dependent only on the quality of the external passive elements, R_2 and R_1. If high-quality 0.1% components are employed in this circuit, the gain ($-R_2/R_1$) will be accurate to 0.1%. The same can be said of most operational amplifier circuits operating at d.c. and moderate frequencies. Incidentally, the circuit in Figure 12-2 will provide a precision, variable voltage source if a precise, fixed voltage is available for the input and one of the resistors is variable (e.g., for an initial voltage source in polarography).

Procedure

A variety of signal sources (function generators, d.c. power supplies) and signal measuring devices (oscilloscopes, voltmeters, etc.) are available for testing the operational amplifier circuitry. Their operation is relatively simple. The laboratory instructor will demonstrate their operation.

The maximum output voltage and current of the operational amplifiers to be used are approximately 12–14 volts and 1–2 milliamps, respectively. Make certain the expected output signals (calculated from theoretical amplifier response equations [see equation (15)] do not exceed these limits or the circuits will not function properly.

Inverting Amplifier

The circuit shown in Figure 12-2 is often referred to as an inverting amplifier. The term "inverting" refers to the inversion of signal polarity performed by the amplifier. Using a Philbrick P35 amplifier, construct this circuit with gains (R_2/R_1) of 0.100 and 10.0. Test its behavior with d.c. and a.c. input signals, carefully observing stability and accuracy. Examine the a.c. response as a function of frequency, increasing the frequency until deviations from the expected response are observed.

A simple modification of the inverting amplifier is shown in Figure 12-3. The response of this circuit is given by the relation

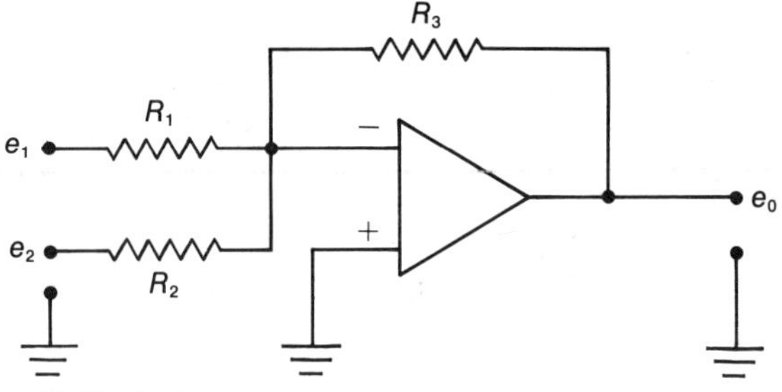

FIGURE 12-3
Inverting amplifier.

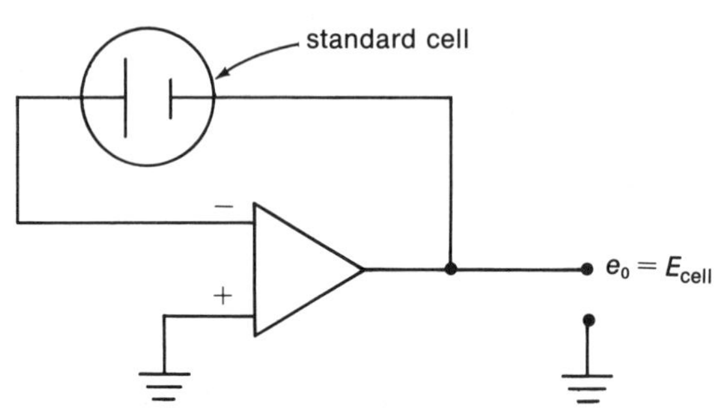

FIGURE 12-4
Precision standard-voltage cell.

$$e_0 = -\left(\frac{R_3}{R_1}e_1 + \frac{R_3}{R_2}e_2\right) \tag{16}$$

Prove equation (16) to your satisfaction, using the simplified approach outlined previously (assume $i_a = 0$ and $e_+ = e_-$). This circuit adds voltages with the inversion of polarity. The coefficients in the addition operation can be made unity by setting $R_3 = R_2 = R_1$. You are not required to test this circuit.

Precision Standard-Voltage Source

The circuit shown in Figure 12-4 delivers a precision d.c. voltage equal to the voltage of the standard cell. The operational amplifier furnishes the current necessary to drive electronic devices, connected to its output, without drawing current from the standard cell. Construct and test this circuit, using the digital voltmeter to measure the d.c. output of the operational amplifier. Use a model P2-A amplifier. Can you explain the response of this circuit? If not, consult your instructor.

Integrator

Employing a condenser as the feedback impedance as shown in Figure 12-5 yields a circuit whose out-

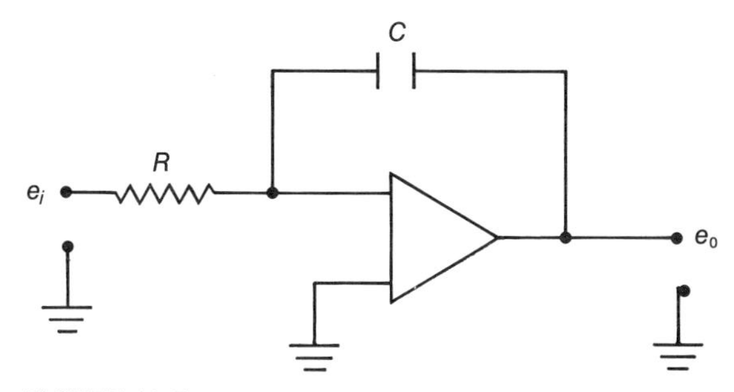

FIGURE 12-5
Integrator.

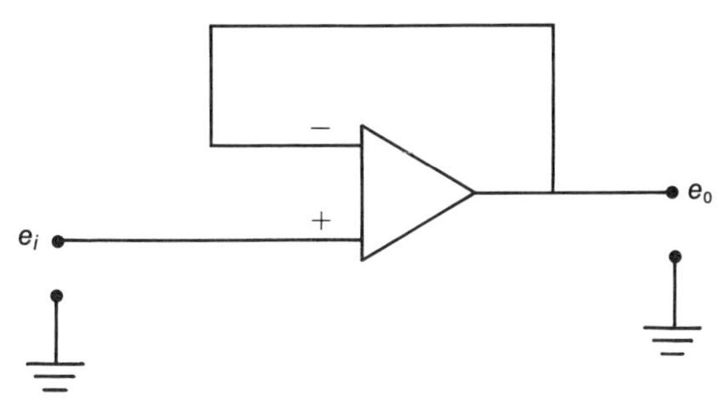

FIGURE 12-6
Follower amplifier.

put is proportional to the time integral of the input voltage. This can be seen by recalling the relations

$$e = \frac{q}{c} = \frac{1}{c} \int idt \qquad (17)$$

where q is the instantaneous charge on the condensor, i the current flowing across the condenser, c the capacitance, and t is time. For the circuit in Figure 12-5, the current across the condenser is equal to the current through the input resistor, the latter being given by equation (2) where $e_- = 0$. Thus,

$$e_0 = -\frac{1}{RC} \int e_i dt \qquad (18)$$

Construct the integrator as in Figure 12-5, using a model P-2A amplifier. Test for the special case where e_i = constant. For this situation

$$e_0 = -\frac{e_i t}{RC} \qquad (19)$$

that is, the output varies linearly with time. Employ a strip-chart recorder to examine the output. Assess whether the observed output is linear as expected and if the observed slope is equal to the calculated slope, e_i/RC. Perform this test at two different values of e_i/RC. Make certain the rate of change of voltage does not exceed the response limit of the recorder.

Test the response of the integrator to a.c. signals. If

$$e_i = A \sin \omega t \qquad (20)$$

then

$$e_0 = \frac{A}{\omega RC} \cos \omega t \qquad (21)$$

Verify equation (21) with respect to both amplitude and phase relationships.

The integrator with a constant input voltage constitutes a "voltage ramp" generator. The linearly increasing output voltage is useful in electrochemistry (e.g., in polarography), as a "clock" in timing circuits and in a variety of other applications. Of course, integration of electrical signals has many uses in chemical instrumentation (e.g., in coulometry to determine the number of coulombs, to determine the area under gas chromatographic peaks for quantitative analysis, etc.)

By switching the positions of the resistor and condensor in Figure 12-5, a differentiator is obtained. You will not be asked to construct and examine this circuit, but take the time to prove through simple mathematical considerations that such a circuit does yield an output that is the time-derivative of the input signal.

Follower Amplifier

Frequently, in chemical instrumentation signal sources with high output impedances are encountered; that is, they can furnish only very small output currents, which are often insufficient to drive various signal conditioning and measuring devices (amplifiers, integrators, recorders, etc.). Common examples are the outputs of photomultiplier tubes and electrolytic cells employing the glass electrode. To make use of such signal sources, a follower amplifier, often called a "buffer amplifier," is frequently employed. This device is characterized by a very high input impedance ($\sim 10^{10}$ ohms or larger) and a low output impedance. In other words, it can respond to a signal source without drawing significant current and can furnish at its output substantial currents to drive other electronic circuitry. The voltage gain of this device is often made to be precisely unity.

A unity-gain follower can be constructed conveniently from an operational amplifier employing the circuit shown in Figure 12-6. Because the

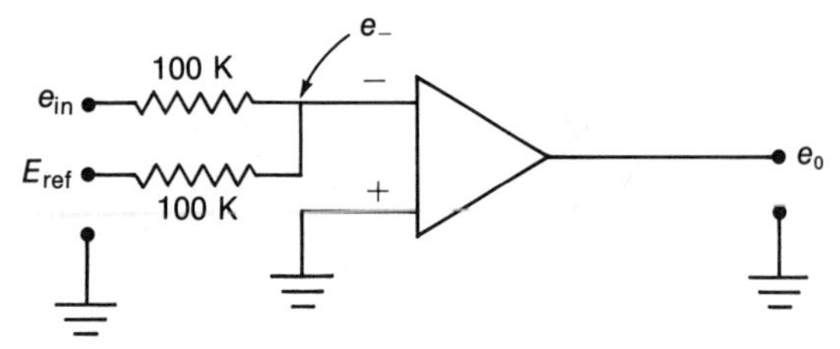

FIGURE 12-7
Trigger or voltage comparator: $e_0 = +12$ for $e_{in} < -E_{ref}$; $e_0 = -12$ for $e_{in} > -E_{ref}$.

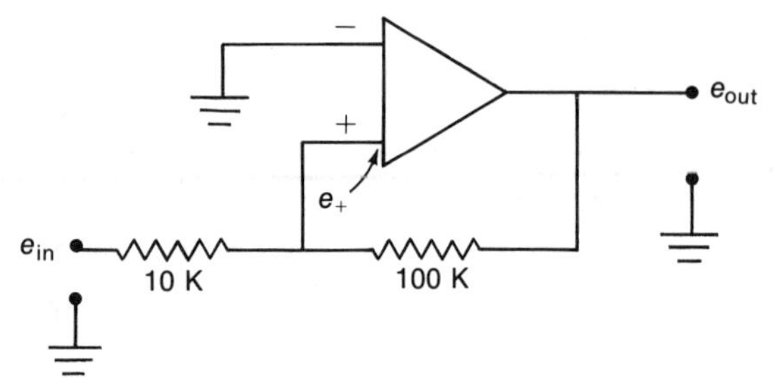

FIGURE 12-8
Trigger with hysteresis. Test this circuit using a P-45 amplifier.

operational amplifier maintains the potential difference between its inputs at zero volts, it is obvious that

$$e_i = e_0 \tag{22}$$

(note that this circuit does not invert the polarity of the signal).

Construct and test this circuit using both d.c. and a.c. signals. Increase the frequency until significant deviations from the expected response are observed. Insert resistors between the input signal source and the follower input and repeat the above measurements (use 1 K, 10 K, 100 K, and 1 meg resistors). Does this have any influence on the output? Explain. Use the P-35 amplifier.

Trigger or Voltage Comparator

The simple circuit shown in Figure 12-7 is useful as a "trigger" (voltage comparator). Because there is no negative feedback loop, the amplifier is unable to "null" the potential at the amplifier input (i.e., make $e_- = e_+ = 0$) so that the open loop gain manifests itself in a more direct manner in this circuit. Assuming that the amplifier draws negligible current, simple Ohm's Law considerations indicate that

$$e_- = \frac{e_{in} + E_{ref}}{2} \tag{23}$$

Recalling equation (1), the following is obtained:

$$e_0 = -\frac{\alpha}{2}(e_{in} + E_{ref}) \tag{24}$$

Because α is so large, an essentially infinitesimal difference between e_{in} and $-E_{ref}$ will cause the amplifier output to "limit," that is, to reach the maximum output voltage attainable for the amplifier in question. The amplifiers you will be using limit at approximately ± 12 volts (the sign depends on the sign of the input voltage; see the relations in Figure 12-7). This limiting voltage will vary somewhat from amplifier to amplifier and should be determined by the student. It will be designated as $\pm E_L$ in the remainder of this discussion. The trigger circuit simply indicates whether a voltage is above or below a certain predetermined level. Among the applications are the conversion of a sine wave to a square wave, the construction of timing circuits of various types, and the construction of safety devices. In the last application a voltage, e_{in}, exceeding a "safe" level, $-E_{ref}$, will cause the trigger to "change state" (i.e., e_0 goes from $+E_L$ volts to $-E_L$ volts), activating an alarm, a relay, and so forth.

Test the circuit shown in Figure 12-7 using a Model P-35 or P-45 amplifier. How precisely can the switching point be located? This test can be effected by using a low-frequency triangular wave signal from the function generator. Keep in mind that because of the lack of negative feedback, nonideality in amplifier characteristics is not suppressed in this application. Thus, the results will be less ideal than in previous examples. Do your results support this statement? How? Here is an opportunity to use the scope camera.

Trigger with Hysteresis

Figure 12-8 shows an interesting variation of a trigger circuit. As in the previous example, the lack of a negative feedback loop yields a circuit that can exist in one of two states, $+E_L$ or $-E_L$ volts. The addition of positive feedback (the 100K resistor) produces a hysteresis effect, that is, the value of e_{in} at which the circuit changes state depends on whether the change in state occurs as e_{in} is increasing or decreasing (i.e., on whether e_0 is $+E_L$ or $-E_L$ volts). This can be seen by the fact that the potential at the + input of the amplifier obeys the relationship

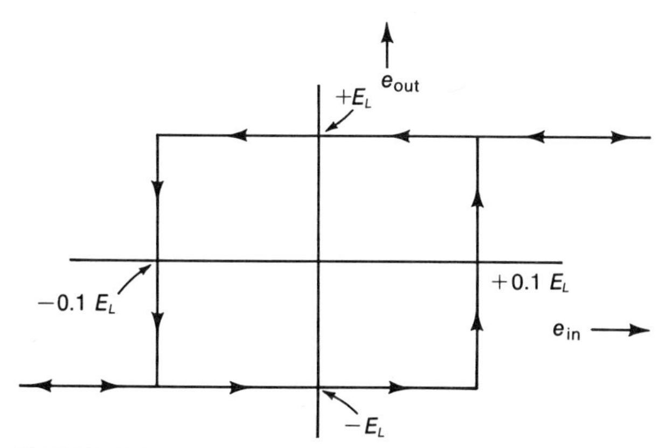

FIGURE 12-9
Hysteresis loop generated by the circuit shown in Figure 12-8.

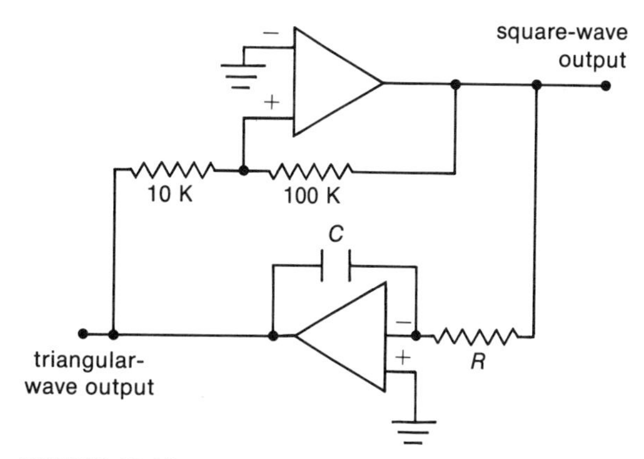

FIGURE 12-10
Triangular-wave–square-wave oscillator.

$$e_+ = \frac{10e_{\text{in}} + e_{\text{out}}}{11} \tag{25}$$

so that

$$e_{\text{out}} = \alpha\left(\frac{10e_{\text{in}} + e_{\text{out}}}{11}\right) \tag{26}$$

Thus, the point at which the circuit changes state (when $10e_{\text{in}} = -e_{\text{out}}$) depends on the value of e_{out}. For example, if e_{out} is $+E_L$ volts, as long as e_{in} is greater than $-0.1E_L$ volts no change in state will occur. However, when e_{in} becomes a few microvolts smaller than $-0.1E_L$ volts, the potential, e_+, will become less than zero and the circuit will change state to $e_{\text{out}} = -E_L$ volts. By the same reasoning, to change the state of the circuit back to $+E_L$ volts, e_{in} must exceed $+0.1E_L$ volts (*not* $-0.1E_L$ volts). Thus, the response of this circuit describes a hysteresis loop as shown in Figure 12-9.

Triangular-Wave – Square-Wave Oscillator

A combination of the integrator (Figure 12-5) and the trigger (Figure 12-8) produces an oscillator giving both triangular- and square-wave outputs. The circuit is shown in Figure 12-10.

This circuit operates in the following way. If the trigger output is $+E_L$ volts, the integrator gives a voltage ramp scanning in a negative direction. This continues until the integrator output decreases to $-0.1E_L$ volts. At this point the trigger changes state to $-E_L$ volts. The integrator output then proceeds to scan in a positive direction until it reaches $+0.1E_L$ volts where the trigger changes state to $+E_L$ volts, and the process repeats itself, ad infinitum. The frequency of the oscillation is controlled by the components R and C. Why? How would you control the amplitude of the triangular wave?

Test this circuit using a model P-45 amplifier as the trigger and a model P-35 as the integrator. Test the response with two significantly different values of R and C. Do the wave form and frequency of the oscillations correspond to expectations? Use the scope camera to obtain a permanent record of wave forms.

Other Circuits

Operational amplifier handbooks are given in the References. They contain diagrams of numerous operational amplifier-based circuits. Refer to Reilly (1962), the *Handbook of Operational Amplifier Amplifications* (1963), and *Applications Manual for Computing Amplifiers* (1966), and try to test one or two of the circuits described therein if time permits. Report the results, together with the data on the "required" circuits.

C. CURRENT-VOLTAGE CURVES AND ELECTROMETRIC TITRATIONS

In this experiment a high quality polarograph will be constructed from operational amplifiers. The instrument is designed for use with a three-electrode polarographic cell. The combination of operational amplifier potential control circuitry and the three-electrode cell provides automatic compensation for most of the IR voltage losses in the electrolytic solution and external circuitry. The polarograph will be employed to obtain current-voltage curves at various stages in the titration

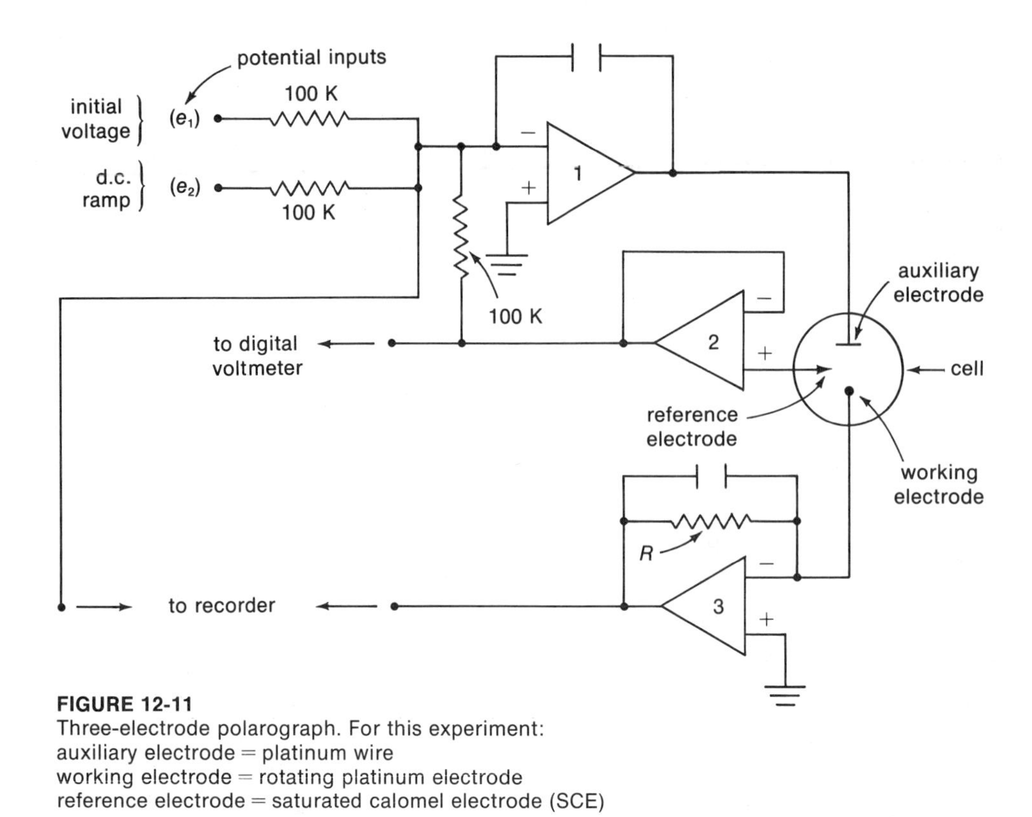

FIGURE 12-11
Three-electrode polarograph. For this experiment:
auxiliary electrode = platinum wire
working electrode = rotating platinum electrode
reference electrode = saturated calomel electrode (SCE)

of iodide ion in hydrochloric acid solution with bromate. These current-voltage curves are useful in the selection of electrometric end points and for the interpretation of the titration curves. The form of titration curves under selected conditions will be predicted, and for comparison amperometric titration curves will be obtained with the polarograph. An operational amplifier will be used to measure the cell potential (as in Section B of this exercise) in obtaining potentiometric titration curves.

Construction of the Polarograph

The three-electrode cell is composed of a rotating platinum working electrode, a platinum wire as an auxiliary electrode, and a standard calomel (SCE) reference electrode. The auxiliary and the reference electrode are variable, while the working electrode is always at instrument ground. The reference electrode does not draw any of the current that flows between the auxiliary and the working electrode.

Since the potential is measured between the reference and the working electrode, only that part of the current which flows through the impedance between the tip of the reference electrode and the working electrode can disturb the potential control process. Effects of the major part of the bulk ohmic resistance are negated if the working electrode and reference electrode are placed next to each other in the solution. The interfacial impedance of the auxiliary electrode is also excluded and the reference electrode does not contribute since it does not experience any current. These concepts are developed in detail in the discussion that follows. A rotating electrode is employed to provide effective stirring of the solution, which decreases the dimensions of the diffusion layer around the working electrode where the solution is depleted of the electroactive ion. Accordingly, the magnitude of the diffusion current increases and the double-layer charging current contributes proportionally less to the observed value. Favorable conditions for voltammetric investigations are thus established (see Section A of this exercise).

The circuit diagram for the polarograph is given in Figure 12-11. The operation of the circuit in this figure can be best understood by bearing in mind that the operational amplifier reacts in the manner required to maintain zero potential difference between its inputs. Thus, the stable state for the loop consisting of amplifiers 1 and 2 corresponds to the output of amplifier 2 being equal in magnitude, but opposite in polarity, to the sum of the input voltages, e_1 and e_2. Since amplifier 2 is in the follower configuration, its input will also equal $-(e_1 + e_2)$ relative to ground potential. Amplifier 3 serves to maintain the working electrode at ground potential (same potential as + input) and provides an output proportional to the polarographic current. Because the reference electrode is maintained at $-(e_1 + e_2)$ volts and the working electrode at zero volts, it is apparent that the potential of the working electrode relative to the reference electrode is maintained at $(e_1 + e_2)$ volts, that is, equal to the sum of the input potentials. To achieve this stable state when the potentials are applied, amplifier 1 emits a voltage leading to current flow between the auxiliary and working electrode. This current polarizes the working electrode until its potential relative to the reference electrode is $(e_1 + e_2)$. A current flow is maintained that is sufficient to keep the working electrode at the stable state potential. The time required to achieve the stable state upon application of the potential input (the transient response) is less than a millisecond. In other words, the control of d.c. potential is essentially instantaneous. Amplifier 3 maintains the working electrode at ground potential by emitting an output voltage equal to the cell current times the feedback resistance R, thus furnishing a signal proportional to the cell current.

Note that this circuit effects potential control without current flow through the reference electrode. This is very important for two reasons. First, problems associated with polarization of the reference electrode due to current flow and the attendant drift in reference electrode potential are negated. Second, some major sources of ohmic resistance in the current path are eliminated. With current flow a potential drop (iR drop) is associated with solution ohmic resistances. This causes an error in potential control because the effective applied potential differs from the external applied potential by the iR drop. Numerical correction for this effect can be accomplished, but it is tedious. A major source of ohmic resistance in two-electrode polarographic setups is the frit or asbestos fiber separating the reference and working electrode compartments. This ohmic resistance cannot contribute with the three-electrode system employed in this experiment because no current flows across it. For this reason the relatively convenient, leak-free, commercial calomel reference electrode used for this experiment will work fine with the circuit of Figure 12-11, whereas it would render inoperative a conventional two-electrode polarograph because of the high impedance of the asbestos fiber used to make contact with the external solution. The only sources of iR drop that are operative with the three-electrode cell are the solution resistance between the tip of the reference electrode and the working electrode and the resistance in the working electrode itself. These contribute a negligible iR drop in this experiment. If a significant iR drop did result from these sources (e.g., as when a high-resistance organic solvent is employed), a minor modification of the circuit in Figure 12-11 could be introduced to effect compensation. Can you envision how this would be done?

Construct the circuit shown in Figure 12-11 using a 10K ohm resistor as the feedback impedance of amplifier 3. This is a tentative value to use as a "starter." If deemed advisable later, the sensitivity (magnitude of output per unit of cell current) of the circuit may be altered by changing this resistor to a more appropriate value. How can the initial d.c. voltage source be constructed from an operational amplifier? An operational amplifier integrator circuit with a constant input voltage is the most convenient source of the d.c. ramp generator. Such a device was constructed in Section B of this exercise. Select components to give a scan rate of approximately 0.10 volt/minute.

Because only d.c. signals will be of interest, it is suggested that you "damp" the circuit response by placing 1 μF capacitors across amplifiers 1 and 3. What effect does this have and why? Apply power to the amplifiers with the working electrode disconnected and all electrodes immersed in the polarographic solution. Connect the working electrode after you are certain the amplifiers are behaving properly. Consult your instructor for the initial test of your circuit.

Current-Voltage Curves

A series of current-voltage curves taken at various stages during the course of a titration provides a complete picture of the relationships between current, voltage, and volume of titrant added for the particular electrode system used. These relationships permit the selection of optimum conditions for the electrometric location of the equivalent point in the titration and the prediction of the general features of the titration curve.

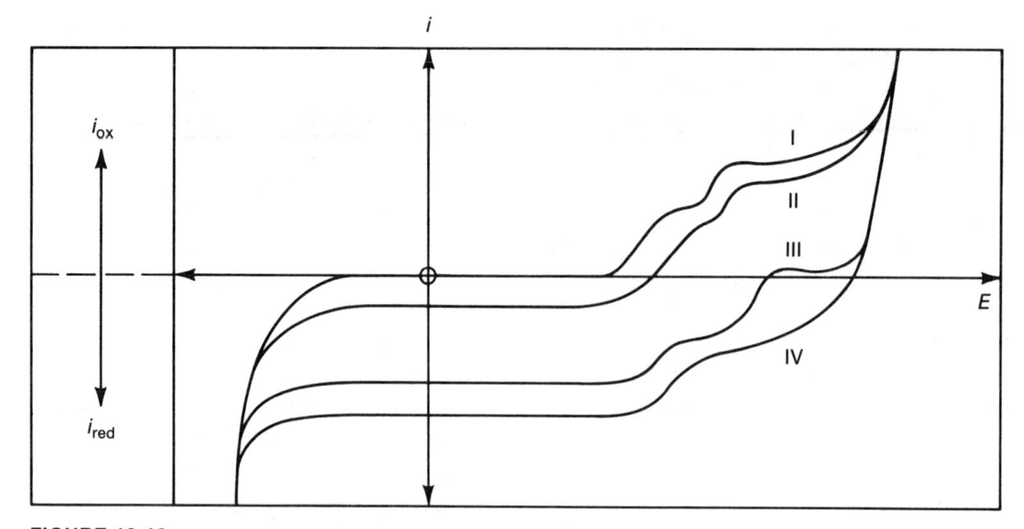

FIGURE 12-12
Current-voltage curves at different stages of titration: curve I is recorded before any titrant is added; curve IV results from titration beyond the equivalence point.

The system used in this experiment is one involving the titration of iodide ion in hydrochloric acid solution with bromate. The titrate solution also contains a small amount of bromide ion. The important reactions involved are:

$$BrO_3^- + 5Br^- + 6H^+ \longrightarrow 3Br_2 + 3H_2O$$

$$Br_2 + 2I^- \longrightarrow I_2 + 2Br^-$$

$$I_2 + Br_2 + 2Br^- \longrightarrow 2IBr_2^-$$

Basically, the titration consists of the oxidation of the iodide ion, first to iodine and then to I^+ by bromine.

The most desirable electrometric end point indication for a certain system must be chosen from the following titration methods (choose the one that produces the sharpest change in the measured property—potential or current—at the equivalence point):

1. Potentiometry with zero impressed current
2a. Potentiometry with constant impressed current and only one polarizable electrode
2b. Potentiometry employing two polarized electrodes at constant impressed current
3a. Amperometric titration with one polarized electrode at constant potential
3b. Anperometric titration with two polarized electrodes at constant potential

The range in which current voltage curves are obtained in this experiment is situated between the potential at which H^+ is reduced and that at which Br^- is oxidized. No limiting current is obtained for these two species, because their concentration is too high. The diffusion currents to be observed can be interpreted by taking into account that the half-wave potential is characteristic of the nature of the depolarizing ion, whereas the magnitude of the diffusion current is proportional to the bulk concentration of the depolarizing ion [see Delahay (1954)]. Since the only oxidizable ion contained in the sample before the titrant is added is I^-, the observed oxidation waves in the first current-voltage curve must originate from the oxidation of I^-, first to $I^0(I_2)$, and then to $I^+(IBr_2^-)$. The Cl^- ion is not oxidized at the applied potentials. Since the amount of oxidized material produced at the electrode is negligible, the first reduction waves (I^+–I–I^-) can be observed only after the titrant has been added. As the equivalence point is reached, the I^- oxidation waves cease to exist and as BrO_3^- is added beyond the equivalence point, free Br_2 appears and is characterized by its reduction wave. Figure 12-12 is a schematic presentation of current-voltage curves similar to the ones you will observe. Keep in mind, however, that in an actual experiment a small residual current is observed at regions where no limiting currents are present.

To choose the optimal end-point method, it is necessary to predict the form of the titration curve for methods 1–3b given previously. These predictions will be checked later by the actual performance of the titrations. A discussion on how to obtain titration curves from current-voltage curves follows.

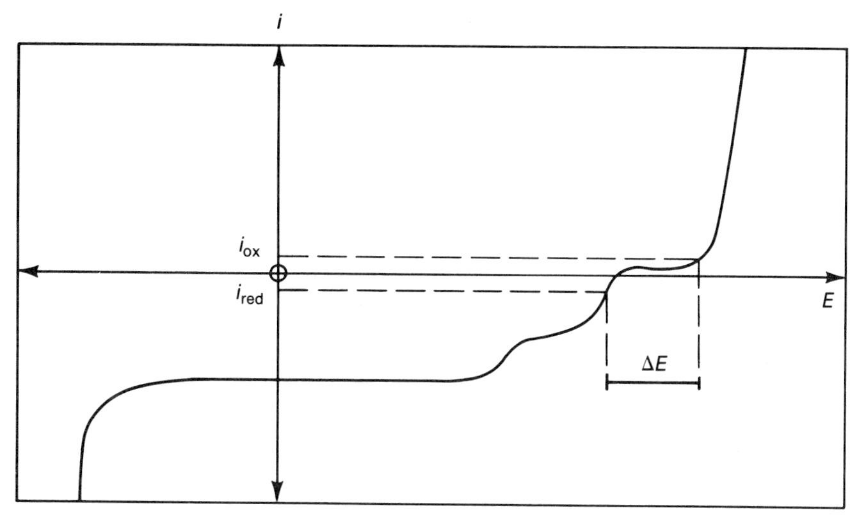

FIGURE 12-13
Determination of observed current in amperometric titration with two polarized electrodes. ΔE is the constant potential between electrodes.

1. *Potentiometric titration with zero impressed current.* In this case the potential difference between electrodes is observed under conditions where current flows. Thus the cell potential is found at those points at which the current-voltage curves cross the voltage axis. If these potentials are plotted against the quantity of titrant added until the respective current-voltage curve was recorded, the predicted titration curve is obtained.

2a. *Potentiometric titration with constant impressed current and one polarized electrode.* Since the current-voltage curves were obtained with one polarized elctrode, those points at which the curves pass through the desired current value are selected. The potentials at these points are plotted against the quantity of titrant added.

2b. *Potentiometric titration with constant impressed current and two polarized electrodes.* The desired amount of current passes into the solution from one electrode and is drawn from the solution (in the same amount) by the other electrode. The point on the current-voltage curve at which the electrode drew the current from the solution and the one at which it passed the same amount back to the solution must be found. The distance on the voltage axis between these two points yields the potential measured between two polarized electrodes on which the selected amount of current is impressed. A plot of this potential against the quantity of titrant added yields the titration curve.

3a. *Amperometric titration with one polarized electrode.* In this case the current values that correspond to the selected voltage are read from the current-voltage curves. Titration curves are obtained in the same way as before.

3b. *Amperometric titration with two polarized electrodes.* This situation is similar to that in 2b. The potential difference remains at a preselected value, ΔE, while one electrode draws the same amount of current from solution as the other one adds. To determine the amount of current, two points must be found on the current-voltage curve that are situated the potential difference ΔE apart, but possess the same current value (oxidizing on one point, reducing on the other). See Figure 12-13 for a clarification of the procedure. The titration curve is obtained by plotting the current against the quantity of titrant added until the curve was recorded.

Apparatus

Philbrick RP Manifold, operational amplifiers, and circuit elements required to construct the circuit in Figure 11-20
X-Y recorder
Digital voltohmmeter
Oscilloscope
Electrodes and synchronous motor for rotating platinum electrode

Chemicals

2 *M* hydrochloric acid solution (dilute 165 ml of the concentrated reagent to 1 liter)

Standard 0.010 *M* potassium iodide solution (prepare a fresh solution by dissolving 1.6 g of the reagent salt in water and diluting to 1 liter)
Standard 0.0050 *M* potassium bromate solution (dissolve 0.84 g of the reagent salt in water and dilute to one liter)
0.10 *M* sodium bromide (dissolve 1.0 g of the reagent salt in 100 ml of water)

Procedure

Construct a polarograph according to the instructions given at the beginning of this section. Employ a rotating platinum electrode as the working electrode, a saturated calomel as the reference, and a platinum wire as the auxiliary electrode. The platinum working electrode should be pretreated before use. The objective is to obtain a reproducible, but not necessarily known, surface state. A simple and usually adequate approach for this experiment is to evolve hydrogen gas from the platinum wire for a few minutes. The polarograph may be used for this purpose, using the hydrochloric acid solution as a hydrogen ion source. This procedure is necessary whenever the platinum electrode has been allowed to stand in contact with air for a few hours, or longer.

Transfer 200 ml of the hydrochloric acid solution and 10 ml of the sodium bromide solution to a 400-ml beaker. Add 10 ml of the standard iodide solution and mix thoroughly. Remove dissolved oxygen from the solution by bubbling nitrogen through it for 10–15 minutes. After this "degassing" process is finished, a stream of nitrogen should be passed over the surface of the solution for the duration of the experiment.

Set the y-axis (pen) zero of the recorder to the center of the chart. Adjust the y-axis sensitivity so that the current signal of about +0.8 volts causes a deflection of 3–4 inches. The x-axis (carriage) of the recorder may be controlled by either the follower or ramp generator outputs. Adjust the x-axis sensitivity and zero so that applied potentials between about −0.2 volts (H_2 evolution) and +1.2 volts (Br_2 evolution) remain on scale.

Record the current-voltage curve of the initial solution starting at about +1.2 volts (i.e., where the Br_2 evolution current drives the pen off scale) and ending at about −0.2 volts (where the H_2 evolution current drives the pen off scale). Add 1 ml of standard bromate solution to the cell, mix thoroughly, and record the current-voltage curve again. Repeat until a total of 9 ml of the bromate have been added.

Predictions

From the recorded current-voltage curves, predict the general shapes of the voltammetric titration curves that would be obtained in each of the following:

1. Potentiometric titration, platinum-calomel electrodes, no impressed current
2. Potentiometric titration, platinum-calomel electrodes, 1 μA of impressed current
3. Potentiometric titration, two platinum electrodes, 1 μA of impressed current
4. Amperometric titration, platinum-calomel electrode, platinum +0.75 volts versus calomel
5. Amperometric titration, two platinum electrodes, 0.1 volt applied

Electrometric Titrations

In addition to the chemicals and apparatus required for obtaining current-voltage curves, an automatic buret[1] with a capacity of 10 milliliters is needed to obtain titration curves. Note the delivery rate of your automatic buret and the time required to deliver the full 10 milliliters. Adjust the internal time base of your x-y recorder so that it will drive the x axis full-scale in about the amount of time required for the buret to deliver 10 milliliters. It is recommended that you calibrate the internal time-base drive of the x-y recorder in terms of the volume of titrant delivered by the automatic buret. How is this done?

Prepare an initial solution of hydrochloric acid, sodium bromide, and standard iodide for titration as before and titrate with the potassium bromate solution, using one of the methods listed under Predictions. Initiate the titration by starting the automatic buret and the x-y recorder sweep as simultaneously as possible. The recorder y axis should be connected to the cell current or the potential signal, whichever is appropriate. Repeat this process, using one or two of the other methods listed. How do the results compare with expectations based on examination of the current-voltage curves?

The polarograph is used for amperometric titrations and the circuits shown in Figures 12-14 to

[1] A buret delivering at a constant rate is described by J. M. G. Baredo and J. K. Taylor in *Trans. Electrochem. Soc.* **92**, 437 (1947).

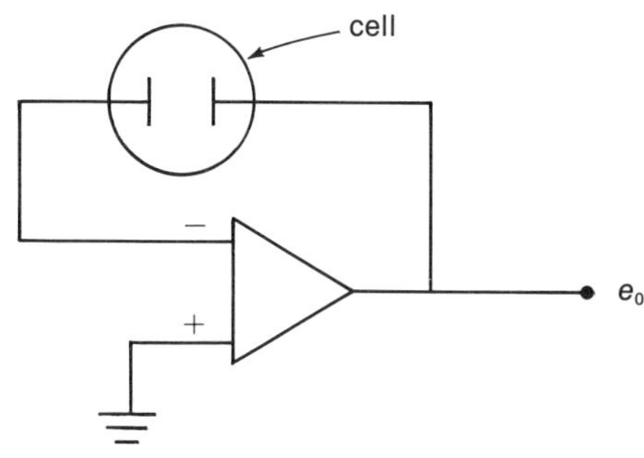

FIGURE 12-14
Circuit for potentiometric titrations with zero impressed current: e_0 = cell potential.

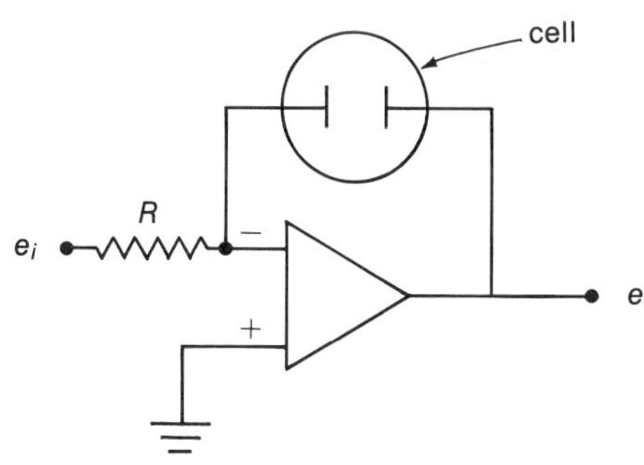

FIGURE 12-15
Circuit for potentiometric titrations with constant impressed current and two polarizable electrodes: e_0 = cell potential + iR; e_i/R = impressed current.

12-16 are used for potentiometric titrations. The circuit in Figure 12-14 is to be constructed for potentiometric titrations with no impressed current. This circuit was examined in Section B of this exercise [refer to Reilly (1962) for further discussion]. The circuit in Figure 12-15 is used for potentiometric titrations with impressed current, provided a high ohmic resistance is not operative, and is ideal for the third method of titration but inappropriate for the second method because of the high resistance of the calomel electrode. The circuit in Figure 12-16 is designed for potentiometric titrations with impressed current where a reference electrode of high resistance is employed. Be prepared to discuss how this circuit works. However, note that the potential output indicated in Figure 12-16 is inappropriate for a potentiometric titration with two polarizable electrodes (method 3). Why?

Construct and use as many of the circuits shown in Figures 12-14 to 12-16 as possible. Compare predicted titration curves with those actually obtained. Select the optimal electrometric titration method for this chemical system.

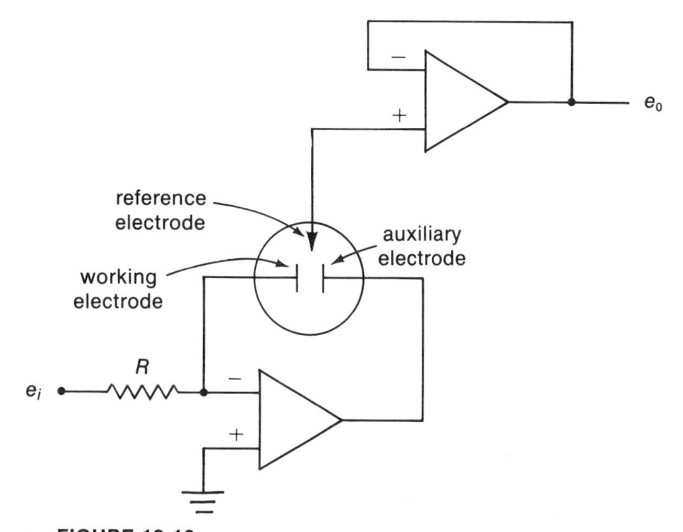

FIGURE 12-16
Circuit for potentiometric titrations with constant impressed current and one polarized electrode: e_0 = potential of reference electrode versus working electrode; e_i/R = impressed current.

D. CHARACTERIZATION OF ELECTRODE REACTION MECHANISMS BY DIRECT-CURRENT POLAROGRAPHY AND CYCLIC VOLTAMMETRY

In this experiment the polarograph constructed from operational amplifiers is applied to d.c. polarographic and cyclic voltammetric measurements using the dropping mercury electrode as the working electrode. Three important basic types of electrode processes (mechanistic schemes) will be examined: (1) the diffusion-controlled or "reversible" process, (2) the process controlled by the rate of chemical reoxidation of the product of electrolytic reduction (the "catalytic" process), and (3) the process controlled by the rate of a chemical reaction preceding the electrochemical charge transfer step (the "kinetic" process). The electrolytic reduction of Ti(IV) to Ti(III) at a dropping mercury electrode is employed. Each of the foregoing types of processes can be obtained with the titanium system through the proper choice of experimental conditions.

Polarograph

The circuit diagram of the polarograph is presented in Figure 12-11. Its construction and behavior are explained in Section C of this exercise. This experiment differs from that in Section C in that

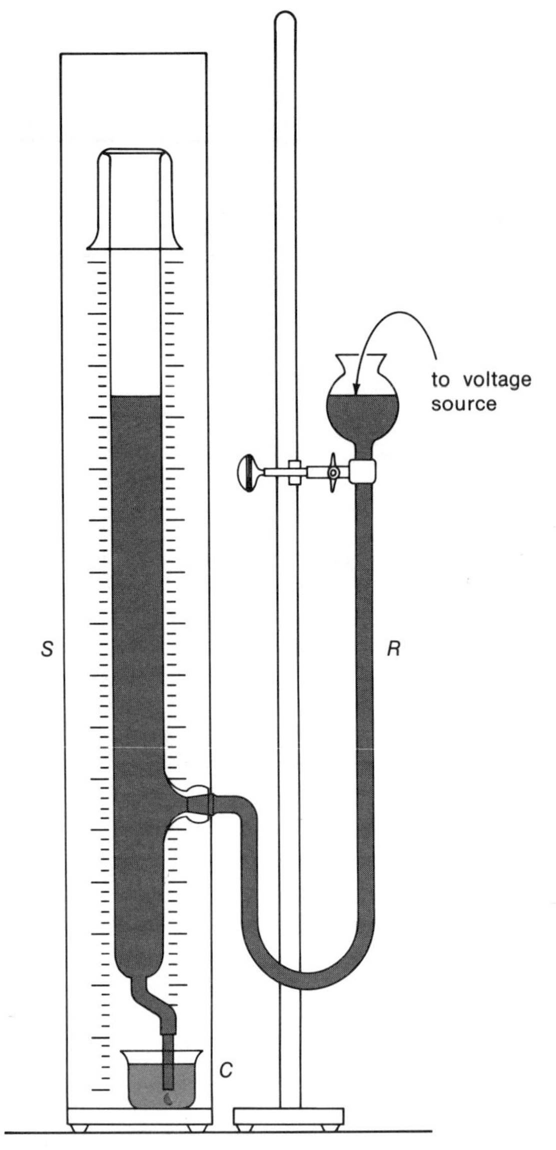

FIGURE 12-17
Apparatus providing constant flow rate for a dropping mercury electrode: S, scale for measuring mercury column height; R, mercury reservoir; C, capillary.

the working electrode in this case is a dropping mercury electrode (DME). Figure 12-17 shows the required apparatus.

As in the previous application, the output of amplifier 3 (cell-current signal) is applied to the y axis of the readout device, which is an x-y recorder in polarography and an oscilloscope in cyclic voltammetry. The x axis of the readout device is driven by the d.c. ramp in polarography and the triangular wave signal in cyclic voltammetry. The experiment described here has been carried out with a Hewlett-Packard Function Generator as the triangular wave signal source.

Chemicals

Oxalic acid dihydrate ($H_2C_2O_4 \cdot 2H_2O$)
Potassium titanium oxalate [$K_2TiO(C_2O_4)_2 \cdot 2H_2O$]
Potassium chlorate ($KClO_3$)
Potassium sulfate (K_2SO_4)

Prepare the following solutions:

1. 0.20 M $H_2C_2O_4$ + 1.00×10^{-3} M Ti(IV)
2. 0.20 M $H_2C_2O_4$ + 1.00×10^{-4} M Ti(IV) + 0.040 M $KClO_3$
3. 0.20 M $H_2C_2O_4$ + 1.00×10^{-3} M Ti(IV) + 0.50 M K_2SO_4

Note: These solutions need not be prepared in precisely these concentrations (±10% is adequate), but their concentrations must be known to the indicated accuracy.

Procedure

Direct-Current Polarography

For each of the solutions, record polarograms with at least three different values of mercury column height (corresponding to drop lives ranging from about 3 to 10 seconds). Use the same set of column height with each solution. A d.c. scan rate of about 0.10 volt per minute is preferred. Initiate the scan with an initial voltage of zero. A recorder x-axis sensitivity of about 50 mv per inch is recommended. Use your own judgment regarding y-axis sensitivity. Record in your notebook all important experimental parameters, such as mercury column height, mercury drop life, potential scan rate, recorder settings, value of feedback resistor R, and so forth. The mercury drop life at potentials in the vicinity of the polarographic wave should be known. Devise a method for obtaining this parameter to reasonable accuracy (±2%).

Cyclic Voltammetry

Obtain cyclic voltammograms for each of the solutions, using triangular wave frequencies of 1, 2, 5, and 10 cycles/second, an amplitude of about 1 volt peak-to-peak, and a fixed mercury column height. The mercury column height should be relatively low to insure a long drop life (10–14 seconds). Adjust the initial potential source so that the rising portion of the cyclic voltammetric wave is observed in approximately the middle of the pattern. Use your judgment in adjusting the oscilloscope x and y sensitivities. Obtain permanent recordings of the cyclic voltammogram by photographing the oscilloscope trace over the life of one mercury drop (time exposure). Consult the instructor for appropriate camera settings. Again, record all relevant experimental data and conditions.

Theory and Calculations

In polarography the current flowing across a polarographic cell is measured as a function of potential between the working and the reference electrode (as in the observation of current-voltage curves). The great advantage in using a DME lies in the fact that each new drop exhibits a fresh electrode surface in exactly the same state as that of the preceding drop. Thus the DME yields very reproducible results. A further convenience is the large overvoltage that H^+ displays on a mercury electrode, which allows a larger voltage range to be scanned in solvents containing hydronium ions.

The voltage drop should be slow enough so that during one drop life the potential remains effectively constant. Thus each drop may be considered a new potentiostatic (constant potential) experiment. It is assumed that in falling each drop stirs the solution enough to provide the same initial concentration conditions as those experienced by the preceding drop. Figure 12-18 shows the current-time curve recorded during one drop life. The highest observed current (at the end of the drop life) is considered to be the relevant current at that potential setting and is used in calculations.

As long as the potential remains too small to allow charge transfer between the electrode and one of the constituents in the solution, only the residual current is observed. It possesses two contributors: the double-layer charging current originating from the increase of drop surface, and faradaic currents due to impurities (e.g., dissolved oxygen). As soon as the decomposition potential of a constituent has been reached, a much larger faradaic current due to this depolarizer adds to the residual current. It increases swiftly with rising potential until the residual current becomes so high that all of the depolarizer ions undergo charge transfer as soon as they reach the electrode. The immediate neighborhood of the electrode becomes completely depleted of depolarizer, and the limiting value of the current does not depend on potential any more. It is governed by the rate at which depolarizer is brought to the electrode. The rate

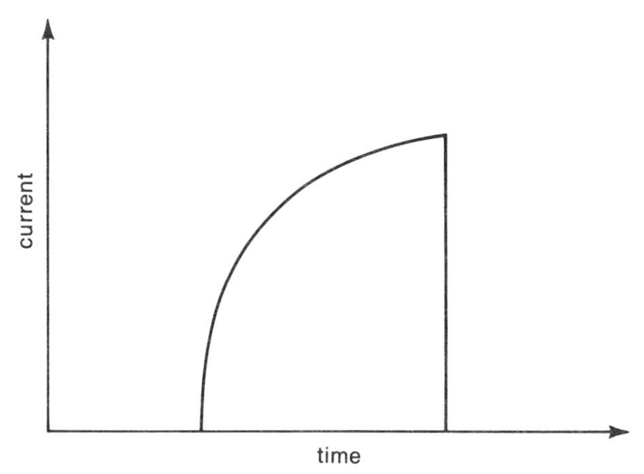

FIGURE 12-18
Current as a function of time during one drop life.

processes determining the height of the limiting current are the subject of this discussion.

The magnitude and shape of the polarographic wave is normally the result of "competition" between several rate processes, one or more of which may be "rate determining" (i.e., contributing significantly to the magnitude and shape of the wave). The possible rate processes to be considered in interpreting polarographic data are: (1) diffusion (mass transfer), (2) the electrochemical charge transfer step, (3) homogeneous chemical reactions coupled with the charge transfer step, (4) heterogeneous (surface) chemical reactions coupled with the charge transfer step, and (5) adsorption. The cathodic polarographic waves of Ti(IV) investigated in this experiment manifest the influence of diffusion and homogeneous chemical reactions on the polarographic wave. Other rate processes are either too rapid (e.g., charge transfer) to exert rate control or nonexistent.

Diffusion-Controlled Wave (Solution 1)

The polarographic wave obtained with solution 1 is an example of a diffusion-controlled or "reversible" wave. The electrode process may be represented by the reaction

$$\mathrm{Ti(IV)} + e \underset{}{\overset{\text{fast}}{\rightleftarrows}} \mathrm{Ti(III)} \qquad \text{(R1)}$$

[the actual Ti(IV) species involved are complex ions of unknown structure]. In such a system any chemical reactions involving the electroactive species are so rapid that chemical equilibrium is maintained at all times. Electrochemical equilibrium is also maintained (the Nernst equation is obeyed) and adsorption plays no significant role. The current is controlled solely by the rate of diffusion of the electroactive species to and from the electrode. An approximate expression for the polarographic wave (disregarding electrode curvature) for such a system is given by

$$i = \frac{nFAC_0^*}{1+e^j}\left(\frac{7D_0}{3\pi t}\right)^{1/2} \quad (1)$$

where

$$j = \frac{nF}{RT}(E_{\text{d.c.}} - E_{1/2}^r) \quad (2)$$

$$E_{1/2}^r = E^0 - \frac{RT}{nF}\ln\left(\frac{f_R}{f_0}\right)\left(\frac{D_0}{D_R}\right)^{1/2} \quad (3)$$

The instantaneous polarographic current is represented by i, t is time, n is the number of electrons transferrred in the electrode reaction, F is Faraday's constant, C_0^* is the concentration of the oxidized form in the bulk of the solution, D_0 and D_R are the diffusion coefficients of the oxidized and reduced forms, respectively, T is absolute temperature, R is the gas constant, $E_{\text{d.c.}}$ is the applied d.c. potential, f_R and f_0 are the activity coefficients of the reduced and oxidized forms, and A is the electrode area. For the dropping mercury electrode, the electrode area is given by

$$A = 0.8515(mt)^{2/3} \quad (4)$$

(at 25°C) where m is the mercury flow rate.

The limiting diffusion current, i_d, (plateau current) occurs when $e^j \ll 1$, that is,

$$i_d = nFAC_0^*\left(\frac{7D_0}{3\pi t}\right)^{1/2} \quad (5)$$

Equations (1) and (5) can be combined to yield the current-potential expression

$$E_{\text{d.c.}} = E_{1/2}^r + \frac{RT}{nF}\ln\frac{i_d - i}{i} \quad (6)$$

or, at 25°C

$$E_{\text{d.c.}} = E_{1/2}^r + \frac{0.0591}{n}\log\frac{i_{d-i}}{i} \quad (7)$$

Equation (7) predicts that a plot of log $\frac{i_d - i}{i}$ against $E_{\text{d.c.}}$ will be a straight line with a slope of $0.0591/n$ volts and an intercept on the potential axis of $E_{1/2}^r$ (the half-wave potential). Thc half-wave potential is very close to the E^0 value since the logarithmic term in equation (3) is often very small. Construct this plot from the data obtained with solution 1, using the currents at the end of drop life. Are the results in agreement with the theory? Is the apparent n value in agreement with reaction (R1)?

The changes in the polarogram with column height actually manifest the time-dependence of the current during each drop life. There are two time-dependent terms in equation (1); $t^{1/2}$ in the denominator and the area term. Equations (1) and (4) indicate that

$$i \propto m^{2/3}t^{1/6} \quad (8)$$

Column height influences both the m and t terms according to the relations

$$m \propto h \quad (9)$$

$$t \propto h^{-1} \quad (10)$$

where h represents mercury column height. Therefore,

$$i \propto m^{2/3}t^{1/6} \propto h^{1/2} \quad (11)$$

that is, the current magnitude at any point on the wave varies directly with the square root of column height with the intercept at the origin.

Plot currents observed with solution 1 (currents at end of drop life) against $h^{1/2}$. Employ data on the diffusion plateau and at the half-wave potential. Do the results agree with the theory?

Catalytic Wave (Solution 2)

The addition of ClO_3^- ion has a rather profound influence on the polarographic wave of Ti(IV). While chlorate is a strong oxidizing agent, it is not reduced electrolytically at the mercury cathode because of a large activation overpotential. However, it does rapidly oxidize the Ti(III) ion produced at the mercury cathode in the polarographic reduction of Ti(IV). The homogeneous redox reaction between Ti(III) and ClO_3^- proceeds according to the stoichiometry

$$6Ti^{3+} + ClO_3^- + 3H_2O \rightarrow 6TiO^{2+} + Cl^- + 6H^+ \quad (R2)$$

Thus, the electrode process may be represented by the scheme

$$\text{Ti(IV)} + e \rightleftarrows \text{Ti(III)} \quad \xleftarrow[k_2]{[ClO_3^-]} \tag{R3}$$

(k = rate constant of the reaction). Hence, Ti(IV) is furnished at the electrode not only by the diffusion process, but by chemical regeneration of the electrolysis product, Ti(III). The net result is a considerable increase in the magnitude of the polarographic current over what is observed in the absence of ClO_3^- (with solution 1). This enhancement of the polarographic wave height can be employed to determine the rate constant for the reaction between Ti(III) and ClO_3^-. The calculation of the chemical rate constant is normally based on the measurement of thc limiting current observed with the catalytic wave i_l (with solution 2), and the limiting current in the absence of the catalytic reaction i_d (with solution 1). For an electrode reaction following the scheme represented by reaction (R3), the ratio i_l/i_d is given by

$$i_l/i_d = \Psi(kt) \tag{12}$$

assuming that identical concentrations, drop life, and so forth, attend the i_l and i_d values; $\Psi(kt)$ is a power series in kt and has been tabulated in the literature. For a first-order or pseudo first-order reaction where $kt \gg 1$ (both conditions apply with solution 2), $\Psi(kt)$ assumes the simple form

$$\Psi(kt) = \left(\frac{3\pi kt}{7}\right)^{1/2} = \frac{i_l}{i_d} \tag{13}$$

Use the experimental results obtained with solutions 1 and 2 to calculate the apparent pseudo first-order chemical rate constant k for the reaction between Ti(IV) and ClO_3^-. The values of i_l and i_d must correspond to the same column height (drop life) and the same concentration—that is, correct for the factor-of-ten difference in Ti(IV) concentrations in two solutions. Calculate the rate constant at each column height used, and average the results. From the rate constant, calculate the half-life of Ti(III) in the presence of 0.04 M ClO_3^-, using the relation

$$\text{half-life} = \frac{0.693}{k} \tag{14}$$

Combining equations (5) and (13) leads to a relation for the limiting current with a catalytic wave given by

$$i_l = nFAC_0^* k^{1/2} D^{1/2} \tag{15}$$

from which

$$i \propto m^{2/3} t^{2/3} \tag{16}$$

and

$$i \propto h^0 \tag{17}$$

are concluded. That is, theory predicts that the current should be independent of column height. What do you observe?

Kinetic Wave (Solution 3)

The addition of sulfate ions to the oxalate solution of Ti(IV) leads to competition between the two anions for positions in the coordination sphere of Ti(IV). As a result, solution 3 contains at least two electroactive complex ions of Ti(IV), one being the species containing only oxalate as a ligand, which was reduced in solutions 1 and 2, and the other containing sulfate as a ligand, possibly a mixed sulfato-oxalato complex. The precise structure of these complex ions is unknown. However, the solution equilibrium between the two forms of Ti(IV) can be represented crudely by

$$\text{Ti(IV) (sulfato)} \underset{k_2}{\overset{k_1}{\rightleftarrows}} \text{Ti(IV) (oxalato)} \tag{R4}$$

Two polarographic waves are observed with solution 3. The first is due to the reduction of the oxalato form and the second (more negative) arises from the sulfato form. The kinetics of the chemical reaction represented by reaction (R4) play an important role in the observed characteristics of the polarogram such as the relative heights of the two waves and their positions on the potential aixs. The very existence of two waves is an immediate indication that the forward and reverse rates of reaction (R4) are not sufficiently rapid to maintain chemical equilibrium in the vicinity of the electrode and that the polarogram is influenced by chemical kinetics. If the reaction rates were sufficiently rapid, only a single wave would be observed. This would occur because if chemical equilibrium is maintained in the vicinity of the electrode, both the sulfato and oxalato forms would be depleted by electrolysis at the first wave, even though only the oxalato form would be reduced electrolytically (sulfato form converts to oxalato form to maintain equilibrium). In effect, both forms would contribute completely to the polarographic current at the potential of the first wave. When the potential became sufficiently negative

TABLE 12-1
F(X) versus X.

X	$F(X)$
0.005	
0.010	0.00441
0.020	0.00880
0.030	0.01748
0.040	0.02604
0.050	0.03447
0.060	0.04281
0.080	0.05102
0.10	0.06712
0.20	0.08279
0.30	0.1551
0.40	0.2189
0.50	0.2749
0.60	0.3245
0.70	0.3688
0.80	0.4086
0.90	0.4440
1.00	0.4761
1.20	0.5050
1.40	0.5552
1.60	0.6326
1.80	0.6623
2.00	0.6879
2.50	0.7391
3.00	0.7730
4.00	0.8250
5.00	0.8577
6.00	0.8803
8.00	0.9093
10.0	0.9268
15.0	0.9508
20.0	0.9629
30.0	0.9752
50.0	0.9851
110.0	0.9932
350.0	0.9979

to reduce the sulfato form, both forms would be electrolyzed but no increase in current would be observed because it makes no difference whether the electron is transferred directly to the sulfato form or to the oxalato form made available by the chemical conversion of the sulfato form. In the actual situation observed with solution 3, chemical equilibrium is not maintained. The chemical reaction is not instantaneous and the conversion of the sulfato form to the oxalato form does not follow the electrochemical depletion of the oxalato form at the first wave. As a result, not all of the Ti(IV) contributes to the current at the potential of the first wave, and the height is diminished (i.e., the first wave is controlled by the kinetics of the chemical reaction). In this case, when the potential becomes sufficiently negative to reduce the sulfato form, a second increase in current is observed because all of the Ti(IV) may now contribute to the current. The total current on the plateau of the second wave corresponds to a diffusion-controlled current.

As indicated previously, the plateau current for the first wave obtained with solution 3 is controlled by the rate of reaction (R4), which is probably pseudo first-order under the conditions of the experiment. For such a simple system, the ratio of the observed limiting current, i_l, to the diffusion controlled current, i_d, is given by

$$\frac{i_l}{i_d} = F(X) \tag{18}$$

where

$$X = \left[\frac{12K(1+K)k_1t}{7}\right]^{1/2}$$

and $K = k_1/k_2$. The significance of k_1 and k_2 is indicated in reaction (R4). The function $F(X)$ is a power series in X. It is tabulated in Table 12-1, which can be used to construct a graph of $F(X)$. The latter is used to obtain the value of X corresponding to the observed $F(X)$, that is, observed i_l/i_d. The value of X permits the calculation of the product $K(1+K)k_1$.

To obtain the individual rate constants, k_1 and k_2, the equilibrium constant K must be known, which, together with knowledge of the composition of the complex ions, can be deduced from polarographic measurements performed over a wide range of concentrations of oxalate and sulfate ions. However, this will not be attempted in this experiment.

From polarographic data on solution 3 determine X, using the height of the first wave as i_l and the combined heights of the first and second waves as i_d. From this value calculate the constant $K(1+K)k_1$. Repeat this calculation at each value of mercury column height employed. Do the results agree? What do you conclude? Plot the currents at the plateau of each wave against $h^{1/2}$. Also plot the ratio of the magnitudes of the first and second waves against $h^{1/2}$. Can you interpret your results qualitatively?

Cyclic Voltammetry and Polarography: Differences and Similarities

In cyclic voltammetry a triangular wave potential is applied to the polarographic cell instead of a ramp voltage, which scans in one direction only. One cycle is much shorter than a drop life; thus many forward and backward scans can be performed with the same drop. The DME therefore behaves like a stationary electrode in an unstirred solution as far as this technique is concerned. Figure 12-19 shows a current-voltage curve registered during one cycle, observed in a solution containing a reducible depolarizer. To aid you in the discussion of your experimental results, the resulting curve will be briefly examined here.

The scan is begun in a cathodic direction (note arrows in Figure 12-19B). At first only a residual current is observed until the potential becomes high enough to allow the depolarizer to become reduced at the electrode. The polarographic wave starts similarly to that in an ordinary polarographic experiment. However, a limiting current is not observed because the solution is not stirred and becomes depleted in depolarizer concentration as the voltage and time variables continue to increase. The observed current begins to sink as a result of the depletion. At the point at which the peak value of the voltage is reached, (t_1), a sharp decrease in current results from the reversal of the double-layer charging current that follows the reversal of the scan. The presence of the double-layer charging current can be clearly observed at this point. The voltage, however, is still high enough to cause the continuation of the reduction wave, which is now retraced until only the residual current remains again. As the potential becomes increasingly anodic, it reaches the point at which an oxidation wave begins to appear. This oxidation wave is caused by the reduction product of the previous electrode reaction. It reaches a peak value and then levels off because the unstirred solution becomes depleted in the reduced species as the scan continues. At the moment the scan is reversed, a sharp drop in anodic current is observed, which again results from the reversal of the double-layer charging current. The current is still anodic, since the working electrode is still positive compared with the reference electrode; but as the applied potential increases, the oxidation wave is retraced to its origin (however, with decreasing magnitude due to depletion). An anodic residual current remains that changes to cathodic residual current, at which point the cycle is closed.

This description of the technique used in cyclic voltammetry is necessarily limited since the objectives of this experiment are rather modest.

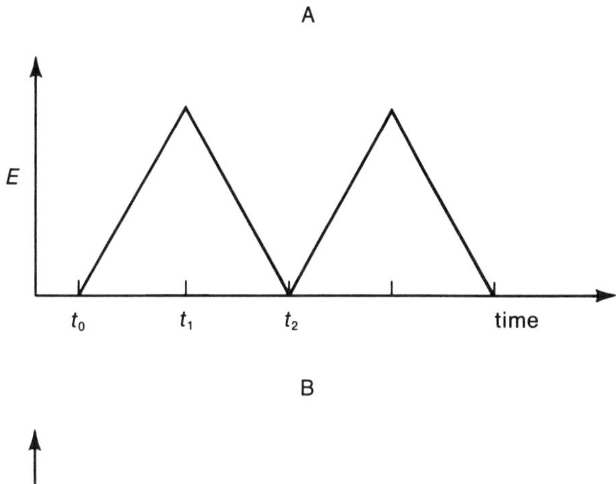

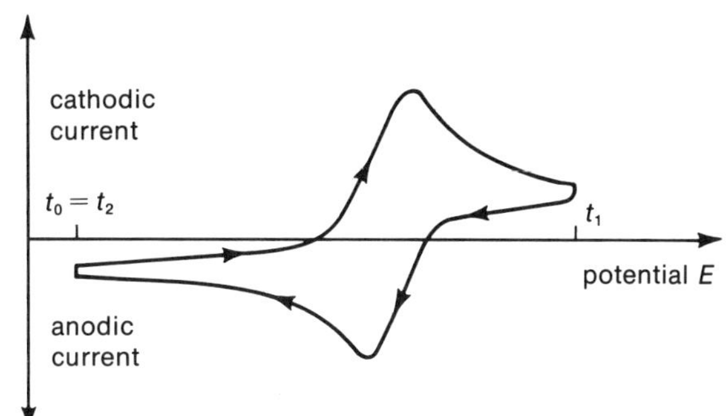

FIGURE 12-19
A. Potential versus time for a triangular-wave potential.
B. Current-voltage curve for a triangular-wave potential. At t_2 the same potential has been reached as at t_0. Figure depicts result after several cycles in a multicycled experiment with a reversible process.

It is, however, important that the student be familiar with this increasingly significant electrochemical technique not only to perform the experiment, but also to know the ways in which the readout is notably sensitive to the characteristics of the electrode reaction. It should be noted that this type of experiment differs substantially from direct-current polarographic measurements and that the information obtained usually supplements knowledge of the electrode reaction obtained from polarography.

In certain respects cyclic voltammetry is similar to polarography. Like polarography, cyclic voltammetry requires that current-voltage curves be recorded under controlled potential conditions. The cell consists of the appropriate electrode materials in a quiescent solution of an electroactive component, a large excess of a supporting electrolyte, and a suitable solvent. The cyclic voltammetric response is influenced by the same types of rate processes as in polarography.

Despite these similarities, the differences between the two techniques are sufficiently profound so that the experimental observables are usually markedly different. First of all, in polarography the normal current-voltage measurement (current at end of drop life) is essentially an observation of the current at a fixed time (the drop life) after the application of an essentially constant potential versus the applied potential. In cyclic voltammetry, on the other hand, the currents observed correspond to elapsed times that are not fixed but depend on the potential. This point can be clarified by recalling that in polarography with the DME each drop is essentially an individual, independent experiment. This is so because when a drop is dislodged the solution at the end of the capillary is sufficiently stirred so that the new drop emerges into nearly the same initial conditions as the preceding drop. Further, the rate of voltage scan is slow enough so that the potential change during a single drop life is negligible. Thus, a polarogram is the result of a series of nearly independent "experiments" involving current measurements at a constant potential, with each successive "experiment" performed at a slightly different potential because of the slow potential scan in progress. Finally, the most readily apparent feature in the qualitative perusal of a polarogram is the variation in current at the end of drop life as a function of the applied potential. This feature, which is also a common observable employed in the quantitative appraisal of polarographic data, corresponds to the current at a specific time (the drop life) after initiation of the experiment (formation of new drop). Since the drop life (time of measurement) is almost independent of potential over the potential range comprising a polarographic wave, the current at the end of drop life versus the potential curve is a fixed time observable. In cyclic voltammetry, on the other hand, the entire potential range is scanned in a short period of time compared with the drop life, so that the drop behaves essentially as a stationary electrode (of course, actual stationary electrodes can also be used), the potential is obviously not constant over the drop life, and each point on the current-potential curve corresponds to an elapsed time that is different from the initiation of the experiment (beginning of potential sweep). Since the rate of diffusion is approximately proportional to $t^{-1/2}$ (t = time), the difference between the mode of control of the time variable in polarography and cyclic voltammetry gives rise to profound differences in the respective current-voltage curves whenever diffusion is at least partially rate-controlling. For example, the diffusion-controlled polarographic reduction wave is S-shaped (sigmoidal), whereas the corresponding cyclic voltammetric wave exhibits an asymmetric peak.

A second distinction between the two techniques is the difference in "time scale" over which the experiment takes place. The time scale in polarography is roughly the drop life, which can be varied from about 2 to 15 seconds. In cyclic voltammetry the time scale corresponds to the period of the triangular wave potential, which can range from a fraction of a millisecond to 15–20 seconds. In other words, a much shorter time scale is feasible and is commonly utilized in cyclic voltammetry. A general principle of chemical kinetics is that the identity of the rate determining steps encountered usually depends on the time scale of the observation. Over a long time scale, slow rate processes are "seen" and vice versa. A process that is not rate-controlling under polarographic conditions may contribute significantly to the cyclic voltammetric response. For example, it is not unusual to observe a reversible (or diffusion-controlled) polarographic wave and find that the cyclic voltammetric response at 10 Hz for the same system is influenced by the rate of the heterogeneous transfer step or a coupled homogeneous chemical reaction.

Finally, a third important difference arises because, unlike polarography, cyclic voltammetry requires a periodic potential scan where the potential range of interest is rapidly traversed in both forward and reverse directions. In the forward (cathodic) sweep, a reduction wave (positive current) is observed and reduced material is generated in the vicinity of the electrode. If the reduced form is chemically stable and if the heterogeneous charge transfer step is reversible, the reduced material will give rise to a reoxidation wave (negative current) during the reverse (anodic) sweep. Thus, the gross reversibility of the electrode reaction is readily determined. In addition, if the primary product of the electrode reaction decomposes, the chemical decomposition products may also give rise to waves at appropriate potentials during the anodic sweep or a later cathodic sweep. (Note that the experiment you perform involves continuous recycling of the triangular wave potential; a single-cycle experiment is also popular but will not be attempted.)

In examining the data from the three solutions employed, make a special effort to find as many similarities and differences between the cyclic voltammetric and polarographic responses as possible. Specific suggestions for data to be examined follow. However, you are expected to go beyond these suggestions, using judgment and curiosity in the selection of additional features of the polarographic and cyclic voltammetric responses that might be worth comparing.

1. *Diffusion-Controlled Wave* (*Solution 1*). If, as in the polarographic experiment, the response is diffusion-controlled, the cyclic voltammetric pattern should exhibit well-formed reduction and oxidation waves whose peaks are separated by $60/n$ (n = number of electrons transferred) millivolts and whose magnitudes vary linearly with the square root of the triangular wave frequency. Do your results agree with these theoretical predictions? If not, explain.

In taking data from the photograph of the oscilloscope trace, which will exhibit an envelope of curves of varying current magnitude, use the outermost curve (largest currents). Consult your instructor for details. Use the same procedure with solutions 2 and 3.

2. *Catalytic Wave* (*Solution 2*). If the cyclic voltammetric response is controlled by the rate of chemical reoxidation of Ti(III) to Ti(IV) by chlorate ion, as in polarography, an S-shaped trace should result in which the current-voltage curve follows precisely the same pattern during the anodic and cathodic sweeps. No oxidation wave should be observed. Do your results agree with this prediction? How do the current levels compare to those obtained in polarography? How do the current levels depend on scan rate?

3. *Kinetic Wave* (*Solution 3*). For this process you are not asked to compare results with any theoretical expectations. Instead, how do the results compare to your expectations based on the polarographic data? What important qualitative conclusions can you form regarding the nature of the electrode reaction?

E. POTENTIOMETRIC TITRATION OF Cu^{2+} WITH ETHYLENEDIAMINE[2]

Chemicals

0.01 *M* copper perchlorate solution
0.10 *M* perchloric acid
0.05 *M* ethylenediamine solution
0.01 *M* barium perchlorate

[2]Adapted from Goldberg (1962).

Apparatus

pH meter with glass and calomel electrodes
200-ml four-necked bottle with gas inlet
Electric stirring motor and glass stirrer
10-ml buret graduated in 0.02 ml
Drying tube filled with ascarite
Thermostat
Presaturator

Introduction

In an aqueous solution Cu^{2+} ions are octahedrally surrounded by water molecules. Upon the addition of ethylenediamine, two neighboring H_2O molecules are replaced by one ethylenediamine. A sufficient increase in ethylenediamine concentration leads to the replacement of still another pair of water molecules. The exchange of ligands is accompanied by a gradual color change, a shift of absorption due to the difference in the coordination sphere. Since the absorption is due to a forbidden *d-d* transition ($t_{2g}^6 e_g^3 \rightarrow t_{2g}^5 e_g^4$), it is very weak and cannot be used as an indicator for the exchange reaction. A sensitive and easily measurable indicator is the pH of the solution. Cu^{2+}, $[Cu(en)]^{2+}$, and H_3O^+ compete for ethylenediamine. The competing reactions are:

$$Cu^{2+} + en \rightleftarrows [Cu(en)]^{2+} \quad (1)$$

$$[Cu(en)]^{2+} + en \rightleftarrows [Cu(en)_2]^{2+} \quad (2)$$

$$en + H^+ \rightleftarrows enH^+ \quad (3)$$

$$enH^+ + H^+ \rightleftarrows enH_2^{2+} \quad (4)$$

Since ethylenediamine participates in each of the reactions, it provides a link between the measurable hydronium ion concentration and the species containing copper. The ethylenediamine concentration can be calculated provided the equilibrium constants of reactions (3) and (4) are known. These can be obtained separately by the titration of perchloric acid with ethylenediamine. A noncoordinating dipositive metal ion (Ba^{2+}) must be added in this case to account for the effect of ionic strength on the equilibrium constants of reactions (3) and (4).

Procedure

Two titrations are to be performed:

1. 10 ml of 0.01 *M* Ba^{2+} solution and 2 ml of 0.1 *M* perchloric acid

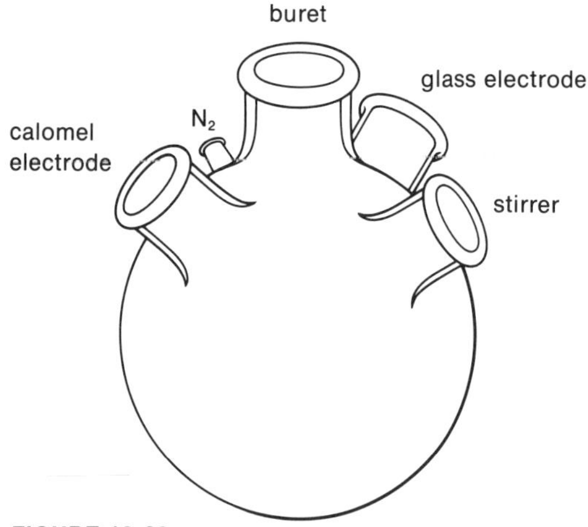

FIGURE 12-20
Four-necked flask for use in the potentiometric titration of Cu^{2+} with ethylenediamine.

2. 10 ml of 0.01 *M* Cu^{2+} solution and 2 ml of 0.1 *M* perchloric acid

(both in 88 ml of water with standard ethylenediamine solution).

This experiment should be done at 35°C by one-half of the students, and at 45°C by the other half. Share your results so that thermodynamic functions can be calculated.

Arrange the four-necked flask, which is shown in Figure 12-20, in the thermostat so that the buret and stirrer can be easily introduced and held in position. Pipet 10 ml of the Cu^{2+} solution and 2 ml of the perchloric acid solution into the flask and add boiled distilled water to bring the total volume to about 100 ml. Note the exact volume. Introduce the buret filled with ethylenediamine solution and the stirrer, and after calibrating the *p*H meter place the electrodes that have been rinsed and dried in the solution. Blow nitrogen presaturated with water at the temperature of the thermostat over the solution and allow it to escape through the stirrer entrance. Add the first small amount of ethylenediamine. Start the stirring, and after enough time has been allowed for equilibration, perform the first *p*H measurement. Report the data. Indicate any observed color changes.

Calculations

General Procedure

Where there is a center atom in equilibrium with n ligands there are n equations relating the n formation constants with the concentrations of species present. In addition there are the equations pertinent to the protonation of the ligand. For $n = 2$ and the system Cu^{2+} and ethylenediamine, reaction equations (1)–(4) given in the introduction apply. The four formation constants are as follows:

$$K_1 = \frac{[MA]\gamma_{2+}}{[M][A]\gamma_{2+} \cdot \gamma_0}$$

$$K_2 = \frac{[MA_2]\gamma_{2+}}{[MA][A]\gamma_{2+} \cdot \gamma_0}$$

$$K_3 = \frac{[AH]\gamma_+}{[A][H]\gamma_+ \cdot \gamma_0} = K_{H1}$$

$$K_4 = \frac{[AH_2]\gamma_{2+}}{[AH][H]\gamma_+^2} = K_{H2}$$

The symbol "M" stands for metal (Cu^{2+}) and "A" for amine (ethylenediamine). Charges are not indicated.

According to the Debye-Hückel theory $\gamma_0 = 1$ and $\gamma_{2+} = (\gamma_+^2)^2$. Bertsch (1955) and McIntyre (1952) state that $\log \gamma_+ = -0.03 \pm 0.005$ for dipositive metal ions in solutions containing ethylenediamine and in the concentration range of this experiment.

In the expressions for K_1, K_2 and K_{H1}, activity coefficients cancel but K_{H2} becomes:

$$K_{H2} = \frac{[AH_2]}{[AH][H]} \cdot \gamma_+^2$$

$$K'_{H2} = \frac{K_{H2}}{\gamma_+^2}$$

The calculation of formation constants becomes lucid and straightforward by the introduction of the "formation function" $\bar{n}$.

$$\bar{n} = \frac{\text{total bound ligand}}{\text{total metal ions}}$$

Titration Curve (1)

In this case ethylenediamine is treated analogous to the metal ion and protons are considered ligands.

$$\bar{n}_H = \frac{[AH] + 2[AH_2]}{[A] + [AH] + [AH_2]} \tag{5}$$

$$= \frac{K_{H1}[H] + 2K_{H1}K'_{H2}[H]^2}{1 + K_{H1}[H] + K_{H1}K'_{H2}[H]^2} \tag{5a}$$

A mass balance for H^+ permits the calculation of $\bar{n}_H$ for any point on the titration curve.

Total bound protons = original protons (from acid)
− free protons
+ protons from dissociation of water.

$$[AH] + 2[AH_2] = C_H - [H] + [OH]$$

In C_H, the acid concentration, the volume change must be taken into account.

The total ethylenediamine C_A is obviously given by the added ethylenediamine and the total volume.

$$\bar{n}_H = \frac{C_H - [H] + \dfrac{K_W}{[H]}}{C_A} \qquad (6)$$

The combination of equations (5a) and (6) and the simultaneous solution for two points on the titration curve leads to K_{H1} and K_{H2}.

That two expressions occur repeatedly during resolution can be verified easily. They are

$$(1 - \bar{n}_H)[H] \equiv J_1 \qquad (7)$$

and

$$(2 - \bar{n}_H)[H]^2 \equiv J_2$$

Variables belonging to one of the points are designated by a prime. Expression (7) refers to point "unprime," and for point "prime" the following is obtained:

$$(1 - \bar{n}'_H)[H]' \equiv J'_1 \qquad (8)$$

and

$$(2 - \bar{n}'_H)[H]'^2 \equiv J'_2$$

Written in these terms, K_{H_1} and K_{H_2} are:

$$K_{H_1} = \frac{\bar{n}_H \cdot J'_2 - \bar{n}'_H \cdot J_2}{J_1 \cdot J'_2 - J'_1 \cdot J_2} \qquad (9)$$

$$K_{H_2} = \frac{\bar{n}'_H \cdot J_1 - \bar{n}_H \cdot J'_1}{\bar{n}_H \cdot J'_2 - \bar{n}'_H \cdot J_2} \cdot \gamma_{\pm}^2 \qquad (10)$$

Since the pH change in curve 1 is solely due to the protonation of ethylenediamine, any two points on the curve should be sufficient for calculation. Of course, the two points that are chosen should be far apart so that both regions of protonation are well covered; furthermore, the error in pH due to incorrect volume of pH readings should be held to a minimum. Regions in which the pH increases rapidly with added volume are therefore unsuitable for calculation.

Titration Curve (2)

This curve is more complicated. Again, $\bar{n}$ is given by:

$$\bar{n} = \frac{\text{total bound amine}}{\text{total metal ion}}$$

or

$$\bar{n} = \frac{K_1[A] + 2K_1K_2[A]^2}{1 + K_1[A] + K_1K_2[A]^2} \qquad (11)$$

Again, the bound ligand can be expressed in terms of a mass balance and the total metal is known from the initial metal adjusted for the total volume at the point of the titration.

bound ligand = total ligand
− free ligand
− protonated ligand
$= C_A - [A] - [AH] - [AH_2]$

and

$$\bar{n} = \frac{C_A - [A] - [AH] - [AH_2]}{C_M} \qquad (12)$$

A mass balance for H^+ yields equation (13).

$$[AH] + 2[AH_2] = C_H - [H] + [OH] \qquad (13)$$

The total protonated amine can now be calculated by the insertion of equations (5) and (13) into equation (12):

$$\bar{n} = \frac{C_A - (C_H - [H] + [OH])/\bar{n}_H}{C_M} \qquad (12a)$$

Note: $\bar{n}_H$ is defined in equation (5a).

A simultaneous solution of equations (11) and (12a) for two points on titration curve (2) leads to K_1 and K_2. The same equations are obtained as for titration curve (1) (equations 9 and 10).

$$K_1 = \frac{\bar{n}J'_2 - \bar{n}'J_2}{J_1J'_2 - J'_1J_2} \qquad (9a)$$

$$K_2 = \frac{\bar{n}'J_1 - \bar{n}J'_1}{\bar{n}J'_2 - \bar{n}'J_2} \qquad (10a)$$

Theoretically, any two points on the titration curve should be sufficient for calculation. In practice points must be selected carefully. The first big change in pH is due to protonation of ethylenediamine. In the region before this change, the pH is insensitive to changes in the coordination of Cu^{2+}, and points should not be selected in this range. Since the H^+ concentration enters the calculation but the logarithm is read, small errors in pH cause large deviations in calculation and can even make calculations impossible. Points should therefore never be taken from regions of steep slopes. The coordination reaction is slow! Allow enough time to be certain that pH readings are taken at equilibrium. Values of $\bar{n}$ should be chosen at 0.5 ± 0.3 and 1.5 ± 0.3.

To calculate ΔG°, ΔH° and ΔS° for the coordination and the protonation reactions, use the results obtained by a student who conducted the experiment at a different temperature.

Additional Work

Take visible spectra of copper perchlorate in water and of the titration solution after the experiment has been finished. Explain the observed band forms and color changes in terms of ligand field theory.

Question

Why is the tris(ethylenediamine) complex not observed?

Note: A FORTRAN program is provided in Exercise 16 to enable students to use their experimental data to a larger extent by basing their calculations on more than two pairs of data in the evaluation of protonation and complexation constants.

F. POTENTIOMETRIC TITRATION OF PHOSPHORUS OXYACIDS

The acid behavior of polyphosphoric acids is characterized by the distinct difference in ease of dissociation between the first proton on each phosphorus group (chain protons) and the ones remaining on each end group (terminal proton). Table 12-2 shows that there are always as many easily dissociable protons as there are phosphorus atoms in an acid. In addition, there are two much less dissociable protons that are identified as the terminal protons. The pK difference between the two terminal protons is much larger than the difference between chain protons in any acid. The close similarity between pK values for terminal protons of different acids also deserves attention. Even the dissociation behavior of orthophosphate fits into the picture, with the exception of the last proton, which cannot be titrated by using a glass electrode. Ring acids, lacking end groups, exhibit just one kind of proton, the pK value for all protons being the same and close to the value found for chain protons. Because of these dissociation phenomena, the estimation of chain length by means of the determination of the ratio of chain to terminal protons becomes possible. It is also feasible to estimate the amount of orthophosphate in a mixture by the ratio of the two terminal protons.

Apparatus

pH meter with glass and calomel electrodes
50-ml buret with tubing clamp and glass tip for ion exchange
50-ml buret for titration
500-ml beaker
300-ml beaker

Chemicals

Dowex 50 ion-exchange resin, 20–50 mesh
1 *M* hydrochloric acid
0.1 *N* sodium hydroxide
Buffer solutions for standardization of pH meter (pH 5, 7, and 9)
Potassium hydrogen phtalate
Small quantity of silver nitrate

Procedure

Chain-length determination can be performed on commercial polyphosphates or on the tripolyphosphate prepared according to the description in Exercise 2B. If prepared polyphosphate is to be used, any pyrophosphate in the product must be removed or determined separately. (Find out why this must be done by considering the stoichiometry for all possible reactions during preparation.) Pyrophosphate can be evaluated by precipitation with Zn in an acetic acid medium and titration of the filtered precipitate with EDTA to obtain the Zn content. In any case, before titration takes place, the salts must be converted to free acids by ion exchange.

TABLE 12-2
pK values of phosphorus oxyacids.

Acid	Formula	pK_1	pK_2	pK_3	pK_4	pK_5	pK_6
Orthophosphoric	H_3PO_4	2.16	7.21	12.33			
Pyrophosphoric	$H_4P_2O_7$	1.52	2.36	6.60	9.25		
Triphosphoric	$H_5P_3O_{10}$		1.06	2.30	6.50	9.24	
Tetraphosphoric	$H_6P_4O_{13}$			1.36	2.23	7.38	9.11
Trimetaphosphoric	$H_3P_3O_9$	2.05					
Tetrametaphosphoric	$H_4P_4O_{12}$	2.74					

Source: All values are from van Wazer (1958).

Preparation of Polyphosphoric Acids By Ion Exchange[3]

Fit a glass-wool plug to the bottom of a 50-ml buret. Fill to a resin height of 40 ml with a slurry of water and Dowex 50 ion-exchange resin (20–50 mesh). Do not allow the water level to fall below the level of the resin. The resin is converted to its hydrogen form by slowly percolating 500 ml of 1 *M* HCl through the column. The excess HCl is removed by washing the resin until the eluant is free of Cl^- ions (test with $AgNO_3$). Dissolve about 0.25 g of the prepared sodium tripolyphosphate in 20 ml of water and let this solution percolate through the column at a rate of 2 drops per minute. Elute the free acids by adding 200 ml of distilled water to the column immediately after the polyphosphate solution has disappeared into the resin. Collect all of the 220 ml in a 500-ml beaker.

Chain-Length Determination

Standardize the *p*H meter and electrodes with buffers of *p*H 5, 7, and 9. Titrate a 200-ml aliquot from the sample obtained by ion exchange with carbonate-free 0.1 *N* NaOH. It is good practice to standardize the normal solution shortly before use with potassium hydrogen phtalate.

Table 12-2 indicates that the protons of chain polyphosphates dissociate in four discrete steps. After passage through an ion-exchange column, the polyphosphates exist in the mono-ionized form. Thus, the first end point observed on potentiometric titration is that of the protons of *pK* from 2.05 to 2.74. Ring polyphosphates, as well as the chain polyphosphates, exhibit this dissociation. Therefore, the first observed end point allows calculations of the total number of chain protons in a sample. Unless the sample contains only cyclic polyphosphates, two or three additional end points will be observed. These end points result from the dissociation of the chain-terminal protons and provide a quantitative measure of these protons. The ratio of chain protons to terminal protons defines the average chain length.

It is not easy to find the exact location of these end points; thus this type of analysis is sensitive to large amounts of contaminants only. An invaluable aid in the detection of end points is a plot of the first derivative of the titration curve with respect to the volume of base added. The first derivative f' may be defined as the change in *p*H divided by the change in the volume of base between two points: $f' = \Delta pH/\Delta$ (volume of base). Plot f' against the volume of base added at the regions of steep change of the titration curve only.

Since various cancellations between short-chain and longer chain or ring acids are possible, there is not much definite information to be gained from an end-group titration alone. It is much more interesting to combine the results of chromatography with those of end-group titration (see the experiment in Exercise 11C).

[3]For a description of ion-exchange phenomena see Exercise 11B.

G. ELECTROLYTIC CONDUCTIVITY OF AQUEOUS SOLUTIONS OF COMPLEX IONS

This experiment consists of the measurement of the concentration dependence of conductance for two coordination compounds that were prepared in Sections A and B of Exercise 1, namely $[Co(NH_3)_5Cl]Cl_2$ and $[Co(en)_3]Cl_3$. Since the conductance of electrolytes in aqueous solution depends on the number of ions present and the mobilities of the different kinds of ions, it serves to confirm our views of the dissociation of coordination compounds. The two compounds are

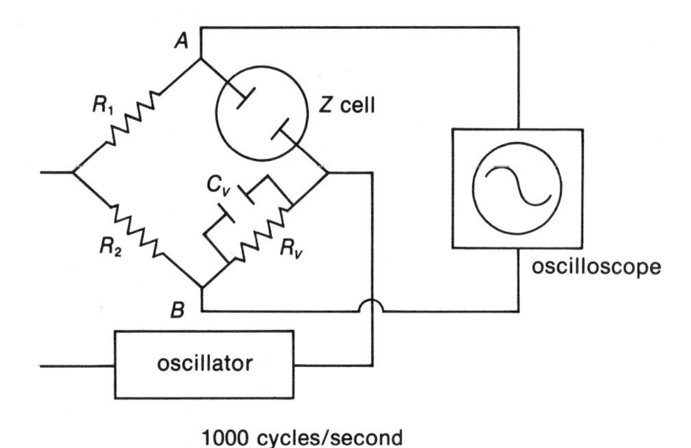

FIGURE 12-21
Diagram of a bridge arrangement.

expected to dissociate completely in the following way:

$$[Co(NH_3)_5Cl]Cl_2 \longrightarrow [Co(NH_3)_5Cl]^{2+} + 2Cl^-$$

$$[Co(en)_3]Cl_3 \longrightarrow [Co(en)_3]^{3+} + 3Cl^-$$

These compounds are regarded as "strong electrolytes." If other ions are released by the complex, for example, by means of replacement of ligands by water, calculations will be meaningless. Solutions, therefore, must be freshly prepared before measurement. The complexes used in this experiment, however, are so stable that their aqueous solutions do not change measureably during an entire laboratory period. The substances also must be pure. Any ionizing impurities present will severely mask the behavior to be observed. This aspect of the experiment serves to evaluate the skill with which the substances were prepared.

Apparatus

Decade resistance box (to 9,999 ohm)
Variable capacitor
0.1 precision resistors (1000, 200, 100 ohm)
Oscilloscope
Conductance cell
Volumetric flasks and pipets
Chassis for bridge connections
Leads (two of them with one stripped end)
Oscillator providing frequency of about 1000 cycles/second

Chemicals

$[Co(en)_3]Cl_3$
$[Co(NH_3)_5Cl]Cl_2$
Conductance water[4]

Procedure

Figure 12-21 shows the arrangement of resistors, capacitor, and cell to be used in this experiment.

The constant precision resistors R_1 and R_2 are chosen in such a way that the range of the decade box R_v is used as effectively as possible. A variable capacitor C_v is placed parallel to R_v to simulate the cell in the branch of the bridge not containing the cell. Points A and B are connected to the vertical reflecting plates of the oscilloscope in such a manner that the difference in potentials is shown on the screen. (Use terminals A and B on a Tektronix 502A dual beam scope with a setting on A–B, or use channels A and B on a Tektronix 561A dual trace scope with a setting on add and channel A inverted.) When the bridge is balanced the potentials at A and B are equal. The cell behaves like a capacitance C_{cell} and a resistance R_{cell} in parallel. Equilibrium for the bridge is therefore given by

$$\frac{R_1}{R_2} = \frac{Z_{cell}}{Z_v} \tag{1}$$

where Z_{cell} and Z_v, the cell and variable impedance, are obtained by adding ohmic and capacitive reactance, R and X, according to Kirchhoff's law.

$$\frac{1}{Z} = \frac{1}{R} + \frac{1}{X} \tag{2}$$

X depends on frequency ν, where $\omega = 2\pi\nu$; and on capacitance C.

$$X = \frac{1}{\omega C}$$

Substitution for X in equation (2) gives

$$Z = \frac{R}{1 + \omega CR} \tag{3}$$

The introduction of equation (3) for cell and variable impedance into equation (1) and the fact

[4]Conductance water is multiple distilled water of a specific conductance of 5×10^{-6} ohm^{-1}cm^{-1} or less. Conductance water can also be prepared by ion exchange.

that ohmic and capacitive reactances must be balanced separately if no phase shift is allowed between potentials A and B leads to the formulation of the equilibrium condition in terms of the measurable quantities R_1, R_2, R_v, and C_v and the desired quantity R_{cell}.

$$\frac{R_1}{R_2} = \frac{R_{cell}}{R_v} = \frac{C_v}{C_{cell}} \tag{4}$$

Without the addition of the capacitor to R_v, perfect balancing of the bridge would be impossible because the part of the current flowing through the capacitance of the cell would be phase shifted. Even with the perfect balance of impedances, however, a residual potential difference may be observed because of "stray reactance" in the bridge. A sensitive test of whether amplitude and phase at points A and B are equal can be performed by connecting A to the vertically deflecting plate and B to the horizontally deflecting plate of the scope. If phase and amplitude are equal, a straight line of 45° slope is observed.

Theoretical Background

The measured resistance of an electrolytic solution in a conductance cell depends on the arrangement of the electrode plates in the cell. The specific resistance is defined as the resistance of a column of solution 1 cm long between plates 1 cm^2 in cross section. Since conductance cells do not usually meet these standards, a conversion factor that depends on the arrangement in any particular cell has to be found. This factor is called cell-constant k and is most easily found by calibration with a solution of known specific conductance κ, where κ is the inverse of the specific resistance:

$$k = \kappa R \text{ cm}^{-1} \tag{5}$$

and R is the measured resistance. The specific conductance obviously depends on the number and charge of ions in the column 1 cm in length and 1 cm^2 in cross section between plates. To avoid this dependence, the equivalent conductance Λ is defined as the conductance of 1 g equivalent of solution placed between plane-parallel electrodes of 1 cm distance. The equivalent conductance Λ and the specific conductance κ are therefore connected by the volume in cm^3 occupied by 1 g equivalent of solute in solution. This volume is $V = 1000/N$, where N, the normality, is the number of gram equivalents per liter.

$$\Lambda = \kappa \frac{1000}{N} \text{ cm}^2 \text{ ohm}^{-1} \text{ equiv}^{-1} \tag{6}$$

Even for strong electrolytes that are completely dissociated, Λ depends on concentration as well as temperature. Kohlrausch found experimentally the following behavior for strong electrolytes at low concentrations:

$$\Lambda = \Lambda_0 - \text{const } c^{1/2} \tag{7}$$

where c is the concentration in moles per liter. A comparison of Λ_0, the limiting equivalent conductance, for different electrolytes containing common ions showed that Λ_0 can be expressed as the sum of the limiting ionic equivalent conductances $(\lambda_-)_0$ and $(\lambda_+)_0$ for anion and cation, respectively.

$$\Lambda_0 = (\lambda_-)_0 + (\lambda_+)_0 \tag{8}$$

The experimental finding of Kohlrausch has been explained and quantitatively expressed by Debye, Hückel and Onsager. In 1923, Debye and Hückel derived an equation of the same form as equation (7) by considering the influence of an atmosphere of counter-ions surrounding each ion in solution. They wrote:

$$\Lambda_0 - (a\Lambda_0 + b)c^{1/2} = \Lambda$$

Debye and Hückel were not able to resolve the expression for a completely, and their expression for b depended on the ionic size. The first term in the expression in parentheses is due to the "relaxation effect" and the second to the "electrophoretic effect." The relaxation effect arises from the fact that as the ion moves through the solution the ionic atmosphere of opposite sign around it must always be newly formed around the new site of the ion and must disintegrate behind it. Since this process requires a certain amount of time, a retarding force is always exerted on the ion, depending on the mobility of the ions, their concentration and valence, the dielectric constant of the medium, and the absolute temperature. The electrophoretic effect depends on the relative velocity of the ions as they travel against the stream of counter ions in their atmospheres and lose momentum by friction; its magnitude is given by valence and concentration of the ions, and the viscosity and dielectric constant of the medium.

In 1927 Onsager refined the theory and was able to derive expressions for the coefficients b and a, which have since been proven to hold very well for 1,1 electrolytes at low concentrations. For ions of higher valence, agreement can be expected only for much lower concentrations. It has been observed that where ion pair formation occurs the limiting curve approaches the limiting tangent [equation (9)] from below, instead of from above as for 1,1 electrolytes. Ion pairs of unsymmetric electrolytes are charged and represent a new conducting species, complicating the situation further.

The Debye-Hückel-Onsager equation predicts the following behavior for strong electrolytes at very low concentrations:

$$\Lambda = \Lambda_0 - \left[\frac{1.981 \times 10^{+6}}{(\epsilon T)^{3/2}} w\Lambda_0 + \frac{29.14(|z_+| + |z_-|)}{\eta(\epsilon T)^{1/2}}\right](|z_+| + |z_-|)^{1/2}N^{1/2} \tag{9}$$

where ϵ is the dielectric constant of the solvent, η the viscosity, and z_+ and z_- are the charges on cation and anion.

$$w = \frac{q|z_+z_-|}{1 + q^{1/2}} \tag{10}$$

with

$$q = \frac{|z_+z_-|\Lambda_0}{[|z_+| + |z_-|][|z_+|(\lambda_-)_0 + |z_-|(\lambda_+)_0]} \tag{11}$$

Equation (9) can be reduced to yield equation (7) for any given electrolyte.

For the purpose of this experiment Λ is expressed in terms of Λ_0 and $(\lambda_-)_0$ for a 3,1 electrolyte and a 2,1 electrolyte containing Cl^- as anion. The limiting equivalent conductance Λ_0 can be obtained from a plot of the measured Λ versus $N^{1/2}$, and $(\lambda_-)_0$ can be taken from the literature. With the aid of the measured value of Λ_0, we can check whether the measured slope of the plot matches the value predicted by the Debye-Hückel-Onsager equation. For this purpose the ratio r is defined as

$$r \equiv \frac{(\lambda_+)_0}{(\lambda_-)_0} \tag{12}$$

and is inserted in equation (8) to get

$$\Lambda_0 = (1 + r)(\lambda_-)_0 \tag{13}$$

The insertion of the appropriate charges and the substitution of Λ_0 and $(\lambda_-)_0$ from equations (12) and (13) in equation (11) for a 3,1 electrolyte (e.g., $[Co(en)_3]Cl_3$) leads to

$$q_{3,1} = \frac{3}{4}\left(1 + \frac{2}{1 + r}\right)^{-1} \tag{14}$$

The quantity $1 + \frac{2}{1 + r} \equiv A$, and from equations (14) and (10) the following is obtained:

$$w_{3,1} = \frac{3}{(4A/3) + (4A/3)^{1/2}} \tag{15}$$

The introduction of equation (15) and appropriate charges into equation (9) yields:

$$\Lambda = \Lambda_0 - \left[\alpha \frac{3\Lambda_0}{(4A/3) + (4A/3)^{1/2}} + 4\beta\right]2 \cdot N^{1/2} \tag{16}$$

where

$$\alpha = \frac{1.981 \times 10^6}{(\epsilon T)^{3/2}}$$

and

$$\beta = \frac{29.14}{\eta(\epsilon T)^{1/2}}$$

From equation (13) and the definition of A the following is derived:

$$A = 1 + \frac{2(\lambda_-)_0}{\Lambda_0}$$

By the same procedure the equation for a 2,1 electrolyte reads:

$$\Lambda = \Lambda_0 - \left[\alpha \frac{2\Lambda_0}{\frac{3}{2}B + (\frac{3}{2}B)^{1/2}} + 3\beta\right]3^{1/2}N^{1/2} \tag{17}$$

where

$$B \equiv 1 + \frac{(\lambda_-)_0}{\Lambda_0}$$

To calculate the slopes, it is assumed that for very dilute solutions the dielectric constant and viscosity are independent of temperature and the values of $\epsilon = 78.6$ and $\eta = 0.00895$ poise at 25°C for aqueous solutions are accepted.

Calculations

Plot Λ_0 against $N^{1/2}$ and determine Λ_0 and the slope of the curve. Check whether the slopes, calculated from equations (16) and (17) with the measured Λ_0, match the measured values. If the curve does not approach linearity, as may be the case for the 3,1 electrolyte, try the empirical equation that follows:

$$\Lambda = \Lambda_0 + Ac^{1/2} + Bc + Dc \log c$$

where A is the slope predicted by the Debye-Hückel-Onsager equation; B and D are empirical constants. Fuoss and Onsager (1957) have calculated B and D for symmetric electrolytes. A plot of $\Lambda - (\Lambda_0 + AN^{1/2})$ against N will show whether the introduction of the term Bc constitutes a sufficient correction.

Instructions

1. Measure the conductance of 0.01 M KCl and 5 solutions of each compound in the concentration range between 1×10^{-2} and 1×10^{-4} M. For $[Co(en)_3]Cl_3$ it is necessary to have most points in the low concentration region.
2. Prepare solutions by the dilution of one initial solution. Keep in mind that successive dilutions result in the compilation of error. All solutions are to be thermostated at 25°C.
3. Begin with the measurement of conductance water so that corrections can be made if necessary. Rinse the cell with conductance water and let it soak repeatedly before the first measurement is done.
4. Start with the lowest concentration. Fill the cell with fresh solution for each measurement (duplicate measurements should be taken in each case; large discrepancies necessitate a third measurement). Rinse the cell repeatedly with each new solution.
5. The recommended procedure for obtaining the bridge balance is first to balance resistances approximately, then adjust the capacitance, and, finally, rebalance the resistance.
6. Return all cells filled with conductance water.

Data

The equivalent conductance of 0.01 M KCl is $\Lambda = 141.27$ cm² ohm⁻¹ equiv⁻¹. The limiting ionic equivalence conductance of Cl^- is $(\lambda_-)_0 = 76.3$ cm² ohm⁻¹ equiv⁻¹. *Note*: Reilley (1962) describes a direct reading conductance meter built from an operational amplifier.

REFERENCES

Bertsch, C. R. 1955. Ph.D. dissertation. Pennsylvania State University.

Bjerrum, J. 1941. *Metal Amine Formation in Aqueous Solution*. Copenhagen: Hase.

Burr-Brown Research Corporation. 1963. *Handbook of Operational Amplifier Applications*. Tucson, Arizona.

Delahay, P. 1954. *New Instrumental Methods in Electrochemistry*. New York: Wiley.

Fuoss, R. M., and Accascina, F. 1959. *Electrolytic Conductance*. New York: Wiley.

Fuoss, R. M., and Onsager, L. 1957. *J. Phys. Chem.* **61**, 668.

Goldberg, D. E. 1962. *J. Chem. Ed.* **39**, 328.

Gucker, F. T., and Seifert, R. L. 1966. *Physical Chemistry*. New York: Norton.

Harned, H. S., and Owen, B. B. 1958. *Physical Chemistry of Electrolytic Solutions*. New York: Reinhold.

Heyrovsky, J., and Kuta, J. 1966. *Progress in Polarography*. New York: Academic Press.

McIntyre, G. H. 1952. Ph.D. dissertation. Pennsylvania State University.

Morrison, Jr., C. F. 1964. *Generalized Instrumentation for Research and Teaching*. Pullman, Washington: Washington State University Press.

Philbrick Research, Inc. 1966. *Applications Manual for Computing Amplifiers, for Modelling, Measuring, Manipulating and Much Else*. Dedham, Massachusetts.

Reilly, C. N. J. 1962. *J. Chem. Ed.* **39**, A853, A933.

Reilly, C. N. J., and Murray, R. W. 1963. In *Treatise on Analytical Chemistry*. Vol. 4, Part 1. Ed. I. M. Kolthoff and P. J. Elving. New York: Wiley, ch. 43.

Robinson, R. A., and Stokes, R. H. 1959. *Electrolyte Solutions*. 2d ed. New York: Academic Press.

Schwarz, W. M., and Shain, I. 1963. *Anal. Chem.* **35**, 1770.

van Wazer, J. R. 1958. *Phosphorus and Its Compounds*. Vol. 1. New York: Wiley.

Vlcek, A. A. 1963. In *Progress in Inorganic Chemistry*. Vol. 5. Ed. F. A. Cotton. New York: Wiley, p. 211.

EXERCISE 13

Analytical Applications of Complex Formation

A. ANALYSIS OF COBALT FERRITE BY TITRATION WITH EDTA

Cobalt and iron form complexes with EDTA (ethylenediaminetetraacetic acid) in a ratio of one to one. In both cases, the compound formed with the trivalent metal is much more stable than the one formed with the divalent metal. The trivalent metal can thus be titrated in acidic solution without interference by the complex formation of EDTA with the divalent metal. The content of the divalent metal is calculated from the titration in basic solution, which gives the total content of the trivalent metal plus the divalent metal. Control of the *p*H is of special importance for Fe^{3+}, where OH^- competes strongly for the coordination sites, which might not be fully available to EDTA even if no precipitation of ferric hydroxides is observed. The *p*H of the solution should never rise above 3 in this case. The total metal content can be derived by the complexation of all metal ions with a surplus of EDTA and back-titration with standard Zn^{2+} solution at a *p*H of about 10. The zinc complex is, of course, less stable than either the iron or the cobalt complex. Because of the properties of Fe^{3+} the complexation must be performed at a *p*H of about 2, after which base and buffer can be added to yield a *p*H of 10, which makes the back-titration with Zn^{2+} feasible. The titration has to be completed within 3 minutes of adding the base because of the release of EDTA from the Fe^{3+} complex by competitive exchange with OH^-. With some planning it is entirely possible to perform the operation in this span of time since it is a simple titration to an indicator end point.

Two standardizations of EDTA solution should be performed. One against an Fe^{3+} solution from reagent iron wire and one against a Zn^{2+} solution of known concentration. Three titrations for each case is suggested; the result should, of course, be the same for both ions. These two methods of standardization should familiarize the student with buffers and indicators used in the actual analysis. The result of the analysis is computed as if all trivalent metals were iron and all divalent metals were cobalt. The sum of Fe_2O_3 and CoO weight calculated on the basis of the determined Fe and Co content should yield the original sample weight.

Apparatus

2 50-ml burets
2 250-ml beakers
2 400-ml beakers
100-ml volumetric flask
250-ml volumetric flask
Magnetic stirrer (optional)
*p*H meter (optional)

Chemicals (All Reagent Grade)

Tiron and Eriochrome Black T indicators
Ammonium chloride
Ammonium hydroxide
Ammonium persulfate
EDTA or disodium salt
Glacial acetic acid
Hydrochloric acid
Iron wire for standardizing
Sodium acetate
Sodium hydroxide
Zinc metal

Prepare the following solutions (I refers to materials used in the titration of $metal^{3+}$; II refers to materials for the back-titration):

Indicator I: 1 g of Tiron (1-2-dihydroxy-5,5-benzenedisulfonic acid) in 20 ml of H_2O. The solution shall be furnished.
Indicator II: 0.2 g of Eriochrome Black T mixed with 100 g of reagent grade NaCl. The mixture shall be furnished.
Buffer I: 250 ml of 10% sodium acetate in glacial acetic acid.
Buffer II: Add 90 ml of concentrated NH_4OH to 14 g of NH_4Cl (reagent grade) dilute to 250 ml with distilled water.
0.01 *M* EDTA: To 2.92 g of EDTA add water to make less than 1 liter. Add the minimum amount of NaOH (solid or concentrated solution) to dissolve the EDTA. Dilute to 1 liter.
0.01 *M* $ZnCl_2$: Weigh as accurately as possible 0.3269 g of reagent grade Zn. (Base the actual molarity of this standard solution on the weight of the Zn sample). Dissolve in 10 ml of 1:1 dilute HCl (warm if necessary). Dilute to 500 ml.

Standardization of EDTA Solution

Method I

Cut the iron wire in small pieces. Wash with benzene and acetone, air-dry for 10 minutes and oven-dry for 5 minutes. Accurately weigh 0.10 g of clean, dry iron wire into a small beaker. Add about 5 ml of concentrated HCl and place over a steam bath until all of the iron wire is dissolved. Transfer to a 100-ml volumetric flask, cool, and dilute with distilled water to the mark.

Transfer 10 ml of the solution to a 250-ml beaker, dilute to about 30 ml, and adjust the *p*H to 2 or 3 (never higher than 3) by adding 5 ml of Buffer I.

Add about 0.05 g of ammonium persulfate to oxidize all of the iron to the ferric state. Shake and add 7 drops of Indicator I; titrate immediately until the solution changes from blue to yellow. The indicator changes color gradually. The first titration might be regarded as a trial run to become familiar with the behavior of the indicator. Add a minute amount of ammonium persulfate and 2–3 drops of indicator; if the solution turns blue again, add more EDTA. Repeat the titration at least three times.

Method II

To 20 ml of EDTA solution in a 250-ml beaker add 25 ml of ethyl alcohol, 5 ml of Buffer II, and enough indicator mixture to color the solution just blue. Titrate with standard Zn^{2+} solution until the first red (purple) color appears. Titrate at least three times.

Sample Preparation

Accurately weigh 0.20 to 0.22 g of the prepared cobalt ferrite into a small beaker. Dissolve the sample in 5 ml of concentrated HCl. Place over a steam bath until dissolved, transfer to a 250-ml volumetric flask, cool, and dilute to the mark. Preparation of the sample should immediately precede the analysis since the oxidation state of the metal ions changes upon exposure to air.

Analysis of Fe^{3+}

Transfer a 25-ml sample to a 400-ml beaker, dilute to about 100 ml and bring the *p*H to 2 or 3 by adding 5 ml of Buffer I. Add 5–7 drops of Indicator I and titrate to a pink color without any blue tint. Repeat at least three times.

Analysis of Total Metal

Transfer a 25-ml sample to a 400-ml beaker. Add 150 ml of ethanol. The solution will appear yellow at this point. Adjust to a *p*H of 2 with HCl. Add 40 ml of standardized EDTA solution; the solution is now colorless or pale pink. Neutralize with NH_4OH (1:3 dilute); the solution turns pink. Add 10 ml of Buffer II and 0.3 mg of Indicator II. Immediately titrate to the first red color with the standard Zn^{2+} solution.

Calculations

From the experimental amount of Fe^{3+} and Co^{2+}, calculate the sample weight based on the sum of the oxides. How does the Fe^{3+} and Co^{2+} content compare to the theoretical content? Has any iron been titrated as Fe^{2+}, together with the Co^{2+}? Check this by the sample weight. Compare these results to those obtained after ion-exchange separation.

B. SPECTROPHOTOMETRIC ANALYSIS OF COBALT FERRITE FOLLOWING ION-EXCHANGE CHROMATOGRAPHIC SEPARATION

The preparation of $CoFe_2O_4$ by the coprecipitation of cobaltous and ferrous oxalate (see Exercise 2B) insures correct stoichiometry provided no ferric oxalate has been formed during precipitation and enough air has been present during the ignition process. Ferric oxalate is soluble and thus leads to a loss of iron in the mixture prepared for ignition, and insufficient access to air during ignition causes the formation of Fe^{2+} instead of Fe^{3+}. (The effect of these complications on the cobalt ferrite obtained can be assessed by reference to the discussion of spinel structure in Exercise 10B). The magnetic properties especially will be seriously affected and since it is these properties that determine the industrial desirability of the product, the detection of traces of Fe^{2+} becomes important. Since it is difficult to find small amounts of Fe^{2+} in the presence of large quantities of Fe^{3+} and Co^{2+}, the three ionic species are determined separately by spectrophotometric analysis following separation by ion-exchange chromatography. The optimization of the separation process was treated in Exercise 11B. If this experiment has been performed, the student can devise a procedure for the separation of Fe^{2+} and Co^{2+}; Fe^{3+} will not interfere since it is retained on the column and is removed with water later. The following procedure is to be followed by students who did not perform the experiment in Exercise 11B.

Apparatus

50-ml buret to serve as ion-exchange column
Glass wool
1-ml pipet
2-ml graduated pipet (0.1 ml)
2 50-ml beakers
4 100-ml volumetric flasks
2 25-ml volumetric flasks
Spectrophotometer with cells

Chemicals

12 *M* HCl purified by ion exchange
6 *M* HCl prepared by dilution of 12 *M* HCl
4 *M* HCl prepared by dilution of 12 *M* HCl
Nitroso-R-salt solution, 1% in water
Hydroxylamine solution, 5% in water (to be prepared immediately before use)
Sodium acetate
Dowex 1 × 8 ion-exchange resin, 100-200 mesh
Stock solutions of 5×10^{-3} *M* Fe^{2+} and 5×10^{-4} *M* Co^{2+}

Procedure

Accurately weigh 0.9–1.0 g of cobalt ferrite and dissolve in 15 ml of concentrated HCl over a steam bath. Transfer quantitatively to a 25-ml volumetric flask and fill to the mark with 6 *M* HCl.

While the sample is dissolving, prepare the column. Weigh about 15 g of the resin and prepare a slurry from it with distilled water. Put some glass wool into the lower end of the column and rapidly pour the slurry into the column while the resin is still suspended in the water. Tap the column gently as the resin settles to insure uniform packing.

The resin should fill about 24 cm of the buret. Drain the water off until it just reaches the resin level. Be careful to avoid letting the liquid level fall below the top of the resin bed at any time during the experiment. Wash the resin with two portions of distilled water and then with two 20-ml portions of 6 *M* HCl. Drain carefully to the resin level and introduce 1 ml of the ferrite solution from a pipet. During application of the sample, the pipet should be inserted almost to the resin level to avoid loss of the sample on the walls of the column. Drain to the resin level again and wash the walls directly above the resin with 1 ml of 6 *M* HCl. Drain again and wait for 10 minutes.

Fill the column with 25 ml of 6 *M* HCl. Do not disturb the resin surface while pouring the liquid. A glass-wool plug can be inserted to protect the surface. Let the eluant flow through the column at a rate of 0.5 cm/minute. Collect the first 15 ml in a 25-ml volumetric flask. This portion should contain all of the Fe^{2+}. Gather the second 10 ml in a 50-ml beaker, and test for iron. If iron is still present, combine both portions. The method of analysis is described in "Spectrophotometric Analysis."

Carefully introduce 25 ml of 4 *M* HCl into the column and let it percolate through at the same rate as before. Gather 25 ml in a 100-ml beaker. This portion should contain all of the Co^{2+}.

Add 50 ml of distilled water to the column in two portions and gather all of the eluant in a 100-ml volumetric flask. Allow a flow rate of 0.7 to 0.8 ml/minute. This portion contains all of the Fe^{3+}.

Spectrophotometric Analysis

To the first portion taken from the ion-exchange column add solid sodium acetate until a *p*H of at least 3 is reached (check with litmus paper). Complex the iron with 1.0 ml of orthophenanthroline solution. A red color due to the complex formed by one Fe^{2+} and three orthophenanthrolines will be observed if any Fe^{2+} is present in measurable amounts. Dilute to the mark with distilled water and measure the absorbance at 5120 Å.

To the 25 ml that contain Co^{2+}, add 60 ml of concentrated sodium acetate solution and 30 ml of 1 % nitroso-R-salt solution. Place over a steam bath for 10 minutes, cool, transfer quantitatively to a 100-ml volumetric flask, and dilute to the mark. The complex formed from three moles of dye with one mole of Co^{2+} is brown-red. Measure the absorbance of a sample after a dilution of 1:100 at 4100 Å and 5100 Å.

To the Fe^{3+} portion, add enough solid sodium acetate to bring the *p*H to higher than 3, add 10.0 ml of 5% hydroxylamine solution, and wait until the yellow color of the Fe^{3+} has disappeared. After all of the iron has been converted to the ferrous state, add 10.0 ml of orthophenanthroline solution. Dilute to the mark with distilled water. Measure the absorbance at 5120 Å after dillution of the sample to 1:100.

Spectrophotometry of Standard Solutions

The stock solutions are best prepared from reagent-grade iron wire and cobalt acetate. Since only a small amount of each is used by each student, it is advantageous to have the stock solutions prepared for the whole class.

The following measurements can conveniently be performed during the preparation of the ferrite since the preparation includes long periods of waiting. Transfer 1 ml of Fe^{2+} stock solution to a 100-ml volumetric flask and another milliliter to a 100-ml beaker. Add a small amount of hydroxylamine to insure complete reduction to Fe^{2+} (Fe^{2+} is oxidized upon exposure to air). Add enough solid sodium acetate to the flask to bring the *p*H to 5 after about 80 ml of distilled water has been introduced. Dilute the contents of the beaker in the same way. Introduce 0.5 ml of orthophenanthroline solution into the flask and 0.3 ml of nitroso-R-salt into the beaker. Place the beaker in the water bath for 10 minutes, cool, transfer the contents quantitatively to a 100-ml flask, and dilute to the mark.

Repeat the procedure with two 1-ml portions of Co^{2+} stock solution. Add 0.9 ml of nitroso-R-salt to the beaker. Check the *p*H before the final dilution and place the beaker in a water bath for 10 minutes before transferring the contents to a 100-ml volumetric flask. Cool, and dilute all solutions to correct volume.

Record the spectra between 350 mμ and 750 mμ. Perform a blank trial containing 0.3 ml of nitroso-R-salt solution without any metal in 100 ml. The spectra of Fe^{2+} with nitroso-R-salt and of Co^{2+} with orthophenanthroline are observed only for reference so that contaminations due to incomplete separation can be detected. (Different batches of ion-exchange resin sometimes perform slightly differently. Changes in the rate of elution or inhomogeneous packing can also lead to incomplete separations.) These spectra can be discarded if separation is complete.

From the spectra of iron with phenanthroline and cobalt with nitroso-R-salt, select wavelengths suitable for analysis and calculate molar absorptivities at these wavelengths. Alter the wavelengths proposed in the preceding discussion according to your discretion and be sure to measure standards and samples under exactly the same conditions. Since your calculations are based on the Lambert-Beer law, check its validity by recording spectra at different concentrations. A plot of absorbance against concentration should yield a straight line. If you find deviations from the Lambert-Beer law, this plot serves as a calibration curve.

Calculations

Calculate the percentage of Co^{2+}, Fe^{2+}, and Fe^{3+} in the ferrite that you analyzed. Report the accurate formula. Does the total amount of recovered metals, together with the appropriate amount of oxygen, yield the weight of the sample that was introduced into the column?

Note: Fe^{2+} can also be analyzed by complexation with nitroso-R-salt. Analysis with ortho-phenanthroline following reduction of the metal with hydroxylamine is preferred, however, because the metal-nitroso-R-salt is destroyed upon standing if excess hydroxylamine remains in solution.

PART IV

CHEMICAL KINETICS

EXERCISE **14**

Rate Measurements

A. MEASUREMENT OF THE RATE OF CONVERSION OF *Cis*-$[Co(en)_2Cl_2]Cl$ TO THE *Trans* FORM

The preparation of *cis*- and *trans*-$[Co(en)_2Cl_2]Cl$ is described in Exercise 1C. The *cis* compound undergoes isomerization to yield the *trans* form at a conveniently measurable rate if dissolved in methanol at 45°C. The rate of the isomerization reaction is to be determined under the assumption that the reaction is of first order and goes to completion. The progress of the isomerization is observed spectrophotometrically.

A first-order reaction follows the rate law

$$\frac{d[\text{cis}]}{dt} = k[\text{cis}] \tag{1}$$

or

$$\ln \frac{[\text{cis}]_t}{[\text{cis}]_0} = kt$$

The concentrations of *cis* and *trans* forms are related to one observable A (absorbance) by the Lambert-Beer law.

$$A_{\text{cis}} = a_{\text{cis}} \cdot b \cdot [\text{cis}] \tag{2}$$

$$A_{\text{trans}} = a_{\text{trans}} \cdot b \cdot [\text{trans}]$$

where a is the absorption coefficient, b the length of the optical path within the solution, and [*cis*] and [*trans*] are the concentrations.

Within the visible region, the spectra of the *cis* and the *trans* forms are distinctly different, but there is not a single maximum, free from overlap with other absorptions, that may be used for spectrophotometric analysis in this region. To be able to follow the reaction progress at one wavelength, the following property of an isomerization (expressed in terms of the present example) is used.

$$[\text{trans}]_t = [\text{cis}]_0 - [\text{cis}]_t \tag{3}$$

From equations (2) and (3) the following relation can be derived:

$$\frac{[\text{cis}]_t}{[\text{cis}]_0} = \frac{A_t - A_\infty}{A_0 - A_\infty} \tag{4}$$

where $[cis]_t$ is the concentration at time t; $[cis]_0$ is the concentration at time $t = 0$; A_t is the absorbance at time t; A_0 the absorbance at time $t = 0$; and A_∞ is the absorbance after the reaction has gone to completion. Absorbances pertain to the reaction mixture composed of *cis* and *trans* forms in solution.

Equation (4) eliminates the necessity of determining extinction coefficients and possible error in the initial concentration of the *cis* compound resulting from initiation of the reaction before the sample completely dissolves.

Procedure

To find a suitable wavelength, record the spectra of the *cis* and the *trans* compounds in the region between 3400 Å and 7000 Å. Optimum conditions are met when a large difference in a_{trans} and a_{cis} is observed, preferably at an absorption maximum of one of the substances. (Why?) Set a spectrograph with a thermostated cell compartment to the desired wavelength and bring the temperature in the cell compartment up to 45°C. Prepare a 1×10^{-3} M solution of the *cis* compound in methanol and thermostat at 45°C for 20 minutes. Fill a 10-cm cell and place it in the cell compartment. (If cells of shorter light path are used, the concentration must be increased accordingly.) Measure the absorption of the reaction mixture for at least three hours in intervals of 15 minutes. (To minimize the time required for dissolution of the weighed samples, grind the sample thoroughly before weighing.)

Take the spectrum at the next laboratory period, since A_∞ is obtained from the spectrum of the reaction mixture after at least 12 hours. The reaction mixture does not have to be thermostated in the meantime.

Evaluation

Use equations (1) and (4) to obtain the rate constant k. A graphical solution is preferred. Do your data agree with a first-order rate law? Do absorbing impurities have an effect on the method used for data evaluation? Can the *trans* spectrum that was used in wavelength selection be employed in the determination of A_∞?

B. INVESTIGATION OF THE KINETICS OF FORMATION OF NITRO- AND NITRITO-PENTAAMMINECOBALT(III) ION

In Exercise 1B nitro- and nitritopentaamminecobalt(III) chloride were prepared from chloropentaamminecobalt(III) chloride and sodium nitrite in a slightly acidic medium. If the preparation was kept cold at all times, the nitrito compound was obtained; otherwise the reaction yielded the nitro compound. It was observed that the nitrito form isomerized to yield the nitro form upon standing at room temperature. The kinetics of this reaction have been investigated by Pearson et al. (1954).

In this experiment, an attempt will be made to determine whether the nitro form can be obtained directly at room temperature [see reaction (R1)]; or is the nitrito compound always formed first, followed by isomerization? If the latter were true, could a reaction mechanism be extracted from observations of the rate law?

At the beginning of the preparative reactions, the use of $[Co(NH_3)_5Cl]^{2+}$ as a reactant is a bit misleading since the actual reaction [as formulated in reactions (R1) and (R2)] starts from $[Co(NH_3)_5H_2O]^{3+}$. The preceding replacement of Cl^- in the coordination sphere by H_2O is very slow and will be not investigated here. In addition, there is an equilibrium between $[Co(NH_3)_5H_2O]^{3+} \leftrightarrow [Co(NH_3)_5OH]^{2+}$ that precedes the reaction of interest to us. The equilibrium constant for this deprotonation reaction is $K_{aq} = 2.0 \times 10^{-6}$. Thus very little hydroxo complex is present, but equilibration is very fast. The starting material in this investigation is therefore $[Co(NH_3)_5H_2O](NO_3)_3$. The complex was prepared as a nitrate for reasons of convenience in the process of synthesis (see Exercise 1D).

The two proposed reactions leading to the formation of the nitro complex are:

$$[Co(NH_3)_5H_2O]^{3+} + NO_2^- \xrightarrow{k_1} [Co(NH_3)_5NO_2]^{2+} \tag{R1}$$

or

$$[Co(NH_3)_5H_2O]^{3+} + NO_2^- \rightarrow [Co(NH_3)_5ONO]^{2+} \rightarrow [Co(NH_3)_5NO_2]^{2+} \tag{R2}$$

The last step in reaction (R2) can be investigated independently, since the nitro, as well as the nitrito, compound is available as the product of Exercise

1B. The isomerization reaction (R3) will thus be the first reaction to be investigated in this experiment.

$$[Co(NH_3)_5ONO]^{2+} \xrightarrow{k_i} [Co(NH_3)_5NO_2]^{2+} \quad (R3)$$

The reaction goes to completion in aqueous solution and the compounds have characteristically different visible spectra providing a convenient means for the observation of the reaction.

The exclusion of the back-reaction in (R3) has some favorable consequences for the investigation of reactions (R1) and (R2). The rate of disappearance of the aquo complex according to reaction (R1) can be written formally as

$$-\frac{d[\text{aquo}]}{dt} = k_1[\text{aquo}]^x[NO_2^-]^y - k_e[\text{nitro}] \quad (1)$$

in which [aquo] stands for the concentration of the aquo complex and [nitro] for the concentration of the nitro complex. The last term in equation (1) results from the possible replacement of NO_2^- in the nitro complex by H_2O.

$$[Co(NH_3)_5NO_2]^{2+} + H_2O \xrightarrow{k_s} [Co(NH_3)_5H_2O]^{3+} + NO_2^- \quad (R4)$$

The substitution reaction is rather slow under the present circumstances; therefore the last term in equation (1) can be disregarded. (It is worthwhile to check on the validity of this operation after measurements have been completed.) Whether or not the last term is disregarded, stoichiometry demands that for each mole of nitro compound formed one mole of aquo compound must disappear. The observation of spectral changes should prove this.

The rate of disappearance of the aquo form according to reaction (R2) assumes the same form as equation (1) except that the last term now refers to the replacement of both ONO^- and NO_2^- by H_2O. Again, this last term can be disregarded. Spectrophotometric investigation of the reacting mixture gives evidence of the intermediate nitrito compound, unless the formation of the nitrito complex is very slow compared with the isomerization reaction. In this case, practically all of the nitrito complex isomerizes immediately after formation. The rate of formation of the nitro compound is then again found to be equal to the rate of disappearance of the aquo compound.

Furthermore, we are interested in the mechanism by which the NO_2^- or the ONO^- group is built into the complex. To this end, the order must be found with respect to the aquo complex [x in equation (1)] and to NO_2^- [y in equation (1)]. This is conveniently done by keeping the concentration of nitrite so high compared with the concentration of the aquo complex that the former virtually does not change during reaction. Thus the reaction order is obtained with respect to the aquo complex. Several kinetic runs under these conditions, but with varying absolute amounts nitrite, indicate the dependence of the obtained rate constant on the concentration of nitrite. Thus the rate with respect to nitrite can be obtained. Keep in mind that the reacting species may include ONONO formed in the following reaction:

$$2NO_2^- + 2H^+ \longrightarrow N_2O_3 + H_2O$$

The results of kinetic investigations on ionic systems can be compared only in a strict sense, if the ionic strength of the media have been kept constant. All runs should therefore be made in solutions of constant ionic strength.

Nitrous acid decomposes in solution if present above certain concentration limits. This difficulty can be circumvented by use of the weak acid properties of HNO_2. At constant pH, a constant ratio of undissociated to dissociated acid will be found in solution, and whichever species is the reacting one will always be present in a concentration proportional to the total nitrite content, $[NO_2^-]_{\text{total}} = [NO_2^-] + [HNO_2] + 2[N_2O_3]$, since equilibrium is rapid. Investigations in a medium of constant pH offer the same advantage for the aquo and hydroxo complex.

Measurements

1. Record the spectra of $[Co(NH_3)_5H_2O](NO_3)_3$, $[Co(NH_3)_5ONO]Cl_2$, and $[Co(NH_3)_5NO_2]Cl_2$ in concentrations of 5×10^{-3} moles per liter in H_2O (immediately after preparation of the solutions) with H_2O in the reference path on a double-beam recording spectrophotometer.

Find the wavelengths at which two of the compounds have the same molar extinction coefficient ϵ. Since observation may be obscured by impurities and deviations in concentrations, these wavelengths are given as follows: $\epsilon_{\text{nitro}} = \epsilon_{\text{nitrito}}$ at 4120 Å and 4840 Å; $\epsilon_{\text{nitrito}} = \epsilon_{\text{aquo}}$ at 5250 Å.

2. Record the spectrum of a 10^{-2} M nitrito compound in a buffer solution of pH 4.1, ionic strength of 1.125, and 0.2 M in $[NO_2^-]_{\text{total}}$. Take the first measurement immediately after preparation, the second 15 minutes later, and the third after another

15 minutes. Further measurements are recorded at intervals of 30 minutes. Record the last spectrum after about 24 hours. The cell in the reference path of the spectrophotometer is filled with buffer solution.

3. Record the optical density of an 8×10^{-3} M aquo complex in buffer solution containing 0.4 M $[NO_2^-]_{total}$ at 4120 Å and 4840 Å immediately after preparation and every 10 minutes thereafter until constant absorption is reached.

4. Measure the extinction of an 8×10^{-3} M solution of aquo complex at 4120 Å in buffers 0.1, 0.2, and 0.3 M in $[NO_2^-]_{total}$ immediately after preparation of the solution and at intervals of 15 minutes. Increase the intervals as the reaction slows down until the extinction does not change any more.

Data

Assume a first-order rate process for the isomerization reaction and analyze the data from part 1 under "Measurements" according to

$$\frac{[\text{nitrito}]}{[\text{nitrito}]_0} = \frac{A - A_\infty}{A_0 - A_\infty}$$

where A is the absorbance of the mixture at a given time and at a suitable wavelength; A_0 is the absorbance at that wavelength at time $t = 0$; and A_∞ is the same after the reaction has gone to completion. Check whether the reaction is first-order and derive k by plotting log $(A - A_\infty)$ against t.

Treat the data obtained under parts 3 and 4 in the same way. Find the rate constant for the disappearance of the aquo compound. Is the rate first-order in the aquo compound? Check the dependence of the rate constant on total nitrite concentration in the buffer and derive the reaction order with respect to nitrate concentration. If your results are not satisfactory, repeat the measurements under part 4 at intermediate concentrations of total nitrite and suitable time intervals.

Compare the rate constant found for the isomerization reaction to that for the disappearance of the aquo compound at 0.4 M total nitrite. What is the rate determining step in the formation of the nitro complex from the aquo compound?

Check the data under part 3. Do the changes confirm the result obtained by comparing rate constants?

Postulate a reaction mechanism that explains all of your observations. Devise additional experiments designed to prove your point.

Reevaluate the preparation method in Exercise 1B in view of your present results.

C. THERMAL CYCLIZATION OF HEXAFLUOROBUTADIENE

Concepts and methods of homogeneous kinetics are demonstrated in the unimolecular isomerization of hexafluorobutadiene:

$$F_2C{=}CF{-}CF{=}CF_2 \rightleftarrows \text{cyclo-}(F_2C{-}CF_2{-}CF{=}CF)$$

The reaction that does not proceed completely to one side was investigated by Schlag and Peatman (1964).

According to Lindemann and Christiansen, a unimolecular reaction proceeds in three steps: activation, deactivation, and reaction to form the product.

$$A + M \xrightarrow{k_1} A^+ + M \tag{1}$$

$$A^+ + M \xrightarrow{k_2} A + M \tag{2}$$

$$A^+ \xrightarrow{k_3} \text{Pr} \tag{3}$$

The reverse reaction, Pr $\rightarrow A$, will not be considered at this point. A^+ represents a molecule that possesses energy equal to or higher than the amount required for reaction; A^+ is an activated molecule. If a steady state has been reached, that is, the production of activated molecules is equal to the destruction of activated molecules by deactivation or reaction, the rate of disappearance of A is given by

$$-\frac{dC_A}{dt} = \frac{k_1 k_3 C_A C_M}{k_2 C_M + k_3}, \tag{4}$$

where C_A is the concentration of A and C_M is that of M.

At sufficiently high pressures, activation and deactivation occur much more frequently than reaction, because of the large number of collisions per unit of time, which only accidentally leaves enough time for A^+ to react between collisions. In other words, $k_2 C_M \gg k_3$. Under this condition, equation (4) becomes

$$-\frac{dC_A}{dt} = \frac{k_1 k_3}{k_2} C_A = k_f \cdot C_A \tag{5}$$

The reaction is observed to be first order.

At low pressure, collisions become infrequent, and the long time between collisions favors reactions rather than deactivation: $k_2C_M \ll k_3$. Thus equation (4) can be written as:

$$-\frac{dC_A}{dt} = k_1 \cdot C_M C_A \tag{6}$$

The reaction rate now appears to be second order. Since $k_f > k_1$, the observed rate of disappearance of A decreases in a complicated manner from its high pressure limit to its low pressure limit, with decreasing total pressure. The pressure at which the rate becomes sensitive to the addition of an inert gas depends on the nature of the molecule A. The time necessary for the redistribution of energy over internal degrees of freedom so that favorable conditions for reaction are obtained is greater if more internal degrees of freedom are available. As a consequence, the high pressure limit is reached at lower and lower pressures as the reacting molecule becomes larger. (For hexafluorobutadiene, the high pressure limit is approached at approximately 4 mm Hg.) Quantitative treatment of the region below the high pressure limit is possible through the use of the Rice-Ramsperger-Kassel model. However, in this experiment only the behavior of the reaction within the high-pressure limit will be considered. The transition state theory is particularly useful for the interpretation of data.

The molecule in the transition state is X^+. Whereas A^+ was any molecule with enough energy for reaction, X^+ is now a molecule out of the pool of all A^+ that has, among the energies of all internal degrees of freedom, those that favor reaction. The reactant A and the transition state species X^+ are in equilibrium.

$$A \rightleftarrows X^+ \rightarrow \text{Pr}$$

The rate constant k_f may then be expressed, with the aid of the equilibrium constant $K^{\ddagger}$, for the reaction $A \leftrightarrow X^+$.

$$k_f = \kappa \frac{kT}{h} \cdot K^{\ddagger} \tag{7}$$

where h = Planck's constant, and κ, the transmission coefficient, is found between 0.5 and 1.0. For details of the derivation of equations (4) to (8) see Eggers et al. (1964).

Thermodynamics yields $K^{\ddagger} = \exp(-\Delta G^{\ddagger}/RT)$ and (7) becomes:

$$k_f = \kappa \frac{kT}{h} \cdot \exp\left(\frac{-\Delta H^{\ddagger}}{RT} + \frac{\Delta S^{\ddagger}}{R}\right) \tag{8}$$

Empirically, the rate constant can be expressed by the Arrhenius equation:

$$k_f = A \exp\left(-\frac{E_a}{RT}\right) \tag{9}$$

where E_a is the energy of activation.

The relation between the two equations becomes apparent, if the following is introduced:

$$\Delta H^{\ddagger} = \Delta E^{\ddagger} + \Delta(PV) = \Delta E^{\ddagger} + \Delta n(RT) \tag{10}$$

If the number of moles does not change during reaction, $\Delta n = 0$ and

$$\Delta H^{\ddagger} = \Delta E^{\ddagger} \tag{11}$$

Thus equation (8) becomes:

$$k_f = \kappa \frac{kT}{h} \exp\left(\frac{\Delta S^{\ddagger}}{R}\right) \exp\left(-\frac{\Delta E^{\ddagger}}{RT}\right) \tag{12}$$

or

$$\ln k_f = \ln\left(\kappa \cdot \frac{kT}{h}\right) + \frac{\Delta S^{\ddagger}}{R} - \frac{\Delta E^{\ddagger}}{RT} \tag{12a}$$

$$= \text{const} - \frac{\Delta E^{\ddagger}}{RT}$$

where const compares to $\ln A$ and $\Delta E^{\ddagger}$ to E_a if only small temperature intervals are considered. From a plot of $\ln k_f$ against $1/T$, the energy of activation, $E_a = \Delta E^{\ddagger}$, and the entropy of activation, $\Delta S^{\ddagger}$, can be obtained. Over a limited temperature range, the first term on the right-hand side of equation (12a) is practically temperature independent, so that E_a can be derived from the slope, whereas the entropy of activation is given by:

$$\Delta S^{\ddagger} = R\left[\text{const} - \ln\left(\kappa \cdot \frac{kT}{h}\right)\right] \tag{13}$$

As an approximation $\kappa = 1$. In addition, if the equilibrium of reactant and product can be measured, the rate of the reverse reaction at the temperature can be calculated.

$$K_{eq} = \frac{C_{Pr}}{C_A} = \frac{k_f}{k_{-f}} \tag{14}$$

where k_{-f} is the rate of the reverse reaction.

$$\text{Pr} \xrightarrow{k_{-f}} A \tag{15}$$

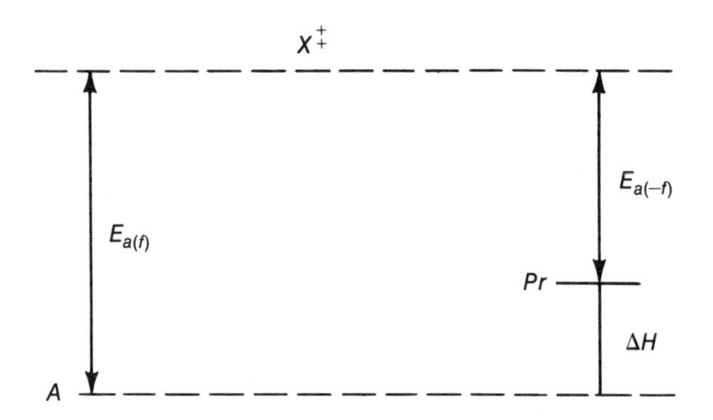

FIGURE 14-1
Relation between energies of activation and enthalpy of reaction for $A \rightleftarrows X^{\ddagger} \rightarrow Pr$.

From the temperature dependence of the equilibrium constant, activation energy and activation entropy can be derived.

$$\ln K_{eq} = -\Delta G/RT = -\frac{\Delta H}{RT} + \frac{\Delta S}{R} \tag{16}$$

also

$$\ln \frac{k_f}{k_{-f}} = \frac{\Delta S_f^{\ddagger} - \Delta S_{-f}^{\ddagger}}{R} + \frac{E_{a(-f)} - E_{a(f)}}{RT} \tag{17}$$

from equations (14) and (12a). Again, $\kappa = 1$ was assumed for forward and reverse reactions. Thus the enthalpy of reaction, as obtained from the temperature dependence of the equilibrium constant, is equal to the difference in activation energies for forward and reverse reactions.

$$E_{a(-f)} - E_{a(f)} = -\Delta H \tag{18}$$

(see Figure 14-1). Similarly, the entropy of reaction is equal to the difference in entropies of activation for forward and reverse reactions.

In principle, it is possible, although difficult, to measure forward and reverse reactions at the same temperatures. Due to the difference in rate constants, one reaction may be too slow or too fast in the region in which the other is easily measurable.

Rate Law

Since forward and reverse reactions occur together at any given temperature and time, the rate of appearance of the product Pr is written as:

$$\frac{dC_{Pr}}{dt} = k_f(C_0 - C_{Pr}) - k_{-f}C_{Pr} \tag{19}$$

where C_0 is the initial concentration of A, and C_{Pr} the concentration of the product formed. At equilibrium, $dC_{Pr}/dt = 0$, because no net reaction occurs. Thus

$$k_{-f} = \frac{k_f(C_0 - C_{Pr}^{eq})}{C_{Pr}^{eq}} \tag{20}$$

with C_{Pr}^{eq} as the equilibrium concentration of the product.

By introducing equation (20) into equation (19) and rearranging it, the following is obtained:

$$\frac{dC_{Pr}}{dt} = k_f \frac{C_0}{C_{Pr}^{eq}} (C_{Pr}^{eq} - C_{Pr}) \tag{21}$$

Integration yields:

$$\int_0^{C_{Pr}} \frac{dC_{Pr}}{(C_{Pr}^{eq} - C_{Pr})} = \int_0^t k_f \frac{C_0}{C_{Pr}^{eq}} dt \tag{22}$$

which leads to:

$$\ln \frac{C_{Pr}^{eq}}{(C_{Pr}^{eq} - C_{Pr})} = \frac{C_0}{C_{Pr}^{eq}} k_f t \tag{23}$$

From equation (20) the following is derived:

$$\frac{C_0}{C_{Pr}^{eq}} k_f = k_{-f} + k_f \equiv k_0 \tag{24}$$

and from equations (23) and (24):

$$\ln \frac{C_{Pr}^{eq}}{(C_{Pr}^{eq} - C_{Pr})} = k_0 t \tag{25}$$

The product concentration in equilibrium is calculated from the equilibrium constant in equation (14).

$$C_{Pr}^{eq} = C_0 \left(\frac{1 + K_{eq}}{K_{eq}}\right) \tag{26}$$

A plot of $-\ln(C_{Pr}^{eq} - C_{Pr})$ against t yields k_0 as the slope. From k_0 the two rate constants k_f and k_{-f} can be obtained with the aid of equation (24). Equations (23) through (26) can of course, also be written in terms of C_A and C_A^{eq}. The selection of the preferable form depends on the accuracy with which C_A or C_{Pr} can be measured.

Proof of Unimolecular Homogeneous Reaction

The previous discussion was based on the assumption that a homogeneous unimolecular gas-phase reaction is being dealt with. Therefore the first task is to establish the validity of this assumption. Experimental proof for homogeneous gas-phase reaction can be obtained through variation of the ratio of surface to volume. If the reaction proceeds only in the gas phase, it should not be affected by this ratio. Variation of the surface to volume ratio is accomplished by filling the reaction vessel with inert material, such as glass wool, glass beads, and so forth, and observing the reaction in this environment. Since these measurements are quite time-consuming, the results of the experiments cited in Schlag and Peatman (1964) will serve the purposes of this discussion and the cyclization of hexafluorobutadiene will be accepted as a truly homogeneous gas-phase reaction. That the reaction is unimolecular remains to be shown. The detection of a region in which the rate is independent of inert gas pressure and is of first order, together with a decrease in rate and increased sensitivity to inert gas pressure as soon as the latter falls below a certain value (high pressure limit), is considered a sufficient criterion for unimolecularity. In addition, the nature of the inert gas is of no consequence. Nitric acid is frequently used as inert gas because of the sensitivity of reactions that involve free radicals towards NO. In this experiment, an attempt will be made to prove the unimolecularity of the reaction.

Apparatus

The experiment is performed on the vacuum line shown in Figure 14-2. The reactant hexafluorobutadiene is introduced into the system from a tank by way of the bubbler manometer to storage volume V_A. Any desired inert gas or the product hexafluorocyclobutene, which is also available in tanks, can be stored in volume V_B. The storage section is evacuated through the stopcock B. Notice that the section between stopcocks C, J, and K is evacuated simultaneously. The volume of this section up to stopcocks A and B and the stopcocks on V_A and V_B is known. The section has access to the manometer through stopcock J, so that by knowing the volume and by measuring the pressure, the number of moles in this volume, V_M, can be calculated. The reaction vessel, V_R, is embedded in an oven and $V_M < V_R$. The volume of V_R between stopcocks K and 2 is also known. Any amount of gas measured in V_M can be transferred into V_R by condensation into the side arm on three-way stopcock K followed by distillation into V_R with K in the proper position. This process, as well as the evacuation of V_R, is slow because of the capillary section between V_R and K. This capillary is necessary to reduce the "dead volume" of V_R, that is, the volume that will not be at the proper temperature. Gas in the dead volume will not react at the same rate as the bulk of the gas in V_R. Care has to be taken that sufficient time is allowed for evacuation of volume V_R whenever the need arises. Samples of the reaction mixture can be extracted from V_R for analysis by opening stopcock 2 to the evacuated known volume V_s between stopcocks 1, 2, and 3. The sample in V_s is transferred by condensation into trap T_1 from which it is later transported into the gas chromatograph, where the analysis is performed. Details of this part of the apparatus are shown in Figure 14-3. If, for some reason, the sample cannot be transferred into the gas chromatograph immediately, it can be stored in any of the four small volumes to the left of stopcock 4. This transfer is done by trap-to-trap distillation. The volume V_s can be evacuated through stopcock 1. Traps T_2 and T_3 serve as containers for emergency recoveries of the reactant and are not used in normal operation. The section between E, F, and G serves the same purpose. It has access to the forepump directly through stopcock G and is arranged so that the reaction section may be evacuated by the forepump if large quantities of inert gas or air have accidentally accumulated. The manometer is connected to the main vacuum line on the left side. Usually, it would be sealed and evacuated at that side, but since the accidental application of pressure higher than atmospheric to the right side is possible, provision has been made for easy evacuation through stopcock H.

Analysis System

For analysis, sample mixtures are separated in a gas chromatograph fitted with a column, 350 cm long with an inner diameter of 0.5 cm, filled with Kel-F 3 oil on firebrick (12% by weight) and kept at 0°C. The detector is a heat conductivity cell, the output of which is read from a chart recorder. Helium is used as a carrier gas. The ratio of reactant to product in the sample is obtained as the ratio of areas under the two peaks (see Exercise 11A for explanations). Figure 14-3 shows a schematic representation of the gas chromatograph and the injection system in detail.

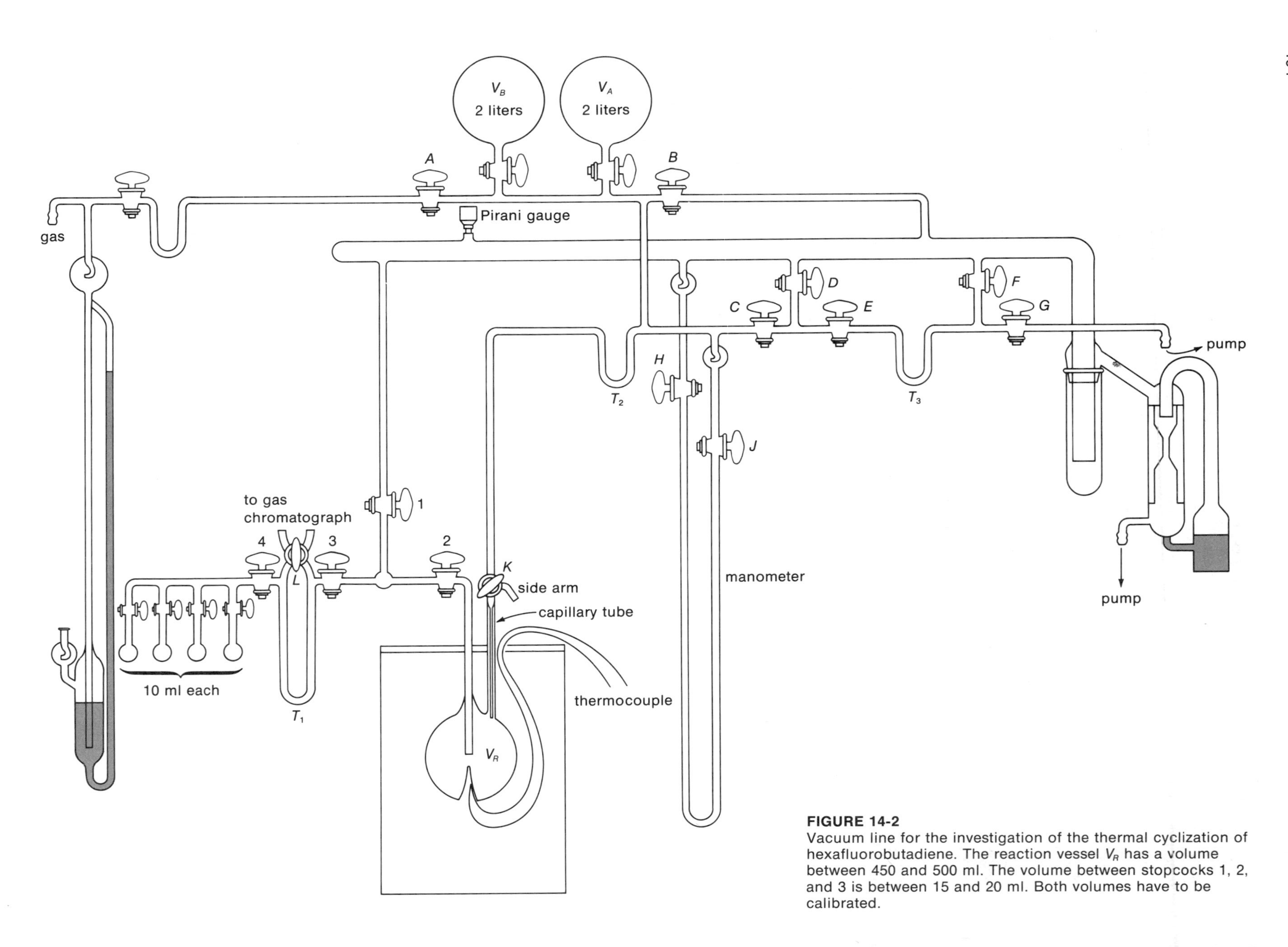

FIGURE 14-2
Vacuum line for the investigation of the thermal cyclization of hexafluorobutadiene. The reaction vessel V_R has a volume between 450 and 500 ml. The volume between stopcocks 1, 2, and 3 is between 15 and 20 ml. Both volumes have to be calibrated.

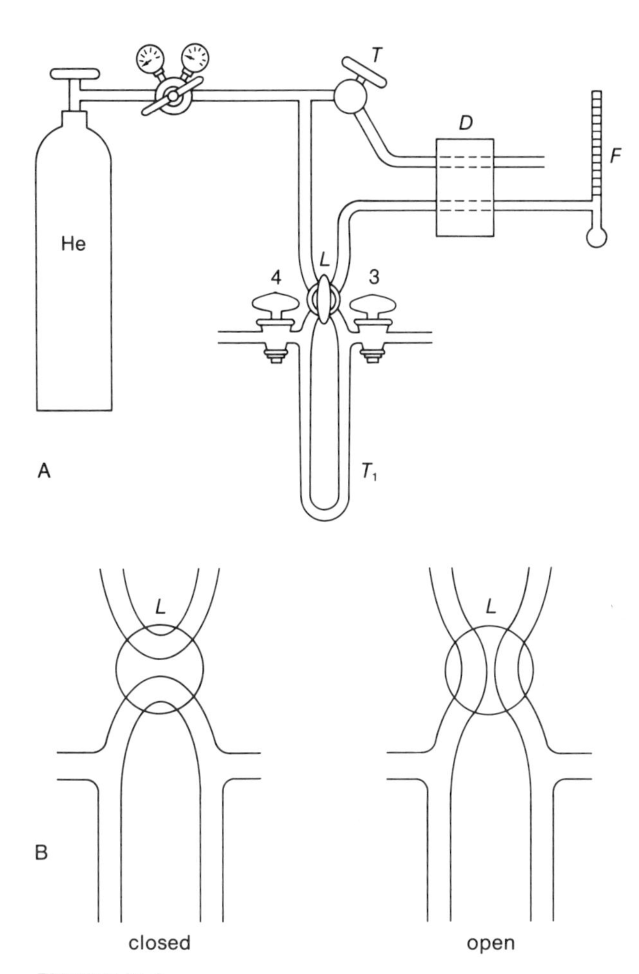

FIGURE 14-3
A. Gas chromatograph: *D*, detector; *F*, flow meter; *T*, throttle to regulate reference flow. B. Details of injection system.

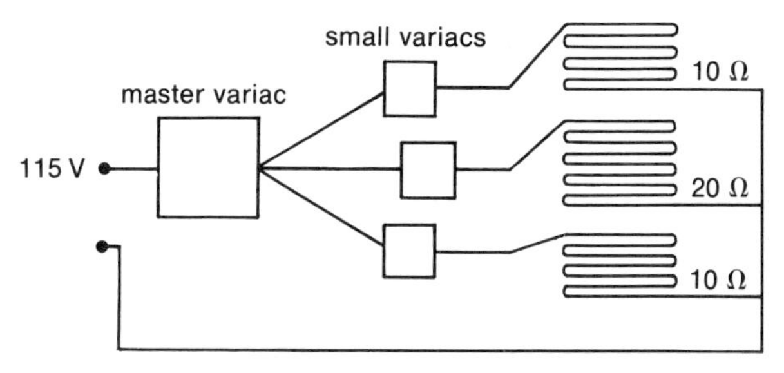

FIGURE 14-4
Wiring diagram of oven.

The flow rate has to be regulated to a constant 95 ccm/minute. Since the flow rate is given by the flow resistance of the column and the pressure drop across it, the flow rate can be regulated with the injection system closed and will not change upon opening *L*. An equal flow rate should be maintained for both the sample and reference. The flow meter can be attached to either one, so that the rate can be read separately during regulation.

Oven

The oven is made from a piece of iron tubing with an inner diameter of 16 cm. Three individual heating coils are wound around a thin layer of asbestos. They are packed in several layers of asbestos paper and then wrapped in material used to insulate steam pipes. Each heating coil is attached to a five amp variac connected to a 20 amp master variac. Figure 14-4 shows the wiring diagram of the oven. Before any experiment is performed, settings that assure a uniform heating of the oven must be found for the three individual variacs. The variacs remain at these settings and should not be altered until, in the course of much more than a semester, deviations from a constant temperature profile are detected. A given temperature for an individual experiment is regulated with the aid of the master variac. A calibration curve showing the approximate temperatures to be reached with different master variac settings should be made available with the apparatus.

The oven is covered with a heavy asbestos lid and is closed by an asbestos bottom. The space remaining around the reaction vessel is packed with glass wool to prevent air currents and assure temperature uniformity. Temperatures surpassing 500°C should be avoided because the glass wool begins to melt and damage to the reaction vessel is incurred.

The oven heating must be switched on several hours before starting the experiment, to allow temperature stabilization.

Temperature Measurement

Measuring the temperature in the reaction vessel is aided by a chromel-alumel thermocouple. It extends into the well on the bottom of the reaction vessel, so that the temperature is measured in the vicinity of the sample capillary. The thermocouple can be calibrated by observation of the cooling curves of Sn (231.9°C) and Pb (327.4°C). This must be done before proceeding with the experiment, or a calibration chart should be made available to students.

Note: Reread Exercises 4A and 11A before starting the experiment.

Procedure

The reaction proceeds at a measurable rate between 190° and 270°C. Below 180°C, hexafluorobutadiene reacts mainly to form dimers and trimers and pure cyclization is found only above 190°C. At temperatures above 270°C, the reaction proceeds too fast to allow the withdrawal of samples at short enough intervals. In fact, even at higher temperatures within the allowable range, samples have to be stored in the small storage volumes prior to analysis, since the time required for separation in the gas chromatograph exceeds the time interval between samplings. The pressure in the reaction vessel can be adjusted to 18 mm Hg of hexafluorobutadiene to insure complete separation.

The furnace should be turned on approximately 3 hours before starting the experiment. Begin with a lower variac setting than required for the temperature you want to reach; correct the variac setting before beginning the final preparations for your experiment. First, evacuate the system with all stopcocks open (except L to the gas chromatograph) with the aid of the forepump. After the pump has quieted down, place a Dewar flask of liquid nitrogen under the main trap in front of the mercury diffusion pump and start the heat and cooling water for the diffusion pump. After a reading of 1 μ or better has been reached on the Pirani gauge, shut off all parts of the line and reopen systematically against the gauge to find out whether the line leaks. If everything is all right, admit hexafluorobutadiene to V_A through the bubbler manometer and stopcock A. Close V_A and evacuate the line. If desired, an inert gas or hexafluorocyclobutene can now be admitted to V_B in the same way. For any kinetics run or for measurement of equilibrium, a certain amount of hexafluorobutadiene must be transferred to the reaction vessel. All kinetics runs are made with hexafluorobutadiene as starting material. For equilibrium measurements, it is advisable to start some of the runs with the cyclo compound, to prove that the equilibrium is reached from both sides without any side reactions. From the volume of V_M, room temperature, the volume V_R, and the furnace temperature, calculate how large the pressure in V_M has to be to reach a pressure between 7 and 18 mm Hg in V_R. Use the ideal gas law for all calculations. Close stopcocks K, H, C, A, and B. Open J and slowly admit enough gas from V_A to reach the calculated pressure in V_M. Close V_A, place a Dewar flask of liquid nitrogen under trap 2, and condense all of the gas in V_M in this trap.[1] Open C and D, and pump on the frozen starting material. By this procedure, all permanent gases that may have accidentally entered V_A during filling are removed. Close C and D. Allow the gas in trap 2 to return to room temperature. Carefully measure room temperature and pressure in V_M. These measurements are used to calculate the pressure in V_R. Place a small Dewar flask of liquid nitrogen under the side arm on K and transfer all gas into the side arm. Now close K to V_M, and open the sidearm to V_R, starting a timer simultaneously. At a temperature of approximately 200°C, take the first sample after 10 minutes and subsequent samples as soon as the chromatograph is free, that is, about every 13 minutes. At high temperatures, start after 2 minutes and take samples as fast as possible, storing them in the small storage volumes. Since the time required depends very much on the skill of the experimenter and, at low temperatures, on the time of the gas chromatographic separation, exact directions cannot be given here. Each student will have to devise his own scheme after a few trial runs. Samples are taken by opening stopcock 2 to the volume enclosed between 1 and 3. Place a Dewar flask of liquid nitrogen under T_1. After a few seconds close 2 and open 3 to the trap. After about 1 minute close 3, remove the Dewar flask so that the sample vaporizes. Open L to the gas chromatograph so that the sample is swept into it by the entering helium flow. As soon as stopcock 3 has been closed, open 1 to the pump to evacuate the system for a new sample. Continue the procedure until the chromatogram shows virtually no change in the ratio of peaks. The equilibrium measurements at low temperature should be taken after several hours. At high temperatures 30 minutes will suffice. Experience with the kinetics runs will give an indication of how long a waiting period is necessary.

At the end of the experiment, leave the line under vacuum and shut off the pumps and furnace heat. If the experiment is to be continued on the following day, the oven may be left at a moderate temperature to save time in the heating process.

[1] Either close J before condensing or correct for the gas volume in the manometer. For a method see Exercise 4A, "Yield Determination."

Make sure that all traps are empty and shut off the helium flow to the gas chromatograph, as well as the voltage to the conductivity cell.

Measurements

Determine k_0 at different temperatures (between 190° and 260°C). Measure the equilibrium composition, preferably starting from the open, as well as from the cyclo compound. At each temperature take runs with different starting pressures (between 6 and 18 mm Hg). Roughly evaluate the data between runs. Add small amounts of inert gas or NO (in the same order of pressure as the starting material) to the reactant to observe independence of rate from inert gas pressure. Since all of these experiments are time-consuming, consult your instructor for the number of runs to be performed at which temperatures. After you know your assignment, write a detailed procedure and discuss it with your instructor before you begin.

Calculations

Since samples are removed from the reactor, the initial concentration would have to be calculated for each point if equation (26) is used to calculate k_0. But since with each removal equivalent parts of both the reactant and the product are removed, the ratio as observed in the gas chromatograph still gives the correct information about the degree of completion of the reaction. From the ratio of peak areas, find the ratios of reactant and product concentration. Express equation (26) in the form of this ratio. From equation (14) the following is obtained:

$$C_{\text{Pr}}^{\text{eq}} = K_{\text{eq}}(C_0 - C_{\text{Pr}}^{\text{eq}}) \tag{27}$$

or

$$C_0 K_{\text{eq}} = (1 + K_{\text{eq}})C_{\text{Pr}}^{\text{eq}} \tag{27a}$$

Multiplication of the numerator and denominator on the left side of equation (25) by $(1 + K_{\text{eq}})$ yields:

$$\ln\left[\frac{C_0 K_{\text{eq}}}{C_0 K_{\text{eq}} - (1 + K_{\text{eq}})C_{\text{Pr}}}\right] = k_0 t \tag{28}$$

Since $C_0 = C_A + C_{\text{Pr}}$ at any time, the following is written after division by C_A

$$-\ln\left(1 - \frac{(C_{\text{Pr}}/C_A)(1 + K_{\text{eq}})}{[(C_{\text{Pr}}/C_A) + 1]K_{\text{eq}}}\right) = k_0 t \tag{29}$$

This treatment has the additional advantage that the peak areas do not have to be calibrated, since only ratios are used.

After k_0 at a certain temperature has been found, calculate k_f and k_{-f}. Check whether the value for K_{eq}, which you measured and used in the calculation of k_0, agrees with the ratio of the two rate constants. From the temperature dependence of k_f and k_{-f}, determine the activation energy for forward and reverse reactions. From the temperature dependence of K_{eq}, obtain the enthalpy and entropy of reaction. The combination of both yields the entropy of activation.

REFERENCES

Basolo, F., and Pearson, R. G. 1967. *Mechanism of Inorganic Reactions*. 2d ed. New York: Wiley, pp. 275–277.

Brönsted, J. N., and Volquartz, K. 1928. *Z. Physik. Chem.* **134**, 97.

Eggers, E. F., Gregory, N. W., Halsey, G. D., and Rabinovitch, B. S. 1964. *Physical Chemistry*. New York: Wiley.

Pearson, R. G., Henry, P. M., Bergmann, J. G., and Basolo, F. 1954. *J. Am. Chem. Soc.* **76**, 5920.

Schlag, E. W., and Peatman, W. B. 1964. *J. Am. Chem. Soc.* **86**, 1676.

PART V

DATA EVALUATION

EXERCISE 15

Treatment of Errors in Experimental Results

Any experimental quantity must be accompanied by a measure of its uncertainty to be of value to the reader. If no error limits are provided, the number of digits in which a value is given provide a measure of uncertainty, insofar as it is understood that only the last figure is unreliable. Too many "significant figures" retained in a result indicate little understanding of the applied method; too few discredit the experimenter. It is therefore very important to be able to calculate or estimate uncertainties in directly measured quantities and in the results derived from them.

Assume that a certain physical quantity, X (e.g., the weight of a glass container), has been measured directly n times by weighing. The result of the ith measurement is called X_i, and the reported result is determined as the mean value $\bar{X}$ of all measurements.

$$\bar{X} = \sum_{i=1}^{n} \left(\frac{X_i}{n} \right)$$

A measure of how nearly the mean approaches the true value, X_{tr}, is the "accuracy," whereas the "precision" refers to the spread of measured data around the mean. Figure 15-1 illustrates the meaning of precision and accuracy.

The precision of an experiment is related to the "random error," for which sign and magnitude are statistically distributed. In an infinitely large number of determinations, subject to random error only, the mean value of the results should be the correct value (i.e., high accuracy). A small number of evaluations of low precision may nevertheless yield a result of low accuracy since the deviations on both sides of the correct result are not balanced. In this case, a single inaccurate value may pull the mean far from the correct result. The experimenter should refer to his notes to find a reason for the deviation of the value in question so that it may be discarded. However, the decision must be made carefully and on the basis of fact. The rejected value may have been the only correct one! The importance of keeping a neat and complete notebook cannot be overstressed.

If the precision of an experiment is sufficient but the accuracy low, a "systematic error" must have occurred. Systematic errors are introduced

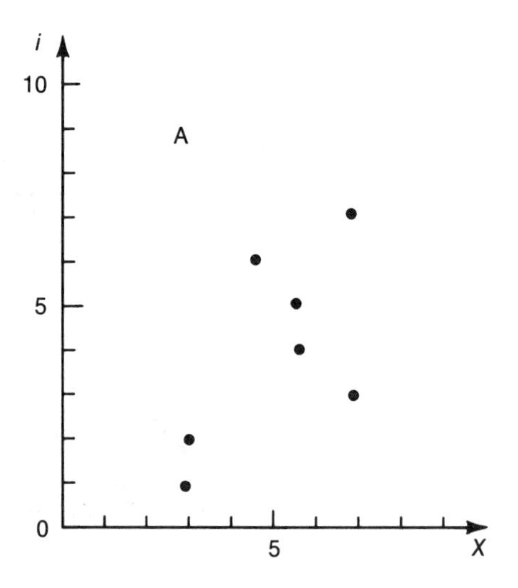

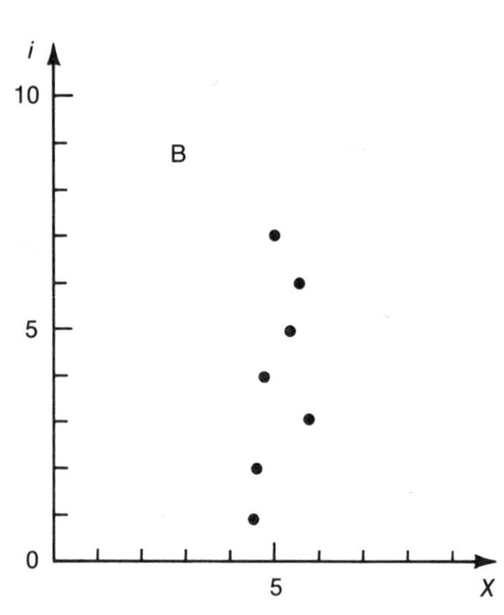

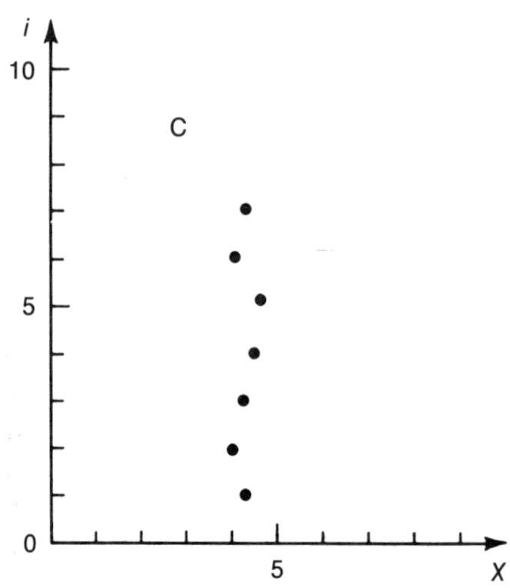

FIGURE 15-1
Spread of experimental results, X, for:
A. High accuracy but low precision, $\bar{X} = X_{tr} = 5.0$
B. High accuracy and high precision, $\bar{X} = X_{tr} = 5.0$
C. High precision but low accuracy, $\bar{X} = 4.2$, $\bar{X}_{tr} = 5.0$
The true value in all cases is $X_{tr} = 5.0$.

by additive or multiplicative constants of fixed sign, which affect each measured value. Faulty positioning of the zero point on a linear scale, for instance, will produce an additive systematic error. Using an incorrect molecular weight when calculating concentrations creates a multiplicative or proportional error. Systematic errors are very hard to detect if comparison with the correct result is impossible. They can be corrected for, once their origin and magnitude has been determined.

A discussion of "error sources" should precede any attempt to estimate or calculate error limits. The only valid error sources are to be found in reading and operating equipment. In derived results approximations in the derivation of the working equation may sometimes introduce deviations from observed behavior that simulate experimental error. All avoidable error sources must be eliminated. The method of calculation, that is, the use of a slide rule or log table, and the choice of scale for graphical presentations, should be adjusted to eliminate rounding-off errors. A few more digits than are to be retained in the result should be carried through the calculation for the same reason. (Millimeter or log paper should be used in all cases in which values will be read from a graph.) Detected systematic errors must be corrected.

Random Errors of Directly Measured Quantities

A simple experiment repeated several times under identical circumstances and performed each time to the limit of sensitivity imposed by apparatus and method will scarcely yield the same result for each measurement. This phenomenon is caused by random error. The following discussion shall be concerned with the postulation of error limits, if the error distribution follows the Gaussian error curve.

As previously defined, X_i is the result of the ith measurement of the property X, and $\bar{X} = \sum_{i=1}^{n} X_i/n$ is the mean value of the property X, where n is the total number of measurements. Then $\kappa_i = X_i - \bar{X}$ is the residual (referred to here as error) of the ith measurement. The relative number of measurements N subject to an error κ is given by:

$$N(\kappa) = \frac{h}{\sqrt{\pi}} e^{-h^2\kappa^2} \qquad (1)$$

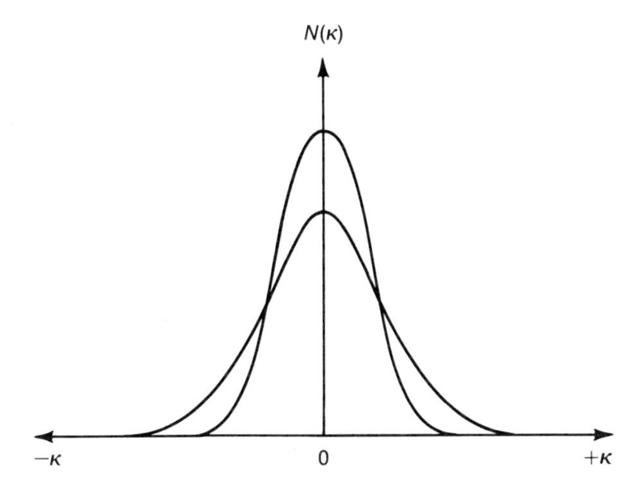

FIGURE 15-2
Gaussian error distribution drawn for two different values of h.

This is the relative number of measurements because the distribution function is normalized so that

$$\int_{-\infty}^{+\infty} \frac{h}{\sqrt{\pi}} e^{-h^2\kappa^2} = 1 \tag{2}$$

The normalization provides an opportunity to calculate the probability, P_i, of finding an error within the interval $d\kappa_i$.

$$P_i = \frac{h}{\sqrt{\pi}} \exp(-h^2\kappa_i^2) d\kappa_i \tag{3}$$

Equation (2) read in this sense means that the probability of finding an error in the interval $-\infty$ to $+\infty$ must be unity.

Figure 15-2 is obtained if N is plotted as a function of κ for two different values of h. It is obvious from equation (2) that the area under both curves must be equal. Since the area under all curves of this type must be equal and the height of the curve depends on h ($N = h/\sqrt{\pi}$ for $\kappa = 0$), the spread of errors must be broader for smaller h. This means the smaller h is, the greater the frequency of errors of large magnitude. It becomes obvious that h is a precision index. The following can be readily shown to hold for a Gaussian error curve:

$$h = \frac{1}{\sqrt{2}} \left(n \Big/ \sum_{i=1}^{n} \kappa_i^2 \right)^{1/2} \tag{4}$$

The quantity $(\sqrt{2} \cdot h)^{-1}$ is called the standard deviation, σ_x, or root-mean-square (rms) error of a single experiment.

$$\sigma_x = \sqrt{\frac{\sum_{i=1}^{n} \kappa_i^2}{n}} \tag{5}$$

It can be shown for an infinite number of experiments, the results of which constitute a Gaussian error curve, that the mean is identical with the true result, $\bar{X} = X_{tr}$.

If the number of measurements is limited, a discrepancy between $\bar{X}$ and X_{tr} should be expected. The standard deviation based on this limited amount of data will thus not coincide with the standard deviation to be found from an infinite number of trials. To correct the situation, instead of dividing by the number of trials, n, divide by the "number of degrees of freedom," $n - n_r$, where n_r is the number of variables to be determined in one trial. For just one variable (which has been the case so far), equation (5) is rewritten as:

$$\sigma_x = \sqrt{\frac{\sum_{i=1}^{n} \kappa_i^2}{n-1}} \tag{6}$$

For a discussion of this correction consult Pugh and Winslow (1966).

We are usually more interested in the standard deviation of the mean value than that of a single measurement. Equation (7) defines the standard deviation for the mean, σ_X.

$$\sigma_X = \sqrt{\frac{\sum_{i=1}^{n} \kappa_i^2}{n(n-1)}} \tag{7}$$

Notice that even though σ acquires a special meaning in the context of the Gaussian error distribution, it contains only experimental values (n, X_i, and $\bar{X}$) and is thus meaningful even if the assumption of a Gaussian distribution cannot be verified. It is therefore preferred over other precision measures, which specifically use features of certain distributions and will not be discussed here.

Systematic Errors of Directly Measured Quantities

In addition to random errors, systematic errors, u, of fixed magnitude and sign occur. The result of the ith measurement must then be written as:

$$X_i = X_{tr} + u + \kappa_i \tag{8}$$

where

$$X_{tr} + u = \bar{X} \tag{9}$$

holds, if the number of measurements approaches infinity and a Gaussian distribution of random errors is found.

The mean is thus displaced from the true result by exactly the same amount because of systematic error regardless of the number of measurements. Systematic errors can therefore be neither detected nor be remedied by repetition of the same measurement. Application of a different method is usually necessary for the detection of systematic errors.

Error Limits of Derived Results

Error Propagation of Systematic Errors

Consider the measured variable R to be a function of the directly measured variables X, Y, and Z.

$$R = F(X,Y,Z) \tag{10}$$

A minute change of the values for X, Y, and Z causes a small deviation dR in R, which may be expressed as follows:

$$dR = \left(\frac{\delta R}{\delta X}\right)_{Z,Y} \cdot dX + \left(\frac{\delta R}{\delta Y}\right)_{X,Z} \cdot dY + \left(\frac{\delta R}{\delta Z}\right)_{X,Y} \cdot dZ \tag{11}$$

If the errors encountered in the measurement of X, Y, and Z are very small (a few percent at the most), then the error in R, called ΔR, may be approximated by:

$$\Delta R = \left(\frac{\delta R}{\delta X}\right)_{Y,Z} \cdot \Delta X + \left(\frac{\delta R}{\delta Y}\right)_{X,Z} \cdot \Delta Y + \left(\frac{\delta R}{\delta Z}\right)_{Y,X} \cdot \Delta Z \tag{12}$$

where the errors ΔX, ΔY, and ΔZ are constant in sign and magnitude. The derivatives are to be taken at the measured points, that is, $X = \bar{X}$, $Y = \bar{Y}$, and $Z = \bar{Z}$.[1] Then ΔR is the error in the derived result due to systematic errors in the variables that were measured directly.

[1] $\bar{X}$ is the mean of several measurements of X within one run.

If the analytical expression from which the result is to be obtained possesses the following form

$$R = aX^k \cdot Y^l \ldots \tag{13}$$

where a, k, and l are constants, then the error in R is derived more conveniently if equation (13) is converted to the logarithmic form.

$$d \ln R = k\, d \ln X + d \ln a + l\, d \ln Y + \ldots \tag{14}$$

or

$$\frac{dR}{R} = k\frac{dX}{X} + l\frac{dY}{Y} + \frac{da}{a} + \ldots \tag{15}$$

For small deviations dX is replaced by ΔX, dY by ΔY, and so forth, and ΔR, ΔX, ΔY, and Δa are identified as the systematic errors in result and variables. If a is a known constant Δa is set equal to zero.

$$\frac{\Delta R}{R} = k\frac{\Delta X}{X} + l\frac{\Delta Y}{Y} + \ldots \tag{16}$$

If R is of the general form $R = aX + bY$, the error in R is expressed as

$$\Delta R = a\Delta X + b\Delta Y \tag{17}$$

For the special case in which one variable has been obtained as the difference between two measurements, this implies that the error in this variable is the sum of the errors in each measurement

$$X = X_1 \,(\pm)\, X_2$$
$$\Delta X = \Delta X_1 \,(\pm)\, \Delta X_2 \tag{18}$$

This applies, for example, to the weight of a sample determined by the difference in two weighings or to the pressure measured by reading two mercury levels in a manometer.

Estimate of Maximal Error. The methods used in the propagation of systematic errors are frequently applied to the propagation of errors that can only be estimated. In the laboratory there are many cases where only one measurement is made of each directly measurable quantity for each run (e.g., a buret has been read only once, not ten times, for the end point of each titration performed). The error in the directly measurable quantity can thus only be estimated and should, in general, be taken from the accuracy with which an instrument can be read or equipment has been calibrated. From observation and experience, the student must decide which element in a direct measurement is the source of greatest error and then assign a value to this error.

The estimated errors are called ΔX, ΔY, and ΔZ. Their contribution to the error in R, ΔR, is contained in equations (12) to (18) with the exception that derivatives and errors are replaced by their absolute magnitudes. This is due to uncertainty about the correct sign of ΔX, ΔY, and ΔZ. Therefore the worst possible case is expected and all contributions are added as if they all changed the result in one direction. The error in the result is thus the maximal estimated error. Equation (12) thus becomes

$$\Delta R = \left|\left(\frac{\delta R}{\delta R}\right)_{Y,Z} \Delta X\right| + \left|\left(\frac{\delta R}{\delta R}\right)_{X,Z} \Delta Y\right| + \left|\left(\frac{\delta R}{\delta R}\right)_{Y,X} \Delta Z\right| \quad (12a)$$

where the vertical bars indicate that the absolute magnitude of the expressions inside are to be taken. Equations (13) to (18) have to be modified in the same way.

A simple example may clarify the situation. Suppose you would like to determine the density, ρ, of a liquid by measuring the weight, m, and the volume, V, of a sample.

$$\rho = \frac{m}{V}$$

The volume error ΔV can be estimated from the manufacturer's specifications for the apparatus (a pyknometer) or from previous calibration with a liquid of known density; Δm can be estimated by knowing the sensitivity of the balance and using equation (18). From equation (12a) the following is obtained:

$$\left|\Delta\rho\right| = \left|\frac{\Delta m}{V}\right| + \left|\frac{m}{V^2}\Delta V\right|$$

where V and m are the measured values. For simplification, divide by ρ; the same result could have been obtained directly from equation (16) by inserting the absolute magnitudes.

$$\left|\frac{\Delta\rho}{\rho}\right| = \left|\frac{\Delta m}{m}\right| + \left|\frac{\Delta V}{V}\right|$$

where $|\Delta\rho/\rho| \equiv$ relative error on density. The density is then to be reported in the following form

$$\rho = \rho_m \pm \Delta\rho \text{ (units)}$$

If more than one sample has been used to determine ρ, the measured value ρ_m in this equation must be replaced by the mean result $\overline{\rho}$. The error limits are calculated for one sample run only. The calculated error in the density serves as a limit within which the standard deviation of the mean experimental result is to be found if the errors have been estimated in an appropriate fashion. This point will be taken up again in the discussion of the propagation of random errors. The following remarks concern error propagation in general.

It is frequently possible to break up complicated equations into parts that can be treated individually and then combined to yield the final error. Assume R to be a function of A and B, where A depends on the variables $X_1, X_2, \ldots$ and B on $Y_1, Y_2, \ldots$

$$R = F(A,B)$$

$$A = f(X_1, X_2, \ldots)$$

$$B = f(Y_1, Y_2, \ldots)$$

then

$$dR = \left(\frac{\delta R}{\delta A}\right)_B dA + \left(\frac{\delta R}{\delta B}\right)_A dB$$

where

$$dA = \left(\frac{\delta A}{\delta X_1}\right)dX_1 + \left(\frac{dA}{\delta X_2}\right)dX_2 + \ldots$$

and

$$dB = \left(\frac{\delta B}{\delta Y_1}\right)dY_1 + \left(\frac{\delta B}{\delta Y_2}\right)dY_2 + \ldots$$

Equations (13) to (18) apply also. Care should be taken to write A and B in terms of completely separated variables. If the expressions for dA and dB contain common variables, all terms have to be combined in dR before numerical values are inserted and the result is calculated. Frequently, the substitution of the result R or intermediate results simplifies the calculation. An example of such simplification was given in the calculation of ρ. It may also be advantageous to watch for terms that yield errors that are negligible compared with those of other terms. The former may then be assumed to be free from error in the course of error propagation and this may greatly reduce the amount of work required. The total error in all of the terms that are dropped should not exceed 20% of the final calculated error.

Propagation of Random Errors

Assume that the measurements of X, Y, and Z are subject to random error only and that n runs have been performed in which X, Y, and Z have been determined. The standard deviation of the result, σ_R, can then be expressed as

$$\sigma_R = \sqrt{\frac{\sum_{i=1}^{n} \Delta R_i^2}{n}} \qquad (19)$$

for n approaching infinity.

The error on the ith run, ΔR_i, is given by equation (12) since the errors $\kappa_{X,i}$, $\kappa_{Y,i}$, and $\kappa_{Z,i}$ are known. Thus n equations of the following form may be written:

$$\Delta R_i = \left(\frac{\delta R}{\delta X}\right)\kappa_{X,i} + \left(\frac{\delta R}{\delta Y}\right)\kappa_{Y,i} + \left(\frac{\delta R}{\delta Z}\right)\kappa_{Z,i} \qquad (20)$$

where ΔX_i is replaced by $\kappa_{X,i}$, and so forth, to return to the previous notation for random errors.

From equations (19) and (20) the square of the resulting standard deviation is obtained as

$$\sigma_R^2 = \left(\frac{\delta R}{\delta X}\right)^2 \frac{\sum \kappa_{X,i}^2}{n} + \left(\frac{\delta R}{\delta Y}\right)^2 \frac{\sum \kappa_{Y,i}^2}{n} + \left(\frac{\delta R}{\delta Z}\right)^2 \frac{\sum \kappa_{Z,i}^2}{n} \qquad (21)$$

in which all terms containing sums over mixed products such as $\sum (\kappa_{X,i} \cdot \kappa_{Y,i})$ become zero because these products are found to be positive as often as negative. Thus summation yields zero. The introduction of standard deviations for the variables X, Y, and Z according to equation (5) leads to

$$\sigma_R^2 = \left(\frac{\delta R}{\delta X}\sigma_X\right)^2 + \left(\frac{\delta R}{\delta Y}\sigma_Y\right)^2 + \left(\frac{\delta R}{\delta Z}\sigma_Z\right)^2 \qquad (22)$$

All sums in equation (21) are to be executed between $i = 1$ and $i = n$.

To obtain the standard deviation of the mean result, $\bar{R}$, that is to be reported, return to the defining equation (7).

$$\sigma_{\bar{R}}^2 = \frac{\sum_{i=1}^{n} \Delta R_i^2}{n^2} \qquad (23)$$

for $n \rightarrow \infty$. For a small number of measurements n^2 should be replaced by $n(n-1)$. Equation (23) written in terms of the standard deviations of the mean values of the variables reads:

$$\sigma_{\bar{R}}^2 = \left(\frac{\delta R}{\delta X}\sigma_{\bar{X}}\right)^2 + \left(\frac{\delta R}{\delta Y}\sigma_{\bar{Y}}\right)^2 + \left(\frac{dR}{\delta Z}\sigma_{\bar{Z}}\right)^2 \qquad (24)$$

In equation (24) the basic equation for propagation of random errors is obtained.

The rules of calculation based on equation (24) are much the same as those used for the propagation of systematic errors. For sums or differences, the squares of the standard deviations are added:

$$R = X_1 \pm X_2 \qquad (25)$$

$$\sigma_{\bar{R}}^2 = \sigma_{\bar{X}_1}^2 + \sigma_{\bar{X}_2}^2$$

For linear functions, the product of the squares of standard deviations for coefficients and variables are added:

$$R = aX + bY \qquad (26)$$

$$\sigma_{\bar{R}}^2 = \sigma_{\bar{a}}^2\sigma_{\bar{X}}^2 + \sigma_{\bar{b}}^2\sigma_{\bar{Y}}^2$$

In product functions, the use of relative errors greatly simplifies the calculation:

$$R = aX^kY^l \qquad (27)$$

$$\left(\frac{\sigma_{\bar{R}}}{\bar{R}}\right)^2 = \left(\frac{\sigma_{\bar{a}}}{\bar{a}}\right)^2 + k^2\left(\frac{\sigma_{\bar{X}}}{\bar{X}}\right)^2 + l^2\left(\frac{\sigma_{\bar{Y}}}{\bar{Y}}\right)^2$$

The error is reported as

$$\pm\sqrt{\sigma_{\bar{R}}^2}$$

Propagation of Estimated Random Error. Frequently, estimated errors in directly measured variables are random rather than systematic, and their propagation should be performed according to equation (24). However, this requires a formulation of standard deviation for an estimated error, for which the distribution of the error must be found. Since nothing is known about this distribution, any error between $-\Delta X$ and $+\Delta X$ is assumed to occur with equal probability. The same assumption is made for Y and Z, resulting in squared distribution functions for the error in all variables. The square of the standard deviation may then be written as:

$$\sigma_X^2 = \frac{1}{X}\int_0^X X^2 dX \qquad (28)$$

and the standard deviation becomes:

$$\sigma_X = \frac{\Delta X}{\sqrt{3}} \tag{29}$$

The estimated rms (root-mean-square) error of the result is thus given by:

$$\sqrt{\sigma_R^2} = \sqrt{\left(\frac{\delta R}{\delta X}\frac{\Delta X}{\sqrt{3}}\right)^2 + \left(\frac{\delta R}{\delta Y}\frac{\Delta Y}{\sqrt{3}}\right)^2 + \left(\frac{\delta R}{\delta Z}\frac{\Delta Z}{\sqrt{3}}\right)^2} \tag{30}$$

This error is much smaller than the error obtained by equation (12a) and represents the true error limits much better. The estimated maximum error according to equation (12a) is frequently so large that inaccurate work is easily covered up.

Error Analysis and the Selection of Experimental Conditions

Besides supplying information about the error incurred in an experiment, the propagation of maximum estimated error serves another, frequently more important, purpose. It furnishes the experimenter with an idea of the magnitude of contributions from different error sources. This enables him to select from the various measurements performed in one experiment those that have to be performed especially carefully or those that need improvement. Selection of the best conditions for a chosen method is more easily facilitated by this kind of reasoning than by experimental trial and error.

For purposes of discussion, the error contributions in the determination of a cubic lattice constant from x-ray scattering, as performed in Exercise 10A, shall serve as an example. The lattice constant is calculated from equation (31):

$$a = \frac{\lambda}{2 \sin \theta} \sqrt{h^2 + k^2 + l^2} \tag{31}$$

where λ is the wavelength of the scattering x-ray and θ the angle of scattering; h, k, and l are integers.

According to equation (12), the following is obtained:

$$\Delta a = -\frac{\lambda\sqrt{h^2 + k^2 + l^2}}{2 \sin^2 \theta} \cos \theta \cdot \Delta\theta \tag{32}$$

under the assumption of negligible error in λ. (Assumptions of this kind are usually fully warranted in a teaching laboratory.) The relative error is then

$$\frac{\Delta a}{a} = -\frac{\cos \theta}{\sin \theta} \Delta\theta = -\operatorname{cotg} \theta \cdot \Delta\theta \tag{33}$$

The relative error on the lattice constant a becomes smaller with increasing θ, which in turn means that the most reliable information is drawn from backscattering. This suggests that the wavelength should be chosen so that backscattering occurs, if this is experimentally possible.

Suppose that the experiment were conducted so that a standard deviation, σ_θ, can be calculated for θ. Propagation of random error yields for the relative error:

$$\left(\frac{\sigma_a}{a}\right)^2 = \operatorname{cotg}^2 \theta \cdot \sigma_\theta^2 \tag{34}$$

The same conclusions concerning reliability of readings are drawn from both equations (33) and (34). Thus for the purpose of investigating error contributions, methods for the propagation of estimated error suffice.

There are additional error sources, which are not explicit in equation (33). To find these the experimenter must rely largely on his knowledge of experimental conditions.

Curve Fitting

Fitting measured data to a given functional relation is a frequently encountered problem. At the same time, it is desirable to obtain information about the error in the result. There are two frequently used methods available for the fitting of data points: (1) the least-squares method or (2) graphical representation. The former provides a measure of precision, although it requires a time-consuming calculation. The latter yields information more rapidly and gives a good visual representation of the functional relation, but it allows for a rather dubious error estimate. It is advantageous to use both methods wherever possible.

Least-Squares Method

Assume that a set of data pairs (X,Y) has been obtained for which the relationship

$$Y = a_0 + a_1X + a_2X^2 + \ldots + a_nX^n$$

holds. Assume that for all data points only random errors following the Gaussian error distribution have been incurred. In this case, the best coefficients $a_0, a_1, a_2, \ldots, a_n$ can be acquired by using the method of least squares.

The error in each point (data pair) can be expressed as the difference between the measured value, Y_i, and the value calculated from X_i and the defining equation.

$$\kappa_i = Y_i - (a_0 + a_1X_i + a_2X_i^2 + \ldots + a_nX_i^n) \quad (35)$$

Since all errors obey the Gaussian distribution, the probability of finding the error on the ith measurement in the region $d\kappa_i$ around κ_i is given by equation (3):

$$P_i = \frac{h}{\sqrt{\pi}} \exp(-h^2\kappa_i^2)d\kappa_i$$

The probability, P, of the simultaneous occurrence of the errors $\kappa_1, \kappa_2, \ldots, \kappa_j$ in the set of j performed measurements ($j > n$) must then be the product of all P_i, $i = 1,2, \ldots, j$, since the measurements were independent.

$$P = \prod_{i=1}^{j} P_i = \left(\frac{h}{\sqrt{\pi}}\right)^j \exp\left(-h^2 \sum_{i=1}^{j} \kappa_i^2\right) d\kappa_1, d\kappa_2, \ldots, d\kappa_j \quad (36)$$

The most probable errors are then those for which P has a maximum. The condition for finding a maximum in P is that $\sum_{i=1}^{j} \kappa_i^2$ be a minimum. For convenience we set $\sum_{i=1}^{j} \kappa_i^2 = s$.

Since all errors κ_i depend on the choice of coefficients, which equation (35) shows, s must be minimized with respect to the coefficients. The coefficients are found by satisfying the following conditions:

$$\frac{ds}{da_0} = \frac{ds}{da_1} = \ldots = \frac{ds}{da_n} = 0$$

This leads to $n + 1$ equations of the following form:

$$\begin{aligned}
\sum Y_i &= ja_0 + a_1\sum X_i + a_2\sum X_i^2 + \ldots + a_n\sum X_i^n \\
\sum Y_iX_i &= a_0\sum X_i + a_1\sum X_i^2 + a_2\sum X_i^3 + \ldots + a_n\sum X_i^{n+1} \\
&\vdots \\
\sum Y_iX_i^n &= a_0\sum X_i^n + a_1\sum X_i^{n+1} + \ldots + a_n\sum X_i^{2n}
\end{aligned}$$

For linear equations, $Y = a_0 + a_1X$, the simultaneous solution of the applicable equations yields:

$$a_1 = \frac{j\sum X_iY_i - \sum X_i\sum Y_i}{j\sum X_i^2 - \left(\sum X_i\right)^2} \quad (37)$$

with

$$a_0 = \frac{\sum X_i^2\sum Y_i - \sum X_i\sum X_iY_i}{j\sum X_i^2 - \left(\sum X_i\right)^2} \quad (38)$$

and for quadratic equations, $Y = a_0 + a_1X + a_2X^2$:

$$a_2 = \frac{\left(j\sum X_i^2Y_i - \sum X_i^2\sum Y_i\right)\left[j\sum X_i^2 - \left(\sum X_i\right)^2\right] - \left(j\sum X_i\sum Y_i - \sum X_i\sum Y_i\right)\left(j\sum X_i^3 - \sum X_i\sum X_i^2\right)}{\left[j\sum X_i^4 - \left(\sum X_i\right)^2\right]\left[j\sum X_i^2 - \left(\sum X_i\right)^2\right] - \left(j\sum X_i^2 - \sum X_i\sum X_i^2\right)}$$

$$a_1 = \frac{j\sum X_iY_i - \sum X_i - \sum Y_i - a_2\left[j\sum X_i^2 - \left(\sum X_i\right)^2\right]}{j\sum X_i^2 - \left(\sum X_i\right)^2}$$

$$a_0 = \frac{\sum Y_i - a_2\sum X_i^2 - a_1\sum X_i}{j}$$

where all summations signs indicate a summation from $i = 1$ to j.

The standard error on any Y (which now carries the error in both X and Y) is then:

$$\sigma_Y = \left(\frac{s}{j-2}\right)^{1/2} \quad (39)$$

Division by $j - 2$ is necessary because the number of trials, j, is small and two variables are measured in each trial.

For a curve of nth order fitted by least-squares minimization, s can be expressed as

$$s = \sum Y_i^2 - a_0\sum Y_i - a_1\sum X_iY_i - a_2\sum X_i^2Y_i - \ldots - a_n\sum X_i^nY_i$$

because the $n + 1$ normal equations have to be satisfied. For linear equations:

$$s = \sum Y_i^2 - a_0\sum Y_i - a_1\sum X_iY_i \qquad (40)$$

and for quadratic equations:

$$s = \sum Y_i^2 - a_0\sum Y_i - a_1\sum X_iY_i - a_2\sum X_i^2Y_i \qquad (41)$$

For linear equations the insertion of equation (39) into equation (22) leads to

$$\sigma_{a_1}^2 = \sum \sigma_Y^2 \left(\frac{\delta a_1}{\delta Y_i}\right)^2 = \frac{s}{j-2}\sum\left(\frac{\delta a_1}{\delta Y_i}\right)^2 \qquad (42)$$

and

$$\sigma_{a_0}^2 = \sum \sigma_Y^2 \left(\frac{\delta a_0}{\delta Y_i}\right)^2 = \frac{s}{j-2}\sum\left(\frac{\delta a_0}{\delta Y_i}\right)^2 \qquad (43)$$

where the summation over i is due to the fact that all Y_i behave as different variables, insofar as the derivative of the a_0 and a_1 are different with respect to each Y_i. Equations (37) and (38) are used to arrive at

$$\sum\left(\frac{\delta a_1}{\delta Y_i}\right)^2 = \frac{j\left[j\sum X_i^2 - \left(\sum X_i\right)^2\right]}{\left[j\sum X_i^2 - \left(\sum X_i\right)^2\right]^2} \qquad (44)$$

and

$$\sum\left(\frac{\delta a_2}{\delta Y_i}\right)^2 = \frac{\sum X_i^2\left[j\sum X_i^2 - \left(\sum X_i\right)^2\right]}{\left[j\sum X_i^2 - \left(\sum X_i\right)^2\right]^2} \qquad (45)$$

The insertion of equations (44) into (42) and (45) into (43) leads to the rms error in a_1 and a_0:

$$\sigma_{a_1} = \pm\sqrt{\frac{js}{(j-2)\left[j\sum X_i^2 - \left(\sum X_i\right)^2\right]}} \qquad (46)$$

and

$$\sigma_{a_0} = \pm\sqrt{\frac{s\sum X_i^2}{(j-2)\left[j\sum X_i^2 - \left(\sum X_i\right)^2\right]}} \qquad (47)$$

The method can be applied to nonlinear equations as well. The standard deviation depends not only on the scatter of experimental points but also on the correct choice of the number of included terms (linear or not linear). See, for example, Exercise 16A, where data fitting on a linear and a quadratic equation is performed.

From equations (37) and (38) it can be seen that the sums over all X_i and Y_i, over the product (X_iY_i), and over all X_i^2 have to be calculated. This creates a great deal of work, especially if there are many points to be calculated. It is therefore of great advantage to use a computer or desk calculator. If the measurements are not very precise, a graphical presentation and solution may be more appropriate.

REFERENCES

Mandel, J. 1961. *The Statistical Analysis of Experimental Data.* New York: Wiley.

Pugh, E. M., and Winslow, G. H. 1966. *The Analysis of Physical Measurements.* Reading, Massachusetts: Addison-Wesley.

Topping, J. 1963. *Errors of Observation and Their Treatment.* London: Chapman Hall.

EXERCISE 16

Reduction of Data by Computer

A. REDUCTION OF DATA FROM EXERCISE 6A

In the reduction of data from the band spectrum of iodine it is necessary to go through a long series of repetitive calculations using four- or five-digit numbers. Work of this kind is more efficiently done by a computer. Therefore a program is furnished here for the most tedious parts of the calculation. Precautions should be taken, however, to avoid mechanical use of the program, which actually inhibits comprehension of the parts of the calculation.

To take advantage of the usefulness of the program without diminishing the learning process, the following procedure is recommended:

1. Calculate manually four points of the Birge-Sponer plot, say for $v' = 10, 11, 12$, and $35, 36, 37$.
2. Include in your report sample calculations of these points and the corresponding computer statements.
3. Plot the four manually computed points, all of the machine-calculated points, and the least-squares line on coordinate paper.
4. Cut up the computer printout and include it in your report.
5. Know the purpose of each statement in the program.

Purpose of the Program

The program calls for data in the form of wavelengths (λ = WAV) in centimeters and vibrational quantum numbers (v' = V). From these data, the computer first calculates wave numbers ($\tilde{\nu}$ = FRQ in cm^{-1}) and then wave number differences [$\Delta\tilde{\nu}$ = DFRQ = $\tilde{\nu}(v'+1) - \tilde{\nu}(v')$].

In the output, the original wavelengths, the vibrational quantum numbers, and the frequencies are printed out in three columns. These are followed by two columns containing the v' and the $\Delta\tilde{\nu}$.

Next, the $\Delta\tilde{\nu}$ are fitted to a linear relationship with v' by means of least-squares calculation:

$$\Delta\tilde{\nu} = A + Bv = \omega_e' - 2(v' + 1)\omega_e'x_e'$$

[See equation (15) in Exercise 6A. Derive this equation according to the instructions in Exercise 6A.] The extrapolated vibrational frequency ω_e' and the first anharmonicity term $\omega_e'x_e$ may be obtained from the coefficients A and B. The computer symbols chosen to represent these are OME (from OMega sub E) and OMEXE (from OMega sub E times X sub E). All of the symbols have a simple mnemonic relationship to the usual Greek letter symbols.

The $\Delta\tilde{\nu}$ are also fitted to a second-order relation to v' by means of a quadratic least-squares fit:

$$\Delta\tilde{\nu} = A' + B'v' + C'v'^2$$

where

$$A' = \omega_e' - 2\omega_e'x_e' + \left(\frac{13}{4}\right)\omega_e'y_e'$$

$$B' = -2\omega_e'x_e' + 6\omega_e'y_e'$$

$$C' = 3\omega_e'y_e'$$

[See equation (17) in Exercise 6A.] (You might derive this equation also. Like the previous equations, it holds only for neighboring lines starting from the same lower state. The program ignores the DFRQ for nonconsecutive points.)

Description of Program

The program is written in standard CDC 6400 FORTRAN, which is a symbolic quasi-mnemonic programming language. The abbreviated description given here is for anyone not acquainted with FORTRAN.

1. The READ, WRITE, and PRINT statements instruct the machine to do just that. For example, the statement READ INPUT TAPE 5, 1, IDENT, M instructs the machine to read from tape 5 (to which the information on your data cards has been automatically transferred) according to FORMAT 1 and the values for IDENT and M. (IDENT identifies the student and M is the number of data points.) In a similar manner, PRINT 901, STDERR instructs the machine to print out, according to FORMAT 901, the value it has computed for the variable STDERR.
2. FORMAT statements simply tell the machine the arrangement in which it should read or write—how long the words or numbers should be, where to put decimal points, and so forth. In general, each READ or WRITE statement must refer to a FORMAT statement.
3. Numbered statements: The numbers on statements can be in any order from 1 to 99999. Statements need not be numbered. The numbers merely serve as references. For example, PRINT 701 tells the machine to print according to the FORMAT numbered 701.
4. A DO statement such as DO 666, K =1,M instructs the computer to perform the calculations between the DO statement and the statement numbered 666 a certain number of times. The value of K is set at 1 when the program first reaches this DO statement. The calculation is performed as far as statement 666, after which the program instructs the computer to return to the statement, increase the value of K from 1 to 2, and once more perform the computation as far as statement 666. The program then returns to the DO, increases K to 3, and cycles to 666 again. This continues until the value of K reaches M, after which the program continues past statement 666. *Note*: Unless otherwise instructed, the program proceeds sequentially downward from statement to statement.
5. CONTINUE and GO TO 4 statements instruct the computer to continue sequentially or switch to statement 4.
6. IF statements instruct the computer to check the truth or falseness of the statement within parentheses and shift control to the indicated (numbered) instruction if this statement is true. For example, IF(NDIF.NE.1) GO TO 87 instructs the computer to check if the variable NDIF (the quantity V(I+1)–V(I)) is not equal (NE) to 1. If NDIF $\neq$ 1, control goes to statement 87. If NIDF = 1, control shifts sequentially downwards.

The quantities SX, SY, SX2, and SXY are

$$\sum_i X_i,\ \sum_i Y_i,\ \sum_i X_i^2,\ \sum_i X_iY_i$$

```
      PROGRAM BIRGEQ (INPUT,OUTPUT, TAPE 60=INPUT)
      QUADRATIC BIRGE SPONER EXTRAPOLATION
      DIMENSION WAV(100),V(100),FRQ(100),DFRQ(100) ,IDENT(14)
       DIMENSION NNV(100)
      PRINT 40
   40 FORMAT(1H1,32X,22HBAND SPECTRA OF IODINE)
   30 READ 1, IDENT,M
      IF (EOF,60)99,13
    1 FORMAT (14A5,I2)
   13 PRINT 7, IDENT,M
    7 FORMAT (///,14A5,I2)
      READ 10,(WAV(L), V(L), L=1,M)
   10 FORMAT(2F20.6)
      DO 20 I=1,M
   20 FRQ(I)=1.0/(WAV(I)*1.E-08)
      M1=M-1
      DO 666  KK=1,M
      NNV(KK)=V(KK)
  666 CONTINUE
      DO 21 I=1,M1
   21 DFRQ(I)=FRQ(I+1)-FRQ(I)
      PRINT 41
   41 FORMAT(1H0,7X,10HWAVELENGTH,10X,11HQUANTUM NO.,9X,9HFREQUENCY,
     111X,11HDELTA FREQ,/,11X,*(ANG)*,33X,*(1/CM)*,17X,*(1/CM)*,//)
      PRINT                  42, (WAV(L),V(L),FRQ(L),DFRQ(L),L=1,M)
   42 FORMAT(9X,F10.5,9X,F5.1,9X,2E20.8)
   50 FN=M-1
      SX=0.0
      SX2=0.0
      SX3=0.0
      SX4=0.0
      SY=0.0
      SXY=0.0
       SY2 = 0.0
      SX2Y=0.0
      DO 4 I=1,M1
      NDIF = NNV(I+1)-NNV(I)
       IF(NDIF.NE.1) GO TO 87
      X=V(I)
      Y=DFRQ(I)
      SX=SX+X
      SX2=SX2+X**2
      SX3=SX3+X**3
      SX4=SX4+X**4
      SY=SY+Y
      SY2 = SY2 + Y*Y
      SXY=SXY+X*Y
      SX2Y = SX2Y + X*X*Y
      GO TO 44
   87  FN = FN -1
```

```
 44  CONTINUE
  4  CONTINUE
     TX2=FN*SX2-SX*SX
     TX3=FN*SX3-SX*SX2
     TX4=FN*SX4-SX2*SX2
     TXY=FN*SXY-SX*SY
     TX2Y=FN*SX2Y-SX2*SY
     UX6=TX4*TX2-TX3*TX3
     UX4Y=TX2Y*TX2-TXY*TX3
     C=UX4Y/UX6
     PRINT 45,    TXY,C,TX3,TX2,FN
 45  FORMAT(/,*  TXY,C,TX3,TX2,FN*,/,5E20.8)
     R=(TXY-C*TX3)/TX2
     A=(SY-C*SX2-B*SX)/FN
     OMEYE=C/3.0
     OMEXE=3.0*OMEYE-0.5*B
     OME=A+2.0*OMEXE-3.25*OMEYE
     PRINT 668
668  FORMAT(*PROGRAM ALLOWS ANY NUMBER OF SKIPPED POINTS  WORKS WITH
    CCONSECUTIVE POINTS*)
     PRINT                  8,A,B,C,OMEYE,OMEXE
   8 FORMAT(1H0,10X,2HA= E20.8  ,10X,2HB= E20.8  ,10X,2HC= E20.8  ,/
    110X,6HOMEYE= E20.8  ,10X,6HOMEXE= E20.8)
     PRINT 9,OME
   9 FORMAT(10X4HOME=  E20.8)
     SD2 = SY2-(A*SY+B*SXY+C*SX2Y)
     STDERR = SQRTF(SD2/FN)
     PRINT 901,STDERR
901  FORMAT(1H0,* THE STANDARD ERROR FOR THE QUADRATIC LEAST SQUARES IS
    C*,E24.8)
     PRINT 701
701  FORMAT(1H3,*THE ABOVE CALCULATION IS FOR A QUADRATIC LEAST SQUARES
    CTO FIT OMEGAEYE  BELOW IS LINEAR LEAST SQUARES TO FIT OMEGAEXE*)
     B = (SY*SX-SXY*FN)/(SX*SX-FN*SX2)
     A = (SY-B*SX)/FN
     OMEXE = -.5*B
     OME = A-B
      SD2 = SY2-A*SY-B*SXY
     STDERR = SQRTF(SD2/FN)
     PRINT                  3,A,B,OMEXE
  3  FORMAT(1H0,10X,2HA= E20.8  ,10X,2HB= E20.8  ,10X
    110X                      ,6HOMEXE= E20.8)
     PRINT 9,OME
     PRINT 902,STDERR
902  FORMAT(1H0,*THE STANDARD ERROR OF THE LINEAR LEAST SQUARES FIT IS
    C*,E24.8)
     GO TO 30
 99  STOP
     END
```

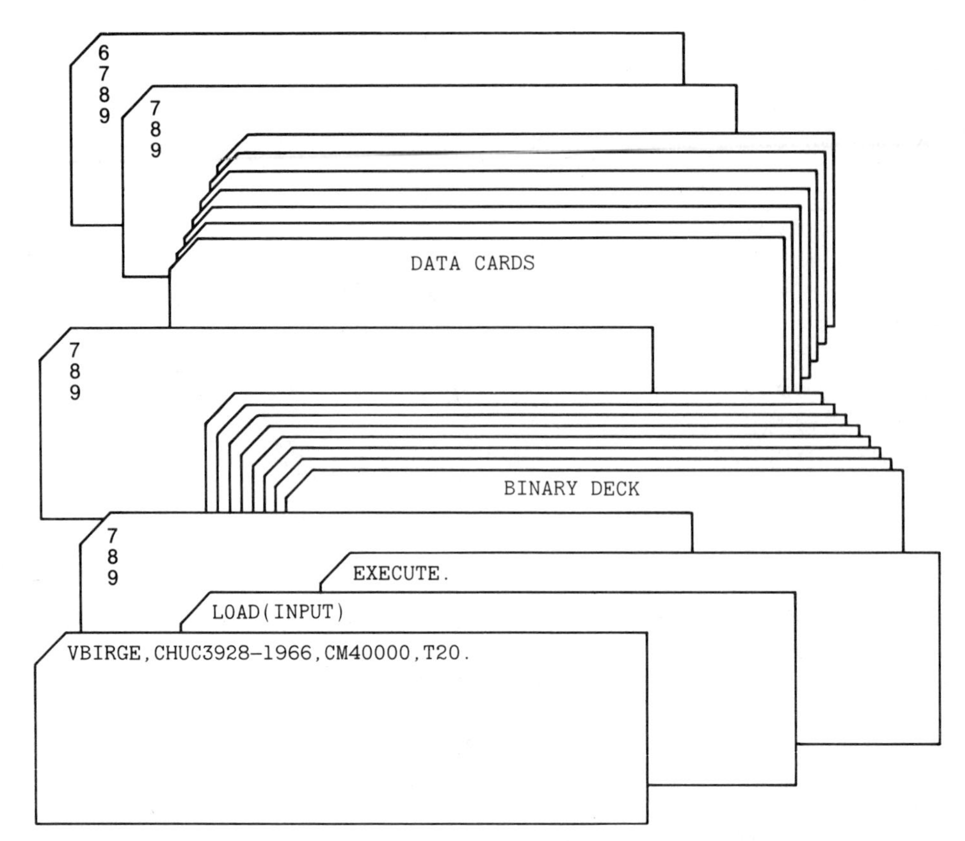

FIGURE 16-1
Deck structure. Arrange data cards in order of increasing v'.

Instructions for Punching Data Cards

Problem Identification Cards

The first data card should be an identification card giving your name and M, the number of observations or data points you have. The first 70 spaces of a white data card may be used to punch out your name, the kind of calculation being performed, and any other desired information, using any spacing you wish. Spaces 71 and 72 are reserved for M. If M is a one-digit number—say, 9—the number must be in space 72 and can be punched "09" using spaces 71 and 72. If desired, spaces 68 and 69 or those preceding can be used to punch M=, but this is not necessary.

Other Data Cards

One value of λ (wavelength in Å), with the corresponding v, is punched on each card. Twenty spaces are reserved for each number. Use "floating point" format, that is, be sure to include the decimal. For example:

5350.2 20.0

Where the number is typed within the spaces allotted is irrelevant. However, only one set of λ and v may be typed on each card.

The number of data cards, excluding the ID card, must equal M as specified previously.

B. FORTRAN PROGRAM FOR EXERCISE 12E

This program has been set up for the calculation of the protonation constants KH1 and KH2 of ethylenediamine and the subsequent evaluation of formation constants K1 and K2 of the ethylenediamine copper(II) and the bisethylenediamine copper(II) complexes. The program is written in standard CDC 6400 FORTRAN. It has been arranged so that average protonation and formation constants can be calculated from more than one pair of data points for each case. Advice on how to select points from the titration curves is given in the experimental procedure in Exercise 12E.

Most of the calculation is done in two nested DO loops. The first inner DO loop begins with the statement DO 2 . . . and leads to the calculation of J_1 and J_2 for protonation from the first pair of data points. In the outer DO loop (DO 7 . . .) K_{H1} and K_{H2} are obtained for all N pairs of data points. The average of the protonation constants is obtained outside of the loops. The calculation of K_1 and K_2 proceeds in much the same way, again employing two nested DO loops to evaluate the average formation constants from M pairs of data points. The number of data pairs that have been selected is given to the computer on the identification card, which is read in statement 31. It is thus especially important that exactly 2N data cards are prepared for the calculation of protonation constants and 2M for the evaluation of formation constants.

Preparation of Data Cards

Identification Card

The identification card is the first data card. Punch your name, title of the experiment, date, and any other pertinent information in the first 70 spaces. The next two spaces are reserved for N, the number of pairs of data points selected from the titration with Ba^{2+} present. Columns 73 and 74 contain M, the number of pairs of data points selected from the titration with Cu^{2+} present.

Data Cards for the Calculation of Protonation Constants

Imagine each card to be divided into eight columns of ten spaces each. In the first column punch the selected *p*H value; in the second, the initial volume; in the third, the volume of base added to reach the selected *p*H; in the fourth, the millimoles of acid that were originally present; in the fifth, the concentration of the standard ethylenediamine solution; and in the sixth, the p*K*w at the correct temperature. Columns 7 and 8 are not punched.

Arrange the 2N cards in N pairs so that each card with a *p*H value, read from the lower region of small slope in the titration curve, is followed by a card containing information from the region of higher *p*H. K_{H1} and K_{H2} are to be calculated from each of these N pairs.

Data Cards for the Calculation of Formation Constants

Again, divide each card into eight columns of ten spaces each and punch information exactly as described for protonation constants until the fifth column is reached. In column 5 punch the number of millimoles of Cu^{2+} present in the solution. Columns 6 and 7 contain the concentration of standard ethylenediamine solution and the correct *p*H, respectively. Again, arrange the 2M cards in M pairs, with the higher *p*H always following the lower one.

The data card structure is shown in Figure 16-2 for one pair of data points for each calculation. The deck structure for Program Amine is shown in Figure 16-3. Table 16-1 provides the key to the symbols used in the program. For further information on FORTRAN programming see the introduction to the program in Section A of this exercise and the references.

TABLE 16-1
Key to the symbols used in Program Amine.

Symbol	Property
V	Initial volume
DV	Added volume
CH	Millimoles of acid in initial volume
CPH	pH $-$ log γ_i
CA	Concentration of standard ethylenediamine solution
VT	Total volume
HP	$[H^+]$
CM	Millimoles of Cu^{2+} present
BHA	$\bar{n}_H$
BLM	$\bar{n}$
QH1	K_{H1}
QH2	K_{H2}
PK1	pK_{H1} } calculated from average protonation constants
PK2	pK_{H2} }
FK1	K_1
FK2	K_n
AK1	pK_1 } from average formation constants
AK2	pK_2 }

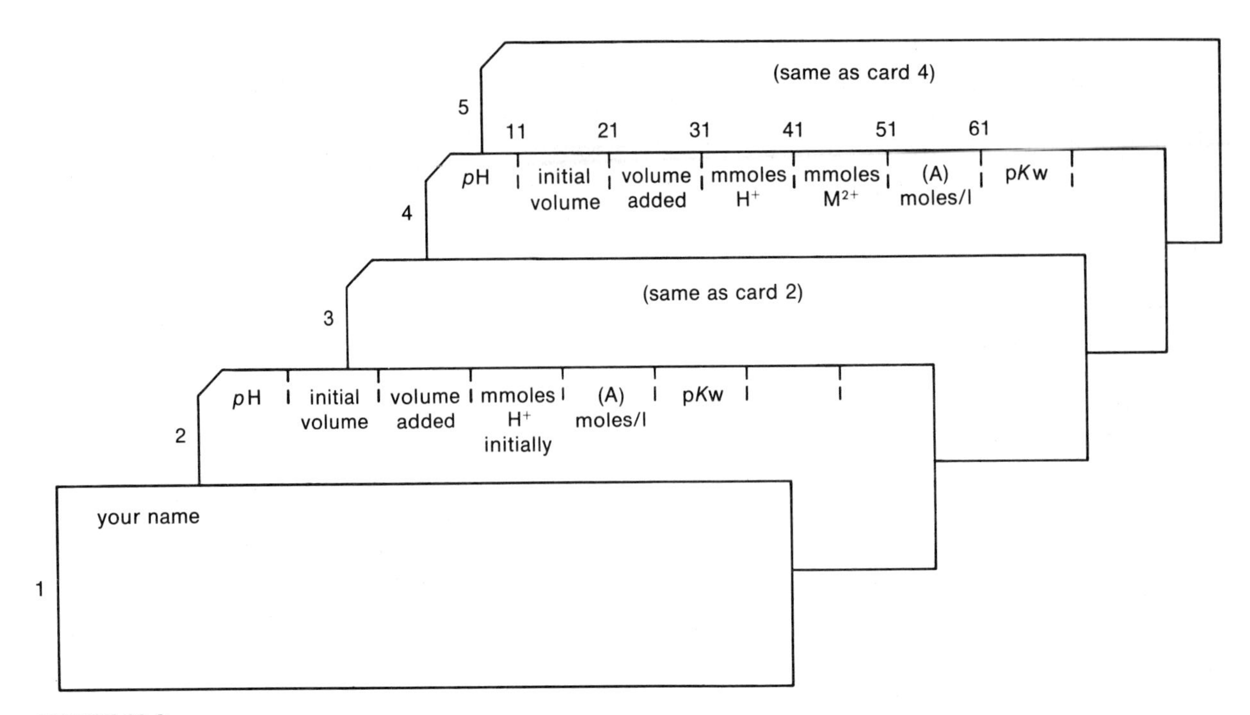

FIGURE 16-2
Data card structure for Program Amine.

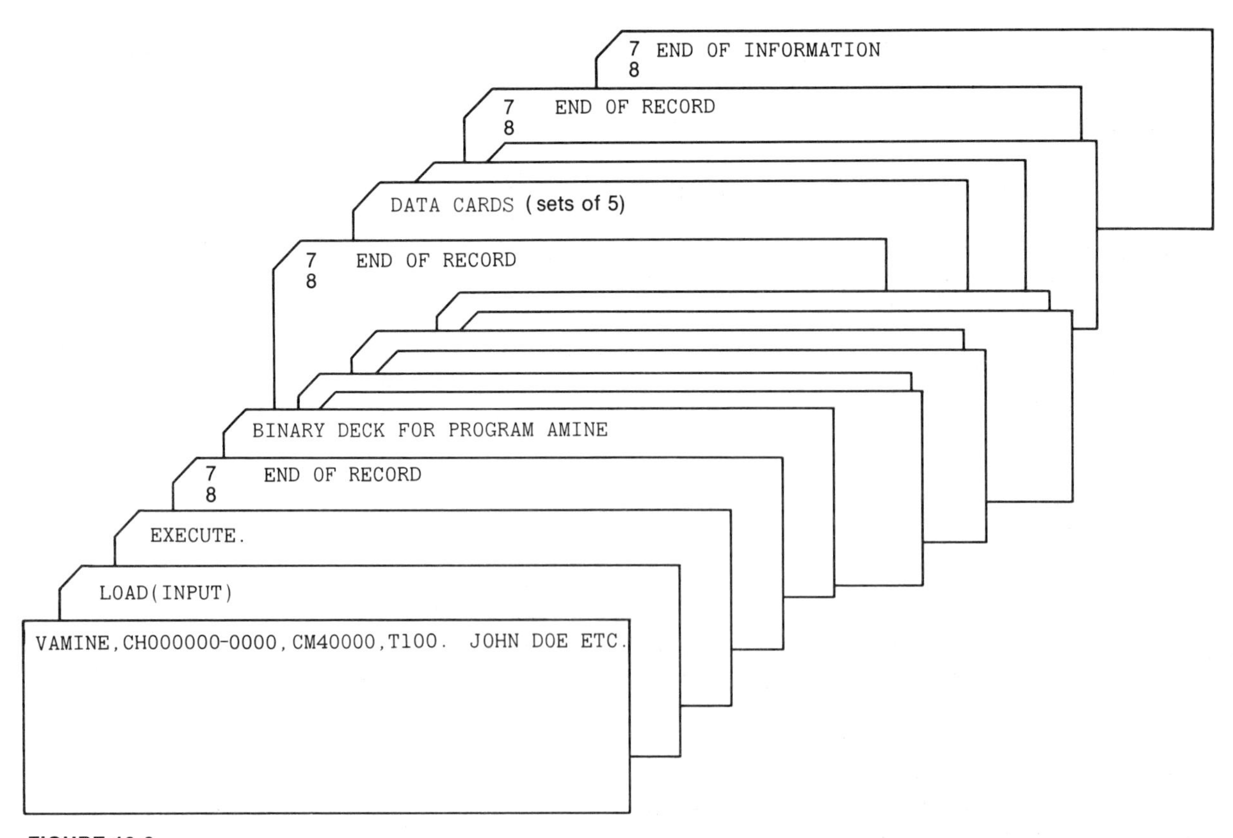

FIGURE 16-3
Deck structure for Program Amine.

```
   PROGRAM AMINE (INPUT, OUTPUT, TAPE 60 = INPUT)
   CALCULATION OF SUCCESSIVE FORMATION CONSTANTS OF A METAL AMINE SYSTEM
   DIMENSION IDENT (14),PH(2),V(2),DV(2),CH(2),CA(2),VT(2),CPH(2),
  1 HP(2),CPOH(2),OH(2),CCH(2),CCA(2),BHA(2),FJ1(2),FJ2(2),CCM(2),
  2 PR1(2),PR2(2),PR3(2),BLM(2),A(2),CM(2)
31 READ 10,IDENT,N,M
10 FORMAT (14A5,I2,I2)
   IF (EOF,60) 20,21
20 STOP
21 PRINT1
 1 FORMAT (1H1 45X *FORMATION CONSTANTS OF A METAL AMINE SYSTEM*//)
   PRINT 11,IDENT
11 FORMAT (1H0 14A5///)
   PRINT 12
12 FORMAT (1H0 25X *CALCULATION OF PROTONATION CONSTANTS FOR THE *
  1 *LIGAND*)
   SQH1 = 0
   SQH2 = 0
   DO 7 II = 1,N
   DO 2 J=1,2
   READ 3,PH(J),V(J),DV(J),CH(J),CA(J),PKW
 3 FORMAT (6F10.0)
   VT(J) = V(J) +DV(J)
   CPH(J) = PH(J) -0.03
   HP(J) = 10.0**(-CPH(J))
   CPOH(J) = PKW - PH(J) - 0.03
   OH(J) = 10.0**(-CPOH(J))
   CCH(J) = CH(J)/VT(J)
   CCA(J) = CA(J)*DV(J)/VT(J)
   BHA(J) = (CCH(J)-HP(J)+OH(J))/CCA(J)
   FJ1(J)  = (1.0 - BHA(J))*HP(J)
   FJ2(J) = (2.0 - BHA(J))*HP(J)*HP(J)
 2 CONTINUE
   QH1 = (BHA(2)*FJ2(1) - BHA(1)*FJ2(2))/(FJ1(2)*FJ2(1) - FJ1(1)
  1*FJ2(2))
   QH2 = (BHA(1)*FJ1(2) - BHA(2)*FJ1(1))/(BHA(2)*FJ2(1) - BHA(1)*
  1FJ2(2))
   PRINT 13
13 FORMAT (1H0 7X 2HPH 7X 2HVT 12X 2HHP 13X 2HOH 13X 2HCH 13X 2HCA
  1 12X 4HNBAR)
   DO 30 J = 1,2
   PRINT 4,PH(J),VT(J),HP(J),OH(J),CCH(J),CCA(J),BHA(J)
 4 FORMAT (1H0 2F10.2,5E15.2)
30 CONTINUE
   PRINT 9,FJ1(2),FJ2(2),FJ1(1),FJ2(1),QH1        ,QH2
 9 FORMAT (1H0 * J1=* E10.2//* J2=* E10.2//* J1P=* E10.2//* J2P=*
  1 E10.2//* QH1=* E10.2,                        *  QH2=* E10.2//)
   SQH1 = SQH1 + QH1
   SQH2 = SQH2 + QH2
 7 CONTINUE
   QH1AV = SQH1/N
   QH2AV = SQH2/N
   PK1 = ALOG10(QH1AV)
   PK2 = ALOG10(QH2AV) - 0.06
   PRINT 19, QH1AV,PK1,QH2AV,PK2
```

```
19 FORMAT (1H0 * QH1AV=* E10.2, 4X *PK1=* F5.2, 8X *QH2AV=* E10.2,
  1 4X *PK2=* F5.2///)
   PRINT 8
 8 FORMAT (1H0 25X*CALCULATION OF K1 AND K2 FOR METAL COMPLEXES*)
   SFK1 = 0
   SFK2 = 0
   DO  18 II = 1,M
   DO 6 J = 1,2
   READ 5,PH(J),V(J),DV(J),CH(J),CM(J),CA(J),PKW
 5 FORMAT (7F10.0)
   VT(J) = V(J) + DV(J)
   CPH(J) = PH(J) - 0.03
   HP(J) = 10.0**(-CPH(J))
   CPOH(J) = PKW - PH(J) - 0.03
   OH(J) = 10.0**(-CPOH(J))
   CCH(J) = CH(J)/VT(J)
   CCM(J) = CM(J)/VT(J)
   CCA(J) = CA(J)*DV(J)/VT(J)
   PR1(J) = QH1AV*HP(J)
   PR2(J) = QH1AV*QH2AV*HP(J)*HP(J)
   PR3(J) = 1.0 +PR1(J) +PR2(J)
   BHA(J) = (PR1(J) + 2.0*PR2(J))/PR3(J)
   BLM(J) = (CCA(J) -(CCH(J) -HP(J) + OH(J))/BHA(J))/CCM(J)
   A(J) = (CCH(J) - HP(J) + OH(J))/(PR1(J) + 2.0*PR2(J))
   FJ1(J) = (1.0 - BLM(J))*A(J)
   FJ2(J) = (2.0 - BLM(J))*A(J)*A(J)
 6 CONTINUE
   FK1 = (BLM(1)*FJ2(2) - BLM(2)*FJ2(1))/(FJ1(1)*FJ2(2) - FJ1(2)
  1*FJ2(1))
   FK2 = (BLM(2)*FJ1(1) - BLM(1)*FJ1(2))/(BLM(1)*FJ2(2) - BLM(2)
  1*FJ2(1))
   PRINT 17
17 FORMAT (1H0 7X 2HPH 7X 2HVT 9X 2HHP 10X 2HOH 10X 2HCH 10X 2HCM 10X
  1 2HCA 9X 5HNHBAR 7X 4HNBAR 10X 1HA)
   DO 15 J = 1,2
   PRINT 14,PH(J),VT(J),HP(J),OH(J),CCH(J),CCM(J),CCA(J),BHA(J),
  1 BLM(J),A(J)
14 FORMAT (1H0 2F10.2,8E12.2)
15 CONTINUE
   PRINT 16,FJ1(J),FJ2(1),FJ1(2),FJ2(2),FK1, FK2
16 FORMAT (1H0  *J1=* E10.2//* J2=* E10.2//* J1P=* E10.2//* J2P=*
  1 E10.2// *  K1=* E10.2, 5X * K2=* E10.2///)
   SFK1 = SFK1 + FK1
   SFK2 = SFK2 + FK2
18 CONTINUE
   FK1AV = SFK1/M
   FK2AV = SFK2/M
   AK1 = ALOG10(FK1AV)
   AK2 = ALOG10(FK2AV)
   PRINT 22, FK1AV,AK1,FK2AV,AK2
22 FORMAT (1H0 * K1=* E10.2, 4X *LK1=* F5.2, 8X *K2=* E10.2, 4X
  1 *LK2=* F5.2)
   GO TO 31
   END
END OF RECORD
```

REFERENCES

McCracken, Daniel D. 1965. *A Guide to FORTRAN IV Programming*. New York: Wiley.

Dickson, T. R. 1968. *The Computer and Chemistry*. San Francisco: W. H. Freeman and Company.

APPENDICES

APPENDIX **A**

Nomenclature of Molecular Electronic States

Electronic states are characterized by their angular momentum values:

$$^{M}\Lambda_{\Omega}$$

where Λ is the component of electronic orbital angular momentum along the internuclear axis and can possess the values $\Lambda = 0, 1, 2, \ldots, L$, where L is the total electronic orbital angular momentum. States of $\Lambda = 0, 1, 2, 3$ are called Σ, Π, Δ, and Φ, respectively. The multiplicity, M, and the resultant spin, S, are related by $M = 2S + 1$. The component of the resultant spin along the internuclear axis can take the values $\Sigma = S, S-1, S-2, \ldots, -S$. The total angular momentum around the internuclear axis is $\Omega = |\Lambda + \Sigma|$.

A molecule of $\Lambda = 2$ and $S = 1$ can adopt the following Σ values: $\Sigma = 1, 0, -1$; and therefore $\Omega = 3, 2, 1$. The three states that result bear the following symbols:

$$^{3}\Delta_{1},\ ^{3}\Delta_{2},\ ^{3}\Delta_{3}$$

In addition, the symbols + and −, as well as g and u, are used. They refer to the symmetry properties of the state. For − states the complete wave function changes sign upon reflection at the origin; for + states it does not. For homonuclear diatomic molecules another symmetry property can be found in whether the electronic wave function changes sign upon reflection at the center of symmetry. Gerade (g) states do not change sign; ungerade (u) states do.

The two states that participate in producing the visible spectrum of iodine are:

$$X = {}^{1}\Sigma^{+}_{0g}$$

and

$$B = {}^{3}\Pi^{+}_{0u}$$

For both states the total angular momentum is $\Omega = 0$. Interaction between electronic and rotational angular momentum for Hund's case a or c yields (c is approached for iodine):

$$F(J) = B[J(J+1) - \Omega^{2}]$$

which reduces to $F(J) = BJ(J+1)$.

Selection Rules

For Hund's cases *a* and *b* the following are found:

$$\Delta\Lambda = 0, \pm 1$$

$$\Delta\Omega = 0, \pm 1$$

$$\Delta S = 0$$

the second of which is also valid for Hund's case *c*. The last is the most frequently violated selection rule, which indeed is violated for iodine.

For symmetry properties the following selection rules exist:

$$+ \leftrightarrow +$$

$$- \leftrightarrow -$$

$$g \leftrightarrow u$$

which are all obeyed in iodine.

Derivation of the Larmor Frequency

A constant and homogeneous magnetic field has no effect on the motion of electrons in an atom except to rotate the motional pattern as a whole around the field axis by a constant frequency,

$$\omega_L = \frac{e \cdot H}{2 \cdot m \cdot c} \tag{1}$$

called the Larmor frequency.

The general proof of this theorem by Larmor is somewhat complicated, but it is very easy to grasp in an elastically bound electron. The forces acting on an electron, bound to the origin of an x,y,z coordinate system, are given by

1. the elastic force with components $-kx, -ky, -kz$
2. the Lorentz force, which is perpendicular to the magnetic field H, which lies along z. The components of the Lorentz force in the velocity directions ($\dot{x}$, $\dot{y}$, $\dot{z}$) are:

$$\frac{e}{m \cdot c} \dot{y} \cdot H, \frac{-e}{m \cdot c} \dot{x} \cdot H, 0.$$

The dot above x, y, and z is an abbreviation for the first derivative with respect to time, $\dot{x} \equiv \frac{dx}{dt}$; two dots symbolize the second derivative. Newton's law, written in x,y,z components, then reads

$$m \cdot \ddot{x} = -kx + \frac{e}{c} \dot{y} H \tag{2}$$

$$m \cdot \ddot{y} = -ky - \frac{e}{c} \dot{x} H$$

$$m \cdot \ddot{z} = -kz$$

The solution of this system of linear differential equations can be easily written for $H = 0$. By setting k/m equal to the square of the electron angular velocity ω_0 (see the harmonic oscillator in Exercise 6B), $k/m = \omega_0^2$, the solutions can be expressed in terms of cos $\omega_0 t$ and sin $\omega_o t$ with real coefficients. If the orbiting electron is viewed along the z axis, the projection of the orbit on the x-y plane is an ellipse.

$$x_0 = a_1 \cos \omega_0 t + b_1 \sin \omega_0 t \tag{3}$$

$$y_0 = a_2 \cos \omega_0 t + b_2 \sin \omega_0 t$$

with arbitrary real coefficients a_1, b_1, a_2, and b_2. The same ellipse can be obtained in the complex plane if the complex number w_0, composed from y_0 and x_0, is defined as

$$w_0 \equiv x_0 + iy_0 \tag{4}$$

Making use of the identity $e^{i\alpha} = \cos \alpha + i \cdot \sin \alpha$,

w_0 can be expressed as

$$w_0(t) = A \cdot \exp(i\omega_0 t) + B \cdot \exp(i\omega_0 t) \quad (5)$$

with complex coefficients A and B.

The general solution of differential equation (6) is represented by $w_0(t)$ and is obtained by multiplying the second equation under (2) by i and adding it to the first equation under (2), all for $H = 0$.

$$m\ddot{w}_0 = -kw_0 \quad (6)$$

Note that $A \cdot \exp(i\omega_0 t)$ describes a vector of length $|A|$ rotating around the origin with constant angular velocity ω_0.

A solution of these differential equations is easily obtained for $H \neq 0$ by the same method used for $H = 0$. The equation for the z component again yields a simple oscillation and for the coupled first two equations the following is found:

$$\ddot{w}(t) = -\omega_0^2 \cdot w - i\frac{e \cdot H}{m \cdot c}\dot{w} \quad (7)$$
$$= -\omega_0^2 \cdot w - 2i\omega_L \cdot \dot{w}$$

The general solution of equation (7) is given by

$$w(t) = A \exp[i(\omega_0 + \omega_L)t] \quad (8)$$
$$+ B \exp[i(-\omega_0 + \omega_L)t]$$

with A and B depending on initial conditions. This solution can be written as

$$w(t) = \exp(i\omega_L t)[A \cdot \exp(i\omega_0 t) \quad (9)$$
$$+ B \cdot \exp(-i\omega_0 t)]$$
$$= \exp(i\omega_L t) \cdot w_0 \cdot t$$

Consequently, the solution for the electronic motion with field H is the solution for $H = 0$, multiplied by a factor $\exp(i \cdot \omega_L t)$. The multiplication of a complex number by a factor $e^{i\alpha}$, however, means a rotation of the representative point in the complex plane by an angle α. Thus, the factor $\exp(i\omega_L t)$ is equivalent to a rotation of the electronic motion described by $w_0(t)$ around the z axis by the Larmor frequency ω_L.